Student Solutions Manual

for use with

Prepared by JILLIAN HATNEAN

MARK VAUGHAN
QUEST UNIVERSITY CANADA

NELSON
EDUCATION
CELEBRATE LIFELONG LEARNING

1914–2014: Nelson Education celebrates 100 years of Canadian publishing

NELSON / EDUCATION

Student Solutions Manual for use with Chemistry, Second Canadian Edition

prepared by Jillian Hatnean and Mark Vaughan

Vice President, Editorial Higher Education:
Anne Williams

Publisher:
Paul Fam

Technical Reviewer:
Rabin Bissessur

Developmental Editor:
Courtney Thorne

Permissions Coordinator:
Lynn McLeod

Senior Production Project Manager:
Imoinda Romain

Copy Editor:
Valerie Adams

Design Director:
Ken Phipps

Managing Designer:
Franca Amore

Cover Design:
Courtney Hellam

Cover Image:
CLAUDIO SANTANA/AFP/ Getty Images

ISBN-13: 978-0-17-672092-6
ISBN-10: 0-17-672092-8

Table of Contents

CHAPTER 1
Human Activity, Chemical Reactivity

REVIEW QUESTIONS

Section 1.2: Harnessing Light Energy and Exciting Oxygen

1.1

PDT or photodynamic therapy requires a photo sensitizer, light, and oxygen. The photo sensitizer is a molecule that readily absorbs light, and transfers its extra energy to neighbouring triplet oxygen (the usual form of oxygen found in tissue and elsewhere). The transferred energy takes triplet oxygen to its singlet state. The very reactive singlet oxygen kills rapidly growing cancer cells.

1.3

The tumour must be located in a place that can be subjected to light. For example, David Dolphin and his collaborators developed porphyrin PDT for the treatment of skin cancers, for which an external light source is sufficient.

1.5

Toxicology is the study of the ill effects (toxicity) of substances on living organisms. Before introducing a porphyrin into the body for PDT, it must be established that the porphyrin, by itself, has little or no significant toxicity.

1.7

Chemotherapy is the use in medicine of substances that are selectively toxic to malignant cells or to a disease-causing virus or bacterium. As such, a vaccine would not be considered chemotherapy. Vaccines expose the body to substances that cause the body to make "antibodies" that fight particular disease-causing viruses or bacteria. The vaccine is administered before exposure to the virus or bacterium, and does not directly attack the disease-causing agent itself. The use of garlic to treat gangrene, on the other hand, is an example of chemotherapy. Garlic is a mild antiseptic, which kills bacteria-infecting tissue leading to gangrene.

1.9

Yes, arsenic is generally considered to be toxic. However, Sec. 6.1 discusses how the toxicity of arsenic varies dramatically depending on the species containing the arsenic atoms. For example, whereas elemental arsenic is toxic, the arsenic-containing species in lobster are not. This is why we can eat lobster with no ill effects.

1.11

The structures of both vitamin B-12 and Visudyne™ are porphyrin-based. Both molecules' porphyrin-backbone make them excellent candidates for photodynamic therapies.

Section 1.3: Where There's Smoke, There's Gavinone

1.13

A natural product is a compound produced by a living organism.

1.15

(a) The Haber process combines hydrogen and nitrogen to make ammonia. Ammonia is used to make fertilizer. It was discovered by Fritz Haber.
(b) In the Bohr model, a hydrogen atom consists of an electron in a circular orbit about a proton. The angular momentum of the electron is a multiple of Planck's constant divided by 2π. It was discovered by Niels Bohr.
(c) A conical flask widely used in chemistry labs to carry out reactions. It was invented by Emil Erlenmeyer.
(d) The van der Waals equation is a relation between the pressure, temperature, and volume of a gas that accounts for the non-zero size of the gas molecules and the attractive forces between them. It was discovered by Johannes Diderik van der Waals.
(e) Gibbs free energy, $G = H - TS$, combines enthalpy and entropy to give a quantity that must decrease for any processes that actually happen. It was discovered by Josiah Willard Gibbs.
(f) Lewisite is a chlorinated organoarsenic compound, which was produced as a chemical weapon that causes blisters and lung irritation. It was discovered by Winford Lee Lewis.
(g) A Lewis base has a lone pair of electrons that it can donate to an electron pair acceptor—a Lewis acid. It was discovered by Gilbert N. Lewis.
(h) Schrödinger's equation determines the wave function that describes the state of an atom. It was discovered by Erwin Schrödinger.

1.17

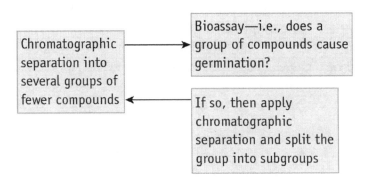

Section 1.4: Chemical Reactivity, Your Activity

1.19

An organic compound consists of molecules made from carbon, hydrogen and oxygen atoms. There can be other atoms (for example, nitrogen in the case of Gavinone), and there may be no oxygen or even hydrogen atoms, as in the case of graphene, graphite, carbon nanotubes, and fullerenes.

1.21

1. Acetylsalicylic acid, the common mild pain reliever, is a derivative of the natural product, salicin, obtained from willow bark.
2. Morphine, obtained from poppies, is a powerful pain reliever.
3. Cocaine, obtained from the coca plant, is also a pain reliever—still used for topical pain relief.
4. Quinine, an anti-malaria agent, was isolated from the bark of cinchona trees.
5. Menthol, isolated from mint leaves, is a topical pain reliever, which also relieves itching.

1.23

Synthetic sources might be cheaper if the natural product is rare. On the other hand, natural products might be cheaper if the drug can be obtained directly from an easily grown plant. Also, natural products require only separation chemistry, so they reduce the need for chemical reagents.

SUMMARY AND CONCEPTUAL QUESTIONS

1.25

Excess production of porphyrin is the cause of porphyria. Porphyria sufferers and patients undergoing PDT treatment both have excess porphyrins in their tissue, making them sensitive to light for the same reason.

1.27

Antimony and arsenic are in the same group in the periodic table—antimony just below arsenic. It is natural that arsenic, having similar properties, might substitute for antimony in its compounds. Since arsenic has the larger natural abundance, it might be expected to constitute a significant impurity in antimony compounds.

1.29

Gavinone has alkane, alkene, ether, and ester functional groups. It would give absorption peaks in the following ranges (in cm^{-1}): 2850–2980 (from the alkane group—note that C–C stretches not shown in Table 3.5 will also cause absorption peaks); 1640–1670 and 3020–3100 (from alkene groups); 1085–1150 (from ether group); and 1000–1300 and 1730–1750 (from ester group).

CHAPTER 2
Building Blocks of Materials

IN-CHAPTER EXERCISES

Exercise 2.1—Classification of Substances

Pure	*Mixture*	
Compounds	*Solutions*	
(h) testosterone (f) sodium chloride	(b) air (c) vinegar (g) athlete's urine sample	(a) mud (e) milk
Elements		
(d) gold		
Homogeneous	*Heterogeneous*	

Exercise 2.3—Names and Symbols of Elements

(a) Na, Cl, and Cr are the symbols for sodium, chlorine and chromium, respectively.
(b) Zn, Ni, and K are the symbols for zinc, nickel and potassium.

Exercise 2.5—Meaning of Subscripts in a Chemical Formula

(a) In $CO(g)$, there is one carbon atom for every oxygen atom (or the ratio of C to O atoms is 1:1).
(b) In $CH_4(g)$, the C to H atom ratio is 1:4.
(c) In $C_2H_2(g)$, the C to H atom ratio is 1:1.
(d) In $C_6H_{12}O_6(s)$, C, H, and O atoms are in proportion to 1:2:1.

Exercise 2.7—Meaning of Chemical Equations in Words

$$O_3(g) + NO(g) \longrightarrow O_2(g) + NO_2(g)$$

Ozone and nitric oxide gases combine to form oxygen and nitrogen dioxide gases. The relative numbers of molecules of the four respective species are 1:1:1:1—i.e., the stoichiometric coefficients.

Exercise 2.9—Atomic Composition

(a) Iron has atomic number 26. Mass number = number of protons + number of neutrons = 26 + 30 = 56.

(b) Zinc has atomic number 30. Therefore, ^{64}Zn has 30 protons.
Number of neutrons = mass number – number of protons = 64 – 30 = 34.
Since it is an atom—rather than an ion—the number of electrons = number of protons = 30.

Exercise 2.13—Mass Ratios of Isotopes

(a) Table 2.2 shows the mass of ^{24}Mg to be 23.9850. The mass of ^{12}C is, by definition, 12.0000.
∴ the ratio of ^{24}Mg to ^{12}C is $\dfrac{23.9850}{12.0000} = 1.99875$

Note that the answer has the right number of significant digits.

(b) The mass of ^{26}Mg is 25.9826
∴ the ratio of ^{26}Mg to ^{12}C is $\dfrac{25.9826}{12.0000} = 2.16522$

(c) The ratio of ^{26}Mg to ^{24}Mg is $\dfrac{25.9826}{23.9850} = 1.08329$

Exercise 2.15—Ratios of Amounts and Numbers of Atoms

It is just the ratio of the numbers of moles, $\dfrac{2\ \cancel{mol}}{0.2\ \cancel{mol}} = 10$

Exercise 2.17—Relative Numbers of Atoms in Samples

Amount, $n(\text{Zn}) = \dfrac{m}{M} = \dfrac{1.00\ \cancel{g}}{65.38\ \cancel{g}\ \text{mol}^{-1}} = 0.0153\ \text{mol}$

We want the mass of 0.0153 mol of copper—same number of moles means same number of atoms.

Mass of 0.0153 mol of copper, $m = n \times M = 0.0153\ \cancel{mol} \times 63.55\ \text{g}\ \cancel{mol^{-1}} = 0.972\ \text{g}$

REVIEW QUESTIONS

Section 2.1: Falsely Positive? The Chemistry of Drugs in Sport

2.25

The isotopic ratio of ^{13}C:^{12}C in endogenous testosterone in athletes is dictated by diet, as it originates from cholesterols in vegetables and in meat. Exogenous testosterone will typically have lower ^{13}C levels since they are synthesized from cholesterol from soybean plants.

2.27

"Heavy" water (^2H_2O or D_2O) contains deuterium, or a heavier form of hydrogen. It is not radioactive. "Light" water is simply H_2O. Scientists can take core samples from glaciers to determine the sea-air temperatures from hundreds of thousands of years ago. Since "heavy" water is denser than liquid water, it is less susceptible to evaporation and the oceans and inner glaciers are enriched with it. Frozen heavy ice would be denser than normal ice, and would form the base of a glacier or ice cap easily. Light water that exists as a vapor can condense on the surface of the glacier in the form of snow and ice. As the temperature has gradually increased over the years, new layers within the glacier may show increased levels of D_2O due to thermally surpassing the barrier to evaporation. The ratio of heavy to light water from core samples at varying depths would provide a picture of temperature change over time. The same argument can be made for light vs. heavy oxygen in water (^{16}O vs. ^{18}O). When we check with e2.10, D_2O is dense and would sink in liquid H_2O, hindering evaporation.

Section 2.6: Chemical Reactions, Chemical Change

2.29

The simplest test might be to apply a flame. Sugar caramelizes when exposed to a flame—it turns brown, then black. Alternatively, you could dissolve the substance in water, and then measure the electrical conductivity of the resulting solution. Only a salt solution conducts electricity. You could also run an infrared (IR) spectrum of the material. Sugar is a molecular substance that absorbs IR light in the 500–3500 cm^{-1} range. Salt does not absorb in this range.

2.31

(a) The change in colour when a T-shirt is bleached is a chemical change, with an accompanying physical change (the colour is a physical property of dyes in the T-shirt that have been chemically changed by the bleach).
(b) The condensation of breath on a cold day is a physical change.
(c) The conversion of carbon dioxide into sugars, by plants, is a chemical change.
(d) The melting of butter in the sun is a physical change.

2.33

(a) "White," "solid," "density," and "gaseous" are physical properties; "reacts readily with an acid" is a chemical property.
(b) "Grey," "powdered," "purple," and "white" are physical properties; "reacts with … iodine" is a chemical property.

Section 2.8: Isotopes of Elements

2.35

(a) mass number = 19 + 20 = 39; the symbol is $^{39}_{19}$K

(b) mass number = 36 + 48 = 84; the symbol is $^{84}_{36}$Kr

(c) mass number = 27 + 33 = 60; the symbol is $^{60}_{27}$Co

2.37

Symbol	^{58}Ni	^{33}S	^{20}Ne	^{55}Mn
Number of protons	28	16	10	25
Number of neutrons	30	17	10	30
Number of electrons in the neutral atom	28	16	10	25
Name of element	nickel	sulfur	neon	manganese

2.39

The atomic weight of potassium is 39.0983. It is much closer to 39 than 41. Therefore, ^{39}K must be the most abundant isotope.

Section 2.9: Relative Atomic Masses of Isotopes and Atomic Mass Units

2.41

Lithium has two isotopes: ^{6}Li and ^{7}Li are 7.50% and 92.50% abundant, respectively. Therefore, the mass spectrum for the element lithium would have 2 peaks at 6.015121 and 7.016003 (from Question 2.45). If we arbitrarily set the most abundant peak (^{7}Li = 7.016003) to 100%, the relative intensity of the peak at 6.015121 is

relative abundance ^{6}Li $= \dfrac{100\%(^6\text{Li})}{(^7\text{Li})} = \dfrac{100\%(7.50\%)}{(92.50\%)} = 8.11\%$

2.43

All % abundances add up to 100%. Therefore, % abundance $^{26}_{12}$Mg = 100.00 − 78.99 − 10.00% = 11.01%.

2.45

Atomic weight of lithium $= \dfrac{7.50}{100} \times 6.015121 + \dfrac{92.50}{100} \times 7.016003 = 6.941$

Section 2.10: Atomic Weights of Elements

2.47

The atomic weight of boron is 10.811028. When the isotopic abundance of ^{11}B is changed to 0.81 (originally 0.801), the atomic weight increases to 10.819995 due to additional contribution from the heavier isotope.

Section 2.11: Amount of Substance and Its Unit of Measurement: The Mole

2.49

$$\text{Molecule ratio} = \text{mole ratio} = \frac{2\ \text{mol}}{1\ \text{mol}} = 2$$

2.51

$$\text{Molecule ratio} = \text{mole ratio} = \frac{0.1\ \text{mol}}{0.1\ \text{mol}} = 1$$

2.53

$$\text{Amount, } n(\text{Ba}) = \frac{m}{M} = \frac{137.3\ \text{g}}{137.3\ \text{g mol}^{-1}} = 1.000\ \text{mol}$$

We want the mass of 5.000 mol of aluminium—five times the number of moles means five times the number of atoms.

$$\text{Mass of 5.000 mol of aluminium, } m = n \times M = 5.000\ \text{mol} \times 26.98\ \text{g mol}^{-1} = 134.9\ \text{g}$$

Section 2.12: The Periodic Table of Elements

2.55

Two periods have 8 elements, two periods have 18 elements, and two periods have 32 elements. However, not all of the last 6 elements in the second 32 element period have yet been synthesized.

2.57

(a) Bk is a radioactive element.
(b) Br is a liquid at room temperature.
(c) B is a metalloid.
(d) Ba is a Group 2 element.
(e) Bi is a Group 15 element.

2.59

(a) Si is a metalloid, while P is a non-metal.
(b) Si has some electrical conductivity, while P does not.
(c) Both Si and P are solids at 25ºC.

Section 2.13: IUPAC Periodic Table of the Isotopes

2.61

Carbon isotope ratio (CIR) testing detects exogenous hormones in urine. The test uses isotope ratio mass spectrometry (IRMS) to determine the ratio of ^{13}C:^{12}C and compares it to an external standard. Synthetic testosterone will have less naturally occurring ^{13}C. A sample with a difference of greater than 3 units from the standard is considered positive for the presence of external sources. This "gold standard" test was invaluable to detect "blood doping" efforts by the Tour de France cycling winners. This same testing can be applied to the debate of volcanoes vs. cars for CO_2 emissions. Volcanic CO_2 is largely ^{13}C depleted. Therefore, detecting high levels of ^{13}C in the atmosphere *could* be attributed to automobile exhaust, but is *not* in any way conclusive.

SUMMARY AND CONCEPTUAL QUESTIONS

2.63

S	N
B	I

2.65

(a) zinc, Zn; cadmium, Cd; mercury, Hg; or Copernicium, Cn
(b) Any from rubidium, Rb, to xenon, Xe
(c) lead, Pb
(d) sulfur, S
(e) sodium, Na
(f) xenon, Xe
(g) selenium, Se, a non-metal
(h) arsenic, As, or antimony, Sb

2.67

The atomic mass of lithium, 6.941, is much closer to 7 than 6. Therefore, ^7Li is the most abundant isotope.

2.69

(a) $(3.79 \times 10^{24}$ atoms$)/(6.022 \times 10^{23}$ atoms mol^{-1}) = 6.29 mol Fe
   (6.29 mol) $\times$ (55.85 g mol^{-1}) = 351 g Fe
(b) (19.921 mol) $\times$ (2 $\times$ 1.008 g mol^{-1}) = 40.16 g H_2
(c) (8.576 mol) $\times$ (12.01 g mol^{-1}) = 103.0 g C
(d) (7.4 mol) $\times$ (28.09 g mol^{-1}) = 208 g Si
   = 210 g with the correct number of significant digits
(e) (9.221 mol) $\times$ (22.99 g mol^{-1}) = 212.0 g Na
(f) $(4.07 \times 10^{24}$ atoms$)/(6.022 \times 10^{23}$ atoms mol^{-1}) = 6.76 mol Al
   (6.76 mol) $\times$ (26.98 g mol^{-1}) = 182 g Al
(g) (9.2 mol) $\times$ (2$\times$35.45 g mol^{-1}) = 650 g Cl_2 (we get 652 g—but the 2 is not a significant digit)

Increasing in mass, these samples have the order, (b), (c), (f), (d), (e), (a) and (g)

2.71

(a) $A_r'(^{16}O)$ =15.9994/1.0078 = 15.8756
   With the convention that the atomic mass of ^1H equals exactly 1, Avogadro's constant would be defined as the number of ^1H atoms in exactly 1.0000 g of ^1H. Since 1.0078 g of ^1H has 6.022 $\times 10^{23}$ atoms, the revised value of the Avogadro constant is
   6.022×10^{23} / 1.0078 = 5.975×10^{23}
(b) $A_r'(^1H)$ = 16 $\times$ 1.0078/15.9994 = 1.0078
   With the convention that the atomic mass of ^{16}O equals exactly 16, Avogadro's constant would be defined as the number of ^{16}O atoms in exactly 16.0000 g of ^{16}O. Since 15.9994 g of ^{16}O has 6.0221$\times 10^{23}$ atoms, the revised value of the Avogadro constant is
   $6.0221 \times 10^{23} \times 16.0000/ 15.9994$ = 6.0223×10^{23}

2.73

$\delta^{13}C$ = 218228/20000002 = 0.0109114 for the first place finisher testosterone
$\delta^{13}C$ = 219135/19999997 = 0.0109568 for the reference steroids
The first place finisher's testosterone $\delta^{13}C$ is consistent with it being endogenous—i.e., he did not take exogenous steroids.

2.75

(a) Two examples of human activity of doing chemistry found in this chapter are GC-MS and IRMS.
(b) Two examples of chemical reactivity found in this chapter are photosynthesis and combustion of methane.

2.77

(a) Barium is expected to react vigorously with water. Since magnesium did not react, whereas calcium did, we see that the trend is increasing reactivity going down the group. Ba is two periods below Ca.

(b) Period = 3 (Mg), 4 (Ca) and 6 (Ba). It appears that reactivity with water increases going down the group.

CHAPTER 3
Models of Structure to Explain Properties

IN-CHAPTER EXERCISES

Exercise 3.1—Formulas of Covalent Network Substances

BN is the formula of boron nitride.

Exercise 3.3—Ions

(a) An S atom has 16 protons and 16 electrons. An S^{2-} ion has 16 protons and 18 electrons, for a net charge of -2.

(b) An Al atom has 13 protons and 13 electrons. An Al^{3+} ion has 13 protons and 10 electrons, for a net charge of $+3$.

(c) An H atom has 1 proton and 1 electron. An H^+ ion has 1 proton and no electrons, for a net charge of $+1$.

Exercise 3.11—Molar Masses of Ionic Compounds

(a) $M(\text{NaOH}) = M(\text{Na}) + M(\text{O}) + M(\text{H}) = 22.99 + 16.000 + 1.008$ g mol^{-1} $= 40.00$ g mol^{-1}

$$\text{Amount, } n(\text{NaOH}) = \frac{m}{M} = \frac{1.00 \text{ g}}{40.00 \text{ g mol}^{-1}} = 0.0250 \text{ mol}$$

(b) $M(\text{Fe(NO}_3)_2) = M(\text{Fe}) + 2 \times (M(\text{N}) + 3 \times M(\text{O})) = 55.85 + 2 \times (14.01 + 3 \times 16.000)$ g mol^{-1} $= 179.87$ g mol^{-1}

$$\text{Amount, } n(\text{Fe(NO}_3)_2) = \frac{m}{M} = \frac{0.0125 \text{ g}}{179.87 \text{ g mol}^{-1}} = 6.95 \times 10^{-5} \text{ mol}$$

(c) $M(\text{CaCO}_3) = M(\text{Ca}) + M(\text{C}) + 3 \times M(\text{O}) = 40.08 + 12.01 + 3 \times 16.000$ g mol^{-1} $= 100.09$ g mol^{-1}

0.500 tonnes $= 500$ kg $= 5.00 \times 10^5$ g

$$\text{Amount, } n(\text{CaCO}_3) = \frac{m}{M} = \frac{5.00 \times 10^5 \text{ g}}{100.09 \text{ g mol}^{-1}} = 5.00 \times 10^3 \text{ mol}$$

Exercise 3.15—Molecular Substances

(b) nitric oxide, NO(g), (d) ammonia, NH_3(g), and (e) sulfur, S_8(s) are all molecular substances.

Magnesium oxide and molten sodium chloride consist of ions, while solid magnesium is a metal consisting of ions and electrons bound in a lattice.

Exercise 3.17—Relative Molecular Mass from High-Resolution Mass Spectrometry Data

^{12}C, ^{16}O and ^{1}H are the most common isotopes of carbon, oxygen and hydrogen. The molar mass of the most abundant isotopologue, $^{12}C_2{}^{1}H_6{}^{16}O$, is $2 \times 12.00000 + 6 \times 1.00783 + 15.99491$ g mol^{-1} = 46.04189 g mol^{-1}. This is not in exact agreement with the experimental value. But, it is much closer to the experimental value than to the average value obtained using natural abundances, 46.06844 g mol^{-1}.

Exercise 3.19—Molar Mass

(a) $M(C_3H_7OH) = 3 \times M(C) + 8 \times M(H) + M(O) = 3 \times 12.01 + 8 \times 1.008 + 16.000$ g mol^{-1} = 60.09 g mol^{-1}

mass of propanol = $m = n(C_3H_7OH) \times M(C_3H_7OH) = (0.0255$ mol$) \times (60.09$ g mol$^{-1}) = 1.53$ g

(b) $M(C_{11}H_{16}O_2) = 11 \times M(C) + 16 \times M(H) + 2 \times M(O) = 11 \times 12.01 + 16 \times 1.008 + 2 \times 16.000$ g mol^{-1} = 180.25 g mol^{-1}

mass of $C_{11}H_{16}O_2 = m = n(C_{11}H_{16}O) \times M(C_{11}H_{16}O) = (0.0255$ mol$) \times (180.25$ g mol$^{-1}) = 4.60$ g

(c) $M(C_9H_8O_4) = 9 \times M(C) + 8 \times M(H) + 4 \times M(O) = 9 \times 12.01 + 8 \times 1.008 + 4 \times 16.000$ g mol^{-1} = 180.15 g mol^{-1}

mass of aspirin = $m = n(C_9H_8O_4) \times M(C_9H_8O_4) = (0.0255$ mol$) \times (180.15$ g mol$^{-1}) = 4.59$ g

Exercise 3.23—Molecular Formula from Mass Spectrum

There are pairs of peaks separated by 2 in *m/z* value at 112 and 114, and 113 and 115. These pairs are consistent with a single chlorine atom with two isotopes ^{35}Cl and ^{37}Cl in roughly 3:1 abundance (75.8% vs. 24.2%). Subtracting 35 from 112 gives 77, which could correspond to C_6H_5. Therefore, a reasonable molecular formula is C_6H_5Cl. Exact mass values corresponding to peaks at 112 and 114 would be given by
$6 \times 12.00000 + 5 \times 1.00783 + 34.968852 = 112.008002$ and
$6 \times 12.00000 + 5 \times 1.00783 + 36.965902 = 114.005052$, respectively.

Exercise 3.25—Interpreting IR Spectra

(a) C=O stretch—ketone, aldehyde or carboxylic acid
(b) C–N–H bend of an amine
(c) C≡N stretch of a nitrile
(d) C=O and O–H stretches of a carboxylic acid
(e) N–H and C=O stretches of an amide

REVIEW QUESTIONS

Section 3.1: Is There Stash on Your Cash?

3.27

Drugs are paid for with cash, which is handled by users and dealers. When these bills re-enter circulation, some cocaine may have transferred to other notes via money sorting machines.

Section 3.2: Classifying Substances by Properties: An Overview

3.29

(a) ionic—high melting, conducts electricity only as a liquid
(b) covalent—a molecular solid—not high melting, does not conduct electricity as solid or liquid
(c) covalent—a molecular solid—sublimes to a molecular gas, does not conduct electricity as solid or liquid
(d) metallic—a good conductor of electricity as a solid or a liquid—this is a low melting metal (some are low melting)

Section 3.3: Covalent Network Substances

3.31

There are equal numbers of Si and C atoms in SiC.

Section 3.4: Ionic Substances

3.33

(a)	Se^{2-}	**(c)**	Fe^{2+} and Fe^{3+}
(b)	F^-	**(d)**	N^{3-}

3.35

(a) An ammonium ion, NH_4^+, has one more proton than electron—hence the +1 charge.
(b) A phosphate ion, PO_4^{3-}, has three more electrons than protons—hence the −3 charge.
(c) A dihydrogenphosphate ion, $H_2PO_4^-$, has one more electron than proton—hence the −1 charge.

3.37

Na_2CO_3 sodium carbonate
$BaCO_3$ barium carbonate
NaI sodium iodide

BaI_2 barium iodide

3.39

(a)	ClF_3 chlorine trifluoride	**(f)**	OF_2 oxygen difluoride
(b)	NCl_3 nitrogen trichloride	**(g)**	KI potassium iodide ionic
(c)	$SrSO_4$ Strontium sulfate ionic	**(h)**	Al_2S_3 aluminum sulfide ionic—but with some covalent character because of the large +3 charge on aluminum and the large polarizability of sulfide (See Chapter 26)
(d)	$Ca(NO_3)_2$ calcium nitrate ionic	**(i)**	PCl_3 Phosphorus trichloride
(e)	XeF_4 xenon tetrafluoride	**(j)**	K_3PO_4 potassium phosphate ionic

3.41

The ionic formulas and number of ions are as follows:

(a) $BeCl_2 \rightarrow Be^{2+} + 2Cl^-$ 1.0 g of $BeCl_2$ (79.912 g/mol) contains 2.3×10^{22} ions.

(b) $MgCl_2 \rightarrow Mg^{2+} + 2Cl^-$ 1.0 g of $MgCl_2$ (95.21 g/mol) contains 1.9×10^{22} ions.

(c) $CaS \rightarrow Ca^{2+} + S^{2-}$ 1.0 g of CaS (72.14 g/mol) contains 1.7×10^{22} ions.

(d) $SrCO_3 \rightarrow Sr^{2+} + CO_3^{2-}$ 1.0 g of $SrCO_3$ (147.63 g/mol) contains 8.2×10^{21} ions.

(e) $BaSO_4 \rightarrow Ba^{2+} + SO_4^{2-}$ 1.0 g of $BaSO_4$ (233.4 g/mol) contains 5.2×10^{21} ions.

Therefore, sample **(a)** has the largest number of ions.

3.43

(a) The atomic number of the element = # protons. Therefore, the atomic number is 20, which is the element calcium (Ca).

(b) Ca^{2+}

(c) This ion is a cation since it has lost 2 electrons (20 electrons – 2 = 18). $Ca \rightarrow Ca^{2+} + 2e^-$

Section 3.5: Metals and Metallic Substances

3.45

The molecules are localized within the solid ice cube, as the water molecules are tightly held together (frozen) through hydrogen bonding and other forces. In liquid water itself, the water molecules are weakly bound to each other but freely move from one molecule to the next with ease (delocalized). At the surface, the localized molecules of the ice cube slowly become delocalized and join the fluid surface as the ice cube melts. We can relate this to magnesium atoms freely giving up electrons (forming ions) that are then delocalized and do not belong to one individual atom, but rather to the whole sample while the Mg ions hold firm positions in the lattice.

Section 3.6: Molecular Substances

3.47

A chemical reaction of oxygen requires breaking the oxygen-oxygen double bond, and generally forming other bonds. As such, the propensity of oxygen to react depends on the strength of the O=O bond, as well as the strength of bonds formed in the reaction. The boiling point of oxygen depends only on the strength of the weak intermolecular forces between neighbouring oxygen molecules.

Section 3.8: Composition and Formula by Mass Spectrometry

3.49

Bromine has two naturally occurring isotopes, ^{79}Br and ^{81}Br, with almost equal abundance and masses 78.918337 and 80.916291, respectively. Molecular ion peaks of the isotopologues, $^{12}C^1H_3^{79}Br$ and $^{12}C^1H_3^{81}Br$, would be most prominent in the spectrum. We would see two peaks of almost equal intensity. Other isotopologue molecular ion peaks would be much smaller.

3.51

Methamphetamine, $C_{10}H_{15}N$ $m/z = 149$
Cocaine, $C_{17}H_{21}NO_4$ $m/z = 303$
Phencyclidine, $C_{17}H_{25}N$ $m/z = 243$
Tetrahydrocannabinol (THC), $C_{21}H_{30}O_2$ $m/z = 314$
Ecstasy, $C_{11}H_{15}O_2N$ $m/z = 193$

3.53

(a) $M(C_{14}H_{10}O_4) = 14 \times M(C) + 10 \times M(H) + 4 \times M(O) = 14 \times 12.01 + 10 \times 1.008 + 4 \times 16.000$ g mol^{-1}
= 242.22 g mol^{-1}
mass of benzoyl peroxide = $n(C_{14}H_{10}O_4) \times M(C_{14}H_{10}O_4) = (0.123 \text{ mol}) \times (242.22 \text{ g mol}^{-1})$
= 29.8 g

(b) $M(Pt(NH_3)_2Cl_2) = M(Pt) + 2 \times (M(N) + 3 \times M(H)) + 2 \times M(Cl)$
= $195.1 + 2 \times (14.01 + 3 \times 1.008) + 2 \times 35.45$ g mol^{-1}
= 300.1 g mol^{-1}
mass of cisplatin, $m = n(Pt(NH_3)_2Cl_2) \times M(Pt(NH_3)_2Cl_2) = (0.123 \text{ mol}) \times (300.1 \text{ g mol}^{-1})$
= 36.9 g

3.55

(a) $M(C_9H_8O_4) = 9 \times M(C) + 8 \times M(H) + 4 \times M(O) = 9 \times 12.01 + 8 \times 1.008 + 4 \times 16.000$ g mol^{-1}
= 180.15 g mol^{-1}
324 mg = 324×10^{-3} g
amount of aspirin, $n = m((CH_3)_2CO) / M((CH_3)_2CO) = (324 \times 10^{-3} \text{ g}) / (180.15 \text{ g mol}^{-1})$
= 1.80×10^{-3} mol

$M(NaHCO_3) = M(Na) + M(H) + M(C) + 3 \times M(O) = 22.99 + 1.008 + 12.01 +$
3×16.000 g mol^{-1}
= 84.01 g mol^{-1}
1904 mg = 1.904 g
amount of sodium hydrogencarbonate, $n = m(NaHCO_3) / M(NaHCO_3) = (1.904 \text{ g}) / (84.01 \text{ g}$
mol^{-1}) = 2.266×10^{-2} mol

$M(C_6H_8O_7) = 6 \times M(C) + 8 \times M(H) + 7 \times M(O) = 6 \times 12.01 + 8 \times 1.008 + 7 \times 16.000$ g mol^{-1}
= 192.12 g mol^{-1}
1000 mg = 1.000 g
amount of citric acid, $n = m(C_6H_8O_7) / M(H_3C_6H_5O_7) = (1.000 \text{ g}) / (192.12 \text{ g mol}^{-1})$
= 5.205×10^{-3} mol

(b) number of molecules of aspirin consumed = $(1.80 \times 10^{-3} \text{ mol}) \times (6.022 \times 10^{23} \text{ molecules}$ mol^{-1})
= 1.08×10^{21} molecules

3.57

$M(FeS_2) = M(Fe) + 2 \times M(S) = 55.85 + 2 \times 32.06 \text{ g mol}^{-1} = 119.97 \text{ g mol}^{-1}$
    $15.8 \text{ kg} = 15.8 \times 10^3 \text{ g}$
Amount of $FeS_2 = n(FeS_2) = \text{mass} / (\text{molar mass}) = (15.8 \times 10^3 \text{ g}) / (119.97 \text{ g mol}^{-1}) = 132 \text{ mol}$
132 mol of FeS_2 contains 132 mol Fe which has
Mass $= (132 \text{ mol}) \times (55.85 \text{ g mol}^{-1}) = 7372 \text{ g} = 7.37 \text{ kg}$

Section 3.11: Connectivity: Evidence from Infrared Spectroscopy

3.59

(a) Visible radiation range is $26\,316 - 12\,821 \text{ cm}^{-1}$ ($380 - 780$ nm).
(b) Infrared radiation range is $12\,821 - 10 \text{ cm}^{-1}$ (780 nm $- 1$ mm) (longest wavelength)
(c) Ultraviolet radiation range is $1 \times 10^6 - 26\,316 \text{ cm}^{-1}$ ($10 - 380$ nm) (highest energy)

3.61

(a) absorption frequencies in cm^{-1}:

2850–2980	due to alkyl C–H stretches
3000–3100	due to aromatic C–H stretches
2700–2850	due to aldehyde C–H stretch
1720–1740	due to aldehyde C=O stretch
1700–1725	due to carboxylic C=O stretch
675–900, 1400–1500, and 1585–1600 due to aromatic C–C bends and stretches	

(b)

3020–3100	due to alkene C–H stretches
2850–2980	due to alkyl C–H stretches
1730–1750	due to ester C=O stretch
1640–1670	due to C=C stretch
1000–1300	due to ester C–O stretch

(c)

3200–3550	due to alcohol O–H stretch
2850–2980	due to alkyl C–H stretches
1705–1725	due to ketone C=O stretch
1000–1260	due to alcohol C–O stretch

3.63

(a) carboxylic acid and aromatic ring

1585–1600, 1400–1500	ring vibrations
675–900	out of plane C–H bend
3000–3100	C–H stretch
1700–1725	C=O stretch
2500–3300	O–H stretch

(b) ester and aromatic ring

1585–1600, 1400–1500	ring vibrations
675–900	out of plane C–H bend
3000–3100	C–H stretch
1730–1750	C=O stretch
1000–1300	C–O stretch

(c) alcohol, nitrile, and aromatic ring

3200–3550	broad peak due to alcohol (phenol) O–H stretch
2210–2260	due to nitrile C≡N stretch

(d) ketone and alkene

3020–3100	due to alkene C–H stretches
2850–2980	due to alkane C–H stretches
1705–1725	due to ketone C=O stretch

(e) ketone and ester

3020–3100	due to alkene C–H stretches
2850–2980	due to alkane C–H stretches
1705–1725	due to ketone C=O stretch
1700–1725	due to ester C=O stretch (would probably overlap with the ketone stretch)
1000–1300	due to ester C–O stretch

SUMMARY AND CONCEPTUAL QUESTIONS

3.65

Measure the electrical conductivity of the solid and the molten liquid. Only the molten liquid conducts electricity if the substance is ionic.

3.67

The structure of cocaine is

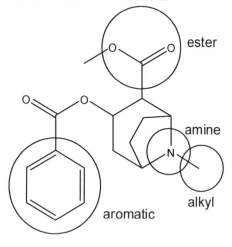

Cocaine contains ester (C=O stretch 1700–1736 cm^{-1}), alkyl (C–H stretch 2850–3000 cm^{-1}), amine (C–N stretch 3000–3100 cm^{-1}) and aromatic (3005 and 704 cm^{-1}) functional groups that are identifiable in the IR spectrum.

3.69

Sodium chloride is ionic. Its solid is held together by the strong attraction between oppositely charged ions. Chlorine is a molecular substance—it consists of Cl_2 molecules. In the solid, the molecules are held together by weak forces between the molecules, independently of the strength of the Cl–Cl bonds within the molecules. Thus, the melting point of chlorine is much lower than that of sodium chloride.

3.71

(a) Silicon dioxide is a covalent network solid that melts at a very high temperature. Carbon dioxide is a molecular substance that melts (sublimes at 1 atm pressure) at a very low temperature because the molecules are held together by weak forces between the molecules.
(b) Sodium sulfide is an ionic substance which melts at a high temperature. Hydrogen sulfide is a molecular substance that melts at a very low temperature.

3.73

(a) Calcium and chloride ions are in a 1:2 ratio in $CaCl_2$ (ionic).
(b) Calcium, carbon, and oxygen atoms are in a 1:1:3 ratio in $CaCO_3$ (ionic).
(c) Nitrogen and hydrogen atoms are in a 1:3 ratio in NH_3 (molecular).
(d) Silicon and carbon atoms are in a 1:1 ratio in SiC (covalent network).
(e) Hydrogen and chlorine atoms are in a 1:1 ratio in HCl (molecular).

3.75

Al^{3+} is most attracted to water because it has the largest magnitude charge.(It is also the smallest, so can get closer to the negatively charged parts of water molecules.)

3.77

(a)

Clostebol has alkyl groups, an alkene, a ketone, a chloride and an alcohol group—the last four groups are circled.

(b) The testosterone molecular ion peak is at 288, whereas the clostebol molecular ion peak is at 322. There are peaks at both these positions, though the peak at 322 is smaller. It would appear that this mass spectrum does not rule out clostebol. However, chlorinated compounds generally show peaks associated with loss of a chlorine atom from the molecular ion. We therefore would expect to see a peak at 287. No such peak is seen. The peak at 322 must be due to some other compound. The data is consistent with the expert analysis.

3.79

Absorption frequencies in cm^{-1}

2850–2980	due to alkyl C–H stretches
2210–2260	due to nitrile C≡N stretch
1705–1725	due to ketone C=O stretch
1640–1670	due to C=C stretch

3.81

(a) Look for the N–H stretch of $CH_3CH_2NHCH_3$—a peak around 3250–3400 cm^{-1} not seen for $(CH_3)_3N$.

(b) Look for the ketone C=O stretch of CH_3COCH_3—a peak around 1705-1725 cm^{-1}—or the OH stretch of $CH_2=CHCH_2OH$—a broad peak around 3200-3550 cm^{-1}.

(c) Look for the aldehyde C-H peaks (2 of them) of CH_3CH_2CHO, around 2700–2850 cm^{-1}—or the ketone C=O stretch of CH_3COCH_3—a peak around 1705–1725 cm^{-1}.

3.83

Many responses may be possible, an example is included. Mass spectroscopy requires fragmentation of the substance in question through ionization, then the fragments are detected and analyzed. This method requires very little sample, is fast and will produce the molecular weight of the unknown substance. The disadvantages of this method are: without high resolution mass spectrometry (HRMS), the fragmentation pattern (m/z peaks) can sometimes be ambiguous, cannot provide structural information and is difficult without volatile compounds. IR spectroscopy detects vibrations due to absorption of radiation. This can be carried out using solid or liquid samples and is a fast method. The disadvantages with this method are: cannot provide the molecular weight of the substance in question or whether you have mixtures of substances.

CHAPTER 4
Carbon Compounds

IN-CHAPTER EXERCISES

Exercise 4.1—Forces of Attraction in Methane

Covalent forces hold the carbon and hydrogen atoms together within a methane molecule.

Exercise 4.3—Reaction of Carbon Compounds with Hydroxyl Radical

The CFCs do not have H atoms bonded to C. Such H atoms readily react with hydroxyl radicals.

Exercise 4.5—"Greenhouse" Gases

Nitrous oxide absorbs strongly in regions of the IR spectrum emitted by the earth that carbon dioxide and water do not absorb. Also, molecules of N_2O have a longer average lifetime in the atmosphere than that of CO_2 molecules.

Exercise 4.7—Carbon Dioxide Storage

Today, I drank from a plastic cup. The plastic material was polymerized from monomers obtained from fractionation and cracking, etc., of petroleum extracted from fossil deposits. I am using a number of plastic products today—for example, the keyboard keys I am typing on, and the computer monitor I am viewing. Another fraction of the same petroleum was used to drive my car. Natural gas—from fossil deposits of mostly methane—is used to heat my house.

Exercise 4.11—Units of Unsaturation

There are 4 units of unsaturation: $14 - 6 = 8 = 4 \times 2$. There is unit of unsaturation for each of the double bonds plus one for the ring.

REVIEW QUESTIONS

Section 4.1: Ice on Fire

4.13

A substance is defined as a single, pure form of matter. Methane clathrate hydrate is a supramolecular compound, in which $CH_4(g)$ is trapped within cages of host water molecules held together by hydrogen bonds to form a symmetrical cavity. Given that methane clathrate hydrate is composed of two species, one encapsulated in the other, by the given definition it could not be considered a substance as it is not purely one species. It is possible to make an argument for the converse.

4.15

Non-covalent bonding refers to an electromagnetic interaction that does not involve sharing electrons. The word *bonding* in chemistry refers to any attraction between atoms, whether strongly (i.e., covalent, ionic) or weakly (dipole-dipole, dispersion, hydrogen bonding). Hydrogen bonding is a perfect example of a non-covalent interaction. Therefore, the word *bonding* should appear in the definition.

4.17

The "clathrate gun hypothesis" is the coined term for the theory that rises in water temperature can cause methane release from clathrates. Methane is a powerful greenhouse gas and contributes to global warming (increased temperatures), which triggers additional methane release from the unstable clathrates on the ocean floor. There has been evidence that there was a ^{13}C isotope excursion during this time, as confirmed by rock and non-surface water samples that contained ^{13}C depleted species. They needed to first prove that warming occurred prior to the ^{13}C isotope excursion, which caused the initial destruction of the clathrates to release the methane that snowballed this effect, which was confirmed in 2002.

4.19

(a) Covalent bonds hold the nitrogen atoms together in the N_2 molecules, inside the clathrate hydrate.

(b) Dispersion forces attract the N_2 and H_2O molecules within the clathrate hydrate cage.

(c) Hydrogen bonding holds neighbouring H_2O molecules together to make the clathrate hydrate cage.

(d) Dispersion forces attract the N_2 and O_2 molecules within the clathrate hydrate cage.

Section 4.3: Climate Change and "Greenhouse Gases"

4.21

A greenhouse prevents convection of heat—a key component of the heat flow from the earth's surface out into the atmosphere. Greenhouse gases do not prevent convection. However, like the greenhouse walls and roof, they prevent some of the radiation emitted by the earth from being lost to space. As shown in the visualization in "Think about it" e4.4, greenhouse gas molecules absorb some of the infrared radiation emitted from earth, causing the molecules to be vibrationally excited. They transfer their energy to surrounding molecules of substances such as N_2 and O_2 that are unable themselves to absorb infrared radiation. This process is called collisional de-excitation.

4.23

$CH_3CH_3(g)$; ozone, $O_3(g)$; and chloroform, $CHCl_3$
Stretching and/or bending bonds in these molecules changes the distribution of charge allowing the molecules to absorb light—an electromagnetic field.
Helium has no vibrations—it is a monatomic gas. Chlorine is a homonuclear diatomic gas—its vibration does not change its distribution of charge.

4.25

The year 1750 is the beginning of the Industrial Revolution. Radiative forcing of our climate was subject only to natural fluctuations before that time.

4.27

Clouds cause cooling by reflecting incoming sunlight back out into space.

4.29

Increasing levels of tropospheric ozone has a positive radiative forcing effect.

4.31

In the troposphere, halocarbons have their greatest impact on radiative forcing. They are important greenhouse gases with a positive forcing.

4.33

The radiative forcing and global warming potential of methane are both enhanced by methane's strong absorption of IR. However, the global warming potential also depends upon how long methane stays in the atmosphere before being eliminated by oxidation and subsequently returning to the surface in rain. The global warming potential of methane is an estimate of how much a given mass of methane will contribute to global warming relative to the same mass of carbon dioxide.

4.35

Consulting the ranking of levels of scientific uncertainty in Figure 2.20 of the 2007 IPCC summary report:
tropospheric carbon dioxide > tropospheric nitrous oxide > stratospheric ozone > jet contrails (we understand that its effect is very small) > solar irradiance

Section 4.4: Capturing, Storing, and Recycling Carbon Compounds

4.37

The advantages of using biofuels are as follows:
(1) They are renewable—we get a new crop each year.
(2) There is no net addition to atmospheric carbon dioxide. Atmospheric CO_2 is consumed as the plant grows. It is returned to the atmosphere when the fuel is combusted.
(3) Biofuel can be produced in most places—provided a suitable crop can be found. Fossil fuels are found only in certain places.
The disadvantages of the biofuel strategy, in practice, are as follows:
(1) Energy is needed to produce the crop and manufacture the biofuel. These energy demands must be minimized to make biofuel production worth the effort.
(2) Only a small portion of corn, for example, can be used to produce biofuels. The cellulosic part of the plant (its bulk), is simply ploughed back into the field. This significantly reduces the viability of biofuel as an alternative energy strategy, as currently produced in North America.
(3) Biofuel production takes land away from food production. This has driven up food prices in North America in recent years.

4.39

The principal difference between the data sets is the size of the annual oscillation in CO_2 level. The CO_2 level at Mauna Loa drops in the spring and summer months due to the growth of vegetation on its land mass and rises again when this growth stops. The Antarctic data oscillates less because of the lack of vegetation in the polar region.

Section 4.5: Alkanes: Saturated Hydrocarbons

4.41

(a)

(b)

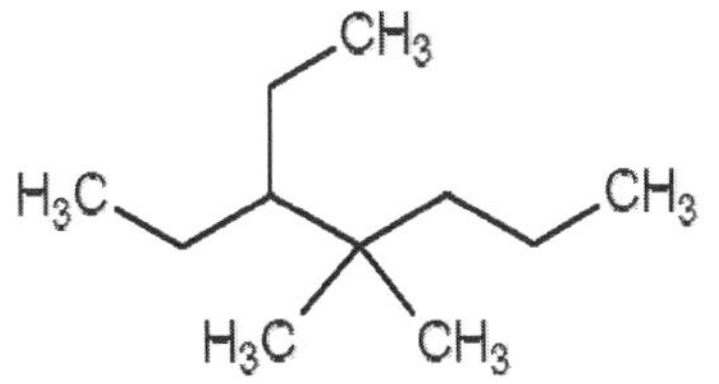

(c)

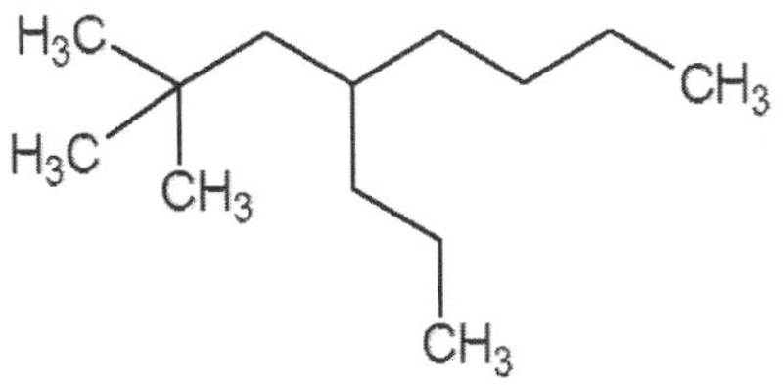

(d)

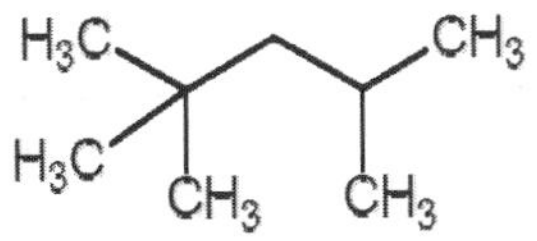

Section 4.6: Polymers and Unsaturated Hydrocarbons

4.43

(a) Hexane has 0 units of unsaturation.
(b) Hept-3-ene has 1 unit of unsaturation—one double bond.
(c) Cyclohex-2-enol has 2 units of unsaturation—a double bond and a ring.
(d) Benzaldehyde has 5 units of unsaturation—4 double bonds and a ring.

4.45

C_4H_8O has 1 unit of unsaturation.
Three possible structures are

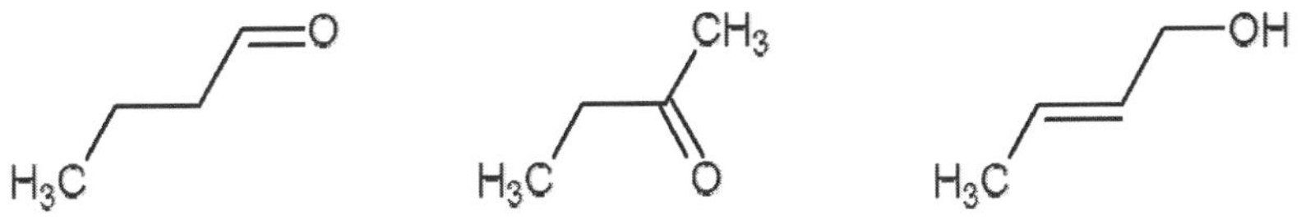

4.47

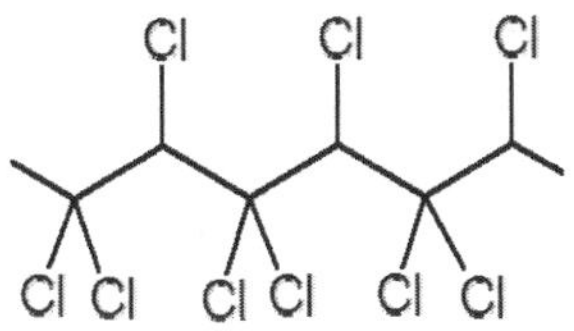

Section 4.7: Where There Is Methane, Is There Life?

4.49

Presence of a strong oxidizing agent on Mars means the environment is hostile to many fragile molecules found on Earth. So the presence of hydrogen peroxide may explain why some expected alkanes besides methane have not been detected. They may have existed, but been rapidly destroyed by oxidation reactions.

SUMMARY AND CONCEPTUAL QUESTIONS

4.51

Force of attraction between carbon and hydrogen atoms in CH_4 molecules > force of attraction between H_2O molecules in methane clathrate hydrate > force of attraction between H_2O and CH_4 molecules in methane clathrate hydrate.
Covalent bonds are stronger than intermolecular forces, and the H-bonds between water molecules (a special type of dipole-dipole interaction) are stronger than the dispersion forces between water and methane molecules (there are also dipole-induced dipole interactions).

4.53

Perfluoropropane has very strong IR absorption and stays in the atmosphere for a very long time—it is quite inert, not readily oxidized and removed from the atmosphere. If there were a source of this gas feeding into the Martian atmosphere, it could accumulate over time until its radiative forcing was sufficient to change the Martian climate. Models suggest it might cause enough initial warming to melt the polar CO_2 ice caps, which would send more CO_2 into the atmosphere, creating a positive feedback loop and perhaps a "runaway" greenhouse effect.

4.55

There are strong absorption peaks in the infrared spectrum of CH_2Cl_2 at frequencies where CO_2 and H_2O do not absorb. As such, CH_2Cl_2 is expected to be an important greenhouse gas.

4.57

Water vapour is a greenhouse gas—generally the most important. However, the amount of water vapour in the atmosphere is not controlled (not directly, in any case) by human activity, except at local scales. Changes in the concentration of water result from climate feedback loops rather than directly from human activity. More water evaporates into the atmosphere in response to global warming. The higher amount of atmospheric water vapour absorbs more IR from Earth, giving a positive feedback cycle. However, as water vapour increases in the atmosphere, more clouds are also formed, which increases Earth's albedo, a negative feedback loop.

4.59

It is certainly important to develop alternative energy schemes to the burning of fossil fuels, in order to reduce the continued accumulation of atmospheric CO_2. However, alternative energy schemes need to be economically viable, and they need to be able to produce large amounts of energy to replace demand currently met by fossil fuels. At present, the large capacity required has not yet been achieved. This problem of capacity is the most important. While reducing demand is possible through conservation and more efficient use of energy, the reduction is not sufficient to allow significant reduction in the use of fossil fuels. A chemist's perspective on fossil fuels must also account for their great importance as feedstocks for the production of many materials we use every day, including plastics and polymers.

4.61

The authors argue that global warming potential has an ambiguous interpretation, and that it relates to effects high up on the cause-effect chain that leads to global warming. They propose an alternative measure of the impact of greenhouse gases that they claim has a less ambiguous interpretation, and uses effects further down the cause–effect chain—that is, more directly linked to global warming.

CHAPTER 5
Chemical Reaction, Chemical Equations

IN-CHAPTER EXERCISES

Exercise 5.1—Balanced Chemical Equations

(a) $Fe_2O_3(s) + 3CO(g) \longrightarrow 2Fe(s) + 3CO_2(g)$

is balanced for electrical charge—there are no charged species among reactants or products. There are 2 mol Fe atoms in 1 mol of Fe_2O_3 on the left, and 2 mol Fe on the right; 3 mol C atoms in 3 mol of CO on the left, and 3 mol C atoms in 3 mol of CO_2 on the right; and $1 \times 3 + 3 \times 1 = 6$ mol O atoms in 1 mol of Fe_2O_3 and 3 mol of CO on the left, and 6 mol O atoms in 3 mol of CO_2 on the right.

Therefore, the reaction is balanced for all relevant atoms.

(b) $CH_3COOH(aq) + C_6H_5CH_2OH(aq) \longrightarrow CH_3COOCH_2C_6H_5(aq) + H_2O(\ell)$

is balanced for electrical charge—there are no charged species among reactants or products. There are $4 + 8 = 12$ mol H atoms in 1 mol of CH_3COOH and 1 mol of $C_6H_5CH_2OH$ on the left, and $10 + 2 = 12$ mol H in 1 mol $CH_3COOCH_2C_6H_5$ and 1 mol H_2O on the right; $2 + 7 = 9$ mol C atoms in 1 mol of CH_3COOH and 1 mol of $C_6H_5CH_2OH$ on the left, and 9 mol C atoms in 1 mol $CH_3COOCH_2C_6H_5$ on the right; and $2 + 1 = 3$ mol O atoms in 1 mol of CH_3COOH and 1 mol of $C_6H_5CH_2OH$ on the left, and $2 + 1 = 3$ mol O atoms in 1 mol $CH_3COOCH_2C_6H_5$ and 1 mol H_2O on the right.

Therefore, the reaction is balanced for all relevant atoms.

(c) $CaCO_3(s) + H_3O^+(aq) \longrightarrow Ca^{2+}(aq) + CO_2(g) + H_2O(\ell)$

is NOT balanced for electrical charge—there is 1 mol of a 1+ species on the left and 1 mol of a 2+ species on the right. There are 1 mol Ca atoms in 1 mol of $CaCO_3$ on the left, and 1 mol Ca (in the form, Ca^{2+}) on the right; 1 mol C atoms in 1 mol of $CaCO_3$ on the left, and 1 mol C atoms in 1 mol of CO_2 on the right; and $3 + 1 = 4$ mol O atoms in 1 mol of $CaCO_3$ and 1 mol of H_3O^+ on the left, and $2 + 1 = 3$ mol O atoms in 1 mol of CO_2 and 1 mol of H_2O on the right.

Therefore, the reaction is balanced for Ca and C, but *not* for O atoms.

Exercise 5.5—Spontaneous Reaction Direction and Relative Stability

(a) A small amount of solid sodium chloride in water has a higher chemical potential than a dilute solution of sodium chloride. Chemical potential is lowered—the spontaneous direction—when the sodium chloride dissolves.

(b) A dilute solution of sodium chloride is more stable than a small amount of solid sodium chloride in water. Dissolution of the solid sodium chloride takes a less stable state to a more stable state.

Exercise 5.7—Relative Masses of Reactants and Products

Always start with the balanced chemical equation. In this case

$$C_6H_{12}O_6(s) + 6O_2(g) \longrightarrow 6CO_2(g) + 6H_2O(\ell)$$

We are given the mass of glucose. To use the balanced equation we need to work in amounts, in moles.

$$\text{Amount of glucose, } n(C_6H_{12}O_6) = \frac{m}{M} = \frac{25.0 \text{ g}}{180.16 \text{ g mol}^{-1}} = 0.139 \text{ mol}$$

The balanced equation tells us how much $O_2(g)$ is consumed, and how much $CO_2(g)$ and $H_2O(\ell)$ is produced. These amounts must be converted to masses to answer the question. The answer is summarized in the following amounts table:

Equation	$C_6H_{12}O_6(s)$	+ $6 O_2(g)$	→ $6 CO_2(g)$	+ $6 H_2O(\ell)$
Initial amount (Initial mass)	0.139 mol (25.0 g)	6 × 0.139 = 0.834 mol (mass of oxygen reacted = 0.834 mol × 32.00 g mol^{-1} = 26.7 g)	0 mol (0 g)	0 mol (0 g)
Change in amount (Change in mass)	−0.139 mol (−25.0 g)	−0. 834 mol (−26.7 g)	+ 6 × 0.139 = 0.834 mol (mass of CO_2 produced = 0.834 mol × 44.01 g mol^{-1} = 36.7 g)	+ 6 × 0.139 = 0.834 mol (mass of H_2O produced = 0.834 mol × 18.02 g mol^{-1} = 15.0 g)
Final amount (Final mass)	0 mol (0 g)	0 mol (0 g)	0.834 mol (36.7 g)	0.834 mol (15.0 g)

Check that the total mass of products formed is the same as the total mass of reactants that react.

Exercise 5.11—Limiting Reactant Situations

(a) The balanced equation is

$$CO(g) + 2 H_2(g) \longrightarrow CH_3OH(\ell)$$

$$\text{Initial amount of CO, } n(CO) = \frac{m}{M} = \frac{356 \text{ g}}{28.01 \text{ g mol}^{-1}} = 12.7 \text{ mol}$$

$$\text{Initial amount of } H_2, n(H_2) = \frac{m}{M} = \frac{65.0 \text{ g}}{2.02 \text{ g mol}^{-1}} = 32.2 \text{ mol}$$

To react all the initial CO requires 2 × 12.7 mol = 25.4 mol of H_2. Note that the 2× factor is the ratio of the stoichiometric coefficients of H_2 and CO. Clearly, there is more than enough initial H_2. So, CO is the **limiting reactant**—i.e., all of the initial CO can react and consume 25.4 mol of H_2. This leaves 32.2 – 25.4 = 6.8 mol of unreacted H_2.

(b) The amount of methanol that can be produced is 12.7 mol (CO and methanol have the same stoichiometric coefficient in the balanced equation). The associated mass of methanol is $m(CH_3OH) = (12.7 \text{ mol}) \times (32.04 \text{ g mol}^{-1}) = 407$ g.

(c) We have already established that 6.8 mol of H_2 is unreacted. This corresponds to the mass, $m(H_2) = (6.8 \text{ mol}) \times (2.02 \text{ g mol}^{-1}) = 14$ g.

The results are summarized in the following amounts table:

Equation	CO(g)	+ 2 H$_2$ (g)	→ CH$_3$OH(ℓ)
Initial amount (Initial mass)	12.7 mol (356 g)	32.2 mol (65 g)	0 mol (0 g)
Change in amount (Change in mass)	−12.7 mol (−356 g)	−2 × 12.7 = −25.4 mol (65 −14 = 51 g) ← not asked for	+12.7 mol (+407 g)
Final amount (Final mass)	0 mol (0 g)	6.8 mol (14 g)	12.7 mol (407 g)

Exercise 5.17—Percent Yield and Theoretical Yield

First, write the balanced chemical equation,

$$CH_3OH(\ell) \longrightarrow 2\,H_2(g) + CO(g)$$

The amount of methanol reacted is $(125 \text{ g}) / (32.04 \text{ g mol}^{-1}) = 3.90$ mol

Theoretical yield of hydrogen $= 2 \times 3.90 \text{ mol} \times 2.02 \text{ g mol}^{-1} = 15.8 \text{ g}$

The actual yield is 13.6 g of hydrogen. Thus,

$$\text{Percent yield of hydrogen} = \frac{13.6 \text{ g}}{15.8 \text{ g}} \times 100\% = 86.1\%$$

Exercise 5.19—Quantitative Analysis Based on Stoichiometry

The key to answering this question is to note that the Ni atoms in the precipitate all came from NiS in the original sample. Since we know the mass of precipitate, we can get the amount of precipitate which equals the amount of Ni^{2+} which is, in turn, equal to the amount of NiS. There are no stoichiometric ratio factors here because Ni^{2+} and $Ni(C_4H_7N_2O_2)_2$ have equal stoichiometric coefficients in the second equation, while NiS and Ni^{2+} have equal coefficients in the first equation.
First, we determine the amount of $Ni(C_4H_7N_2O_2)_2$.

Amount of $Ni(C_4H_7N_2O_2)_2 = m(Ni(C_4H_7N_2O_2)_2) / M(Ni(C_4H_7N_2O_2)_2) = (0.206 \text{ g}) /(288.92 \text{g mol}^{-1}) = 7.13 \times 10^{-4}$ mol

Next, we note that this is the amount of NiS in the sample of millerite. Therefore, the mass of NiS in the sample was

$$\text{mass} = n(\text{NiS}) \times M(\text{NiS}) = (7.13 \times 10^{-4}\ \text{mol}) \times (90.75\ \text{g mol}^{-1}) = 0.0647\ \text{g}$$

Finally, we compare the mass of NiS with the mass of the sample which clearly contains additional non-nickel compounds.

$$\text{Mass percent of NiS} = \frac{0.0647\ \cancel{g}}{0.468\ \cancel{g}} \times 100\% = 13.8\%$$

REVIEW QUESTIONS

Section 5.1: Don't Waste a Single Atom!

5.21

A "loss" of atoms refers to the fact that specific atoms from the reactants fail to end up in the molecules that are intended (target) products. The law of conservation of atoms is still satisfied since these atoms are "wasted" in by-products, not destroyed. However, the calculated atom economy for the formation of the wanted product will be less than 100% if atoms are lost to by-products.

5.23

(a) $CO_2(g) + 3\,H_2(g) \rightarrow CH_3OH(\ell) + H_2O(\ell)$
The atom efficiency is calculated to be:
The C atom efficiency = 100% × (number of C atoms in product)/(number of C atoms in all reactants) = 100% × 1/1 = 100%
The O atom efficiency = 100% × (number of O atoms in product)/(number of O atoms in all reactants) = 100% × 1/2 = 50%
The H atom efficiency = 100% × (number of H atoms in product)/(number of H atoms in all reactants) = 100% × 4/6 = 66.7%
Overall atom efficiency = 100% × $M(CH_3OH)$ / [($M(CH_3OH)$)) + $M(H_2O)$]
$$= 100\% \times 32.04 / [32.04) + 18.02]$$
$$= 64\%$$
The overall atom efficiency is only 64%, indicative that some atoms are "lost" to the $H_2O(\ell)$ by-product.
(b) The balanced equation for the combustion of methanol is
$2\,CH_3OH(\ell) + 3\,O_2(g) \rightarrow 2\,CO_2(g) + 4\,H_2O(g)$
The net result following the addition of reactions is as follows:
$$CO_2(g) + 3\,H_2(g) \rightarrow CH_3OH(\ell) + H_2O(\ell)$$
$$2\,CH_3OH(\ell) + 3\,O_2(g) \rightarrow 2\,CO_2(g) + 4\,H_2O(g)$$
net result: $CH_3OH(\ell) + 3\,O_2(g) + 3\,H_2(g) \rightarrow CO_2(g) + 5\,H_2O(g)$
The atom efficiency for the production of 1 mol of CO_2 is calculated to be:
The C atom efficiency = 100% × (number of C atoms in product)/(number of C atoms in all reactants) = 100% × 1/1 = 100%
The O atom efficiency = 100% × (number of O atoms in product)/(number of O atoms in all reactants) = 100% × 2/7 = 28.6%

Overall atom efficiency = 100% × $M(CO_2)$ / [($M(CO_2)$)) + (5 × $M(H_2O)$)]
= 100% × 44.01 / [(44.01) + (5 × 18.02)]
= 32.8%

The overall atom efficiency for this reaction is only 32.8%. Although methanol can be produced by a variety of methods (four examples provided in text), methanol produced via this method would not be an efficient fuel.

5.25

The effect of a greenhouse gas on the climate depends on three things: concentration, how strongly/where in the IR it absorbs energy, and its atmospheric lifetime. Consulting Table 4.1, while nitrous oxide has a seemingly low concentration when compared to CO_2, its atmospheric lifetime is 114 years (CO_2 is variable). The term "variable" fails to provide much information; however, one outstanding statistic is over 20 years. N_2O has a global warming potential 289 times that of CO_2. Over 100 years, that increases to 298 and finally diminishes to 153 times over 500 years. As well, N_2O strongly absorbs IR radiation over the large range that the Earth emits, particularly at ~2200 and 1300 cm^{-1} where nitrous oxide displays strong bands.

Section 5.2: Chemical Reaction, Chemical Change

5.27

Since all chemical reactions obey the law of conservation of mass, any **balanced** chemical equations will be correct.

5.29

(a) Yes (law of conservation of atoms)
(b) Yes (law of conservation of mass)
(c) No since bonds are broken and new bonds are made

Section 5.3: Chemical Equations: Chemical Accounting

5.31

(a) $C_6H_6(\ell) + 3H_2(g) \longrightarrow C_6H_{12}(\ell)$
is balanced for electrical charge—there are no charged species among reactants or products. There are 6 mol C atoms in 1 mol of C_6H_6 on the left, and 6 mol C atoms in 1 mol of C_6H_{12} on the right; and $1 × 6 + 3 × 2 = 12$ mol H atoms in 1 mol of C_6H_6 and 3 mol of H_2 on the left, and 12 mol H atoms in 1 mol of C_6H_{12} on the right.
Therefore, the reaction is balanced for all relevant atoms.

(b) $MnO_4^-(aq) + 5Fe^{2+}(aq) + 8H^+(aq) \longrightarrow Mn^{2+}(aq) + 5Fe^{3+}(aq) + 4H_2O(\ell)$
has charges, $-1 + 5 × (+2) + 8 × (+1) = +17$ on the left and $+2 + 5 × (+3) = +17$ on the right. It is balanced for electrical charge.
There are ...
1 mol Mn atoms on the left and right

5 mol Fe atoms on the left and right

8 mol H atoms on the left, and 4×2 H atoms on the right; and

4 mol O atoms on the left, and 4 mol O atoms on the right.

Therefore, the reaction is balanced for all relevant atoms.

(c) $Cu(OH_2)_6^{2+}(aq) + 4\,NH_3(aq) \longrightarrow Cu(NH_3)_4^{2+}(aq) + 6\,H_2O(\ell)$

has a charge +2 on both the left and right. It is balanced for electrical charge.

There are ...

1 mol Cu atoms on the left and right

4 mol N atoms on the left and right

$6 \times 2 + 4 \times 3 = 24$ mol H atoms on the left, and $4 \times 3 + 6 \times 2 = 24$ mol H atoms on the right; and 6 mol O atoms on the left and right.

Therefore, the reaction is balanced for all relevant atoms.

5.33

(a) This reaction is atom economical since the product contains all atoms from the reactants. It requires 3 mol H_2 to "fix" dinitrogen, thereby producing 2 mol NH_3. The reaction obeys the law of conservation of mass, so we can determine the amount of NH_3 produced by adding the masses of the reactants:

$$n(N_2) = m(N_2) / M(N_2) = \frac{100000\ \text{g}}{28.02\ \text{g mol}^{-1}} = 3568.9\ \text{mol}$$

$$n(H_2) = 3 \times n(N_2) = 3 \times 3568.9 = 10706.7\ \text{mol}$$

$$m(H_2) = M(H_2) \times n(H_2) = 2.02\ \text{g mol}^{-1} \times 10706.7\ \text{mol} = 21627.5\ \text{g} = 21.63\ \text{kg}$$

$$m(NH_3) = 100\ \text{kg} + 21.63\ \text{kg} = 121.63\ \text{kg}\ (122\ \text{kg significant figures})$$

(b) **(i)** The reactants ($N_2 : H_2$) must be present in the relative ratio of 1 : 3. **(ii)** The absolute amount of gases present in the vessel before the reaction occurred may be closer to equal ratios, as it is difficult to know with certainty the quantity of gas (unless injecting known volumes into a closed vessel). If they are equimolar, at the end of the reaction there would be unreacted N_2 in the vessel as all H_2 would be consumed.

(c)

$$N \equiv N + H_2\ \rightarrow\ HN{=}NH$$

$$HN{=}NH + H_2\ \rightarrow\ H_2N - NH_2$$

$$H_2N - NH_2 + H_2\ \rightarrow\ 2\,NH_3$$

(d) One cannot deduce the rate law expression from the balanced equation of the reaction. The expression of the rate law has to be determined from kinetic data

Section 5.4: Spontaneous Direction of Reaction

5.35

(a) $2\,H_2(g) + O_2(g) \longrightarrow 2\,H_2O(\ell)$ is the spontaneous direction of reaction. However, this is a very slow reaction at ambient temperatures. Reaction can be started by localized heating, e.g., with a spark. There is a barrier to this and many other reactions that require high temperatures. The hydrogen–oxygen reaction requires only a spark since it liberates heat raising neighbouring molecules to the required high temperature.

(b) Liquid water is more stable than a mixture of hydrogen and oxygen, each at 1 bar pressure.

Section 5.6: Masses of Reactants and Products: Stoichiometry

5.37

(a) $C_6H_6(\ell) + \dfrac{15}{2}O_2(g) \longrightarrow 6CO_2(g) + 3H_2O(g)$

Amount of benzene burned, $n = m(\text{benzene}) / M(\text{benzene}) = (16.04 \text{ g}) / (78.11 \text{ g mol}^{-1}) = 0.2053$ mol

Amount of $O_2(g)$ required for complete combustion of benzene $= n(O_2) = \dfrac{15}{2}n(C_6H_6)$

$= 7.5 \times 0.2053 \text{ mol} = 1.540$ mol

Mass of $O_2(g)$ required for complete combustion of benzene, $m = n(O_2) \times M(O_2) = (1.540$ mol$) \times (32.00 \text{ g mol}^{-1})$

$= 49.28$ g

(b) Total mass of products expected = total mass of reactants

We do not need to determine masses of products to answer the question.

Total mass of reactants $= 16.04 + 49.28 \text{ g} = 65.32$ g

5.39

$CH_3COCH_2CO_2H \longrightarrow CH_3COCH_3 + CO_2$

$m(\text{acetoacetic acid}) = 125 \text{ mg} = 0.125$ g

Amount of acetoacetic acid $(CH_3COCH_2COOH) = m(\text{acetoacetic acid}) / M(\text{acetoacetic acid})$

$= (0.125 \text{ g}) / (102.09 \text{ g mol}^{-1}) = 0.00122$ mol

Amount of acetone produced = amount of acetoacetic acid = 0.00122 mol

Mass of acetone produced $= n(\text{acetone}) \times M(\text{acetone}) = (0.00122 \text{ mol}) \times (58.08 \text{ g mol}^{-1}) = 0.07086 \text{ g} \quad = 70.86$ mg

5.41

(a) $2Fe(s) + 3Cl_2(g) \longrightarrow 2FeCl_3(s)$

$m(Fe) = 10.0$ g

Amount of Fe consumed, $n = m(Fe) / M(Fe)$

$= (10.0 \text{ g}) / (55.85 \text{ g mol}^{-1}) = 0.179$ mol

Amount of chlorine required to consume this much iron $= \dfrac{3}{2} \times$ amount of iron $= 1.5 \times 0.179$ mol $= 0.269$ mol

Mass of chlorine required, $m = n(\text{chlorine}) \times M(\text{chlorine}) = (0.269 \text{ mol}) \times (70.90 \text{ g mol}^{-1}) = 19.1$ g

Amount of iron(III) chloride produced, n = amount of iron = 0.179 mol

Mass of iron(III) chloride produced, $m = n(FeCl_3) \times M(FeCl_3) = (0.179 \text{ mol}) \times (162.20 \text{ g mol}^{-1}) = 29.0 \text{ g}$

(b) If only 18.5 g of $FeCl_3(s)$ is obtained from 10.0 g of iron, then
percent yield = 100% × (actual yield) / (theoretical yield) = 100% × (18.5 g) / (29.0 g) = 63.8%
The theoretical yield is the answer to part (a).

(c) If 10.0 g of each of iron and chlorine are combined, then there is not enough chlorine to consume the entire 10.0 g of iron. In part (a), we found 19.1 g of chlorine was required to consume 10.0 g of iron. In this case, chlorine is the limiting reactant.
Amount of chlorine consumed, $n = m(Cl_2) / M(Cl_2)$
$= (10.0 \text{ g}) / (70.90 \text{ g mol}^{-1}) = 0.141 \text{ mol}$

Amount of $FeCl_3$ produced, $n = \dfrac{2}{3} \times$ amount of chlorine $= 0.6667 \times 0.141 \text{ mol} = 0.0940 \text{ mol}$

Mass of iron(III) chloride produced, $m = n(FeCl_3) \times M(FeCl_3) = (0.0940 \text{ mol}) \times (162.20 \text{ g mol}^{-1}) = 15.2 \text{ g}$
= theoretical yield of iron(III) chloride

5.43

(a) $TiCl_4(\ell) + 2 H_2O(\ell) \longrightarrow TiO_2(s) + 4 HCl(g)$
Mass of $TiCl_4(\ell)$ consumed, $m =$ volume × density $= (14.0 \text{ mL}) \times (1.73 \text{ g mL}^{-1}) = 24.2 \text{ g}$
Amount of $TiCl_4$ consumed, $n = m(TiCl_4) / M(TiCl_4) = (24.2 \text{ g}) / (189.67 \text{ g mol}^{-1}) = 0.128 \text{ mol}$
Amount of water required, $n = 2 \times$ (amount of $TiCl_4$ consumed) $= 0.256 \text{ mol}$
Mass of water required for complete reaction, $m = n(H_2O) \times M(H_2O) = (0.256 \text{ mol}) \times (18.02 \text{ g mol}^{-1}) = 4.61 \text{ g}$

(b) Amount of TiO_2 expected, $n =$ amount of $TiCl_4$ consumed $= 0.128 \text{ mol}$
Mass of TiO_2 expected, $m = n(TiO_2) \times M(TiO_2) = (0.128 \text{ mol}) \times (79.87 \text{ g mol}^{-1}) = 10.2 \text{ g}$
Amount of HCl expected, $n = 4 \times$ (amount of $TiCl_4$ consumed) $= 0.512 \text{ mol}$
Mass of HCl expected, $m = n(HCl) \times M(HCl) = (0.512 \text{ mol}) \times (36.46 \text{ g mol}^{-1}) = 18.7 \text{ g}$

Section 5.7: Reactions Limited by the Amount of One Reactant

5.45

(a) The balanced equation for the reaction is
$CH_4(g) + H_2O(g) \longrightarrow CO(g) + 3 H_2(g)$

Initial amount of CH_4, $n(CH_4) = \dfrac{m}{M} = \dfrac{995 \text{ g}}{16.04 \text{ g mol}^{-1}} = 62.0 \text{ mol}$

Initial amount of H_2O, $n(H_2O) = \dfrac{m}{M} = \dfrac{2510 \text{ g}}{18.02 \text{ g mol}^{-1}} = 139.3 \text{ mol}$

ratio $\dfrac{n(H_2O)}{n(CH_4)} = \dfrac{1 \text{ mol } H_2O}{1 \text{ mol } CH_4}$

To react all the initial CH_4 requires 62.0 mol H_2O. Clearly there is more than enough H_2O initially, therefore CH_4 is the **limiting reactant**.

(b) The maximum amount of H_2 that can be produced is 62.0 mol × 3 = 186.0 mol (from the balanced equation 1 mol CH_4 : 3 mol H_2)

$m(H_2) = M(H_2) \times n(H_2) = (2.02 \text{ g mol}^{-1}) \times 186.0 \text{ mol} = 376 \text{ g } H_2(g)$

(c) unreacted H_2O = 139.3 mol – 62.0 mol = 77.3 mol

$m(H_2O) = M(H_2O) \times n(H_2O) = 18.02 \text{ g mol}^{-1} \times 77.3 \text{ mol} = 1393 \text{ g}$

Section 5.8: Theoretical Yield and Percent Yield

5.47

(a) The theoretical yield of $Cu(NH_3)_4SO_4$ is:

$$n(CuSO_4) = \frac{m(CuSO_4)}{M(CuSO_4)} = \frac{10\,g}{159.61\,g\,mol^{-1}} = 0.063\,mol$$

$$n(CuSO_4) = n(Cu(NH_3)_4SO_4) = 0.063\,mol$$

$$m(Cu(NH_3)_4SO_4) = n(Cu(NH_3)_4SO_4) \times M(Cu(NH_3)_4SO_4) = 0.063\,mol \times 227.75\,g\,mol^{-1} = 14.4\,g$$

(b) $\%\,Yield = 100\% \times \dfrac{\text{Actual Yield}}{\text{Theoretical Yield}} = 100\% \times \dfrac{12.6\,g}{14.4\,g} = 87.5\%$

Section 5.9: Stoichiometry and Chemical Analysis

5.49

Mass of $Cu_2S(s)$, m = 1.00 tonne = 1000 kg = 1.00×10^6 g
Amount of Cu_2S, n = $m(Cu_2S) / M(Cu_2S) = (1.00 \times 10^6 \text{ g}) / (159.16 \text{ g mol}^{-1}) = 6.29 \times 10^3$ mol
Amount of H_2SO_4 that can be produced, n = amount of Cu_2S = 6.28×10^3 mol
Each S atom in $Cu_2S(s)$ leads to one molecule of H_2SO_4
Mass of H_2SO_4 that can be produced, m = $n(H_2SO_4) \times M(H_2SO_4) = (6.28 \times 10^3 \text{ mol}) \times (98.08 \text{ g mol}^{-1}) = 6.16 \times 10^5 \text{ g} = 616$ kg.

5.51

Here is the balanced reaction:
$x\,TiO_2(s) + (2x - y)\,H_2(g) \rightarrow Ti_xO_y(s) + (2x - y)\,H_2O(\ell)$
Amount of TiO_2, n = $m(TiO_2) / M(TiO_2) = (1.598 \text{ g}) / (79.87 \text{ g mol}^{-1}) = 0.02001$ mol

Amount of Ti_xO_y, $n = \dfrac{1}{x} \times (1.598 \text{ g}) / (79.87 \text{ g mol}^{-1}) = \dfrac{1}{x} \times (0.02001 \text{ mol})$

$\qquad = m(Ti_xO_y) / M(Ti_xO_y) = (1.438 \text{ g}) / M(Ti_xO_y)$

Therefore, $\dfrac{M\left(Ti_xO_y\right)}{x} = (1.438\text{ g}) /(0.02001\text{ mol}) = 71.86\text{ g mol}^{-1}$

But, we also have $\dfrac{M\left(Ti_xO_y\right)}{x} = M(\text{Ti}) + \dfrac{y}{x} \times M(\text{O}) = 47.867 + \dfrac{y}{x} \times (16.000)$.

So $\dfrac{y}{x} \times (16.000\text{ g mol}^{-1}) = 71.86 - 47.867\text{ g mol}^{-1} = 23.99\text{ g mol}^{-1}$

and $\dfrac{y}{x} = \dfrac{23.99}{16.000} = 1.499 \approx 1.5$

Therefore, $Ti_xO_y = TiO_{1.5}$ or Ti_2O_3.

5.53

Mass of AgCl, $m = 0.1804$ g
Amount of AgCl, $n = m(\text{AgCl}) / M(\text{AgCl}) = (0.1840\text{ g}) / (143.35\text{ g mol}^{-1}) = 0.001284$ mol
2 mol of Cl atoms in 1 mol 2,4-D (2,4-dichlorophenoxyacetic acid) leads to 2 mol of AgCl (one AgCl for each Cl)

Amount of 2,4-D (2,4-dichlorophenoxyacetic acid), $C_8H_6Cl_2O_3 = \dfrac{1}{2}$ (amount of AgCl) =

6.42×10^{-4} mol
Mass of 2,4-D (2,4-dichlorophenoxyacetic acid) in sample, $m = n(C_8H_6Cl_2O_3) \times M(C_8H_6Cl_2O_3)$
$= (6.42 \times 10^{-4}\text{ mol}) \times (221.03\text{ g mol}^{-1}) = 0.142$ g
Mass percent of 2,4-D in sample = $100\% \times m(\text{2,4-D}) / m(\text{sample}) = 100\% \times (0.142\text{ g}) / (1.236\text{ g})$
$= 11.5\%$

5.55

$$CaCO_3(s) \longrightarrow CaO(s) + CO_2(g)$$

Mass $CaCO_3$ in limestone, $m = \dfrac{95\%}{100\%} \times 125\text{ kg} = 119\text{ kg} = 1.19 \times 10^5$ g

Amount of $CaCO_3$, $n = m(CaCO_3) / M(CaCO_3) = (1.19 \times 10^5\text{ g}) / (100.09\text{ g mol}^{-1})$
$= 1.19 \times 10^3$ mol
1 mol of lime, CaO(s), is produced for every 1 mol of $CaCO_3$
    Amount of lime produced, n = amount of $CaCO_3 = 1.19 \times 10^3$ mol
Mass of CaO, $m = n(\text{CaO}) \times M(\text{CaO}) = (1.19 \times 10^3\text{ mol}) \times (56.08\text{ g mol}^{-1}) = 6.67 \times 10^4$ g
$= 66.7$ kg

5.57

Mass of copper in product metal, $m = \dfrac{89.5\%}{100\%} \times 75.4\text{ g} = 67.5$ g

Amount of copper, $n = m(\text{Cu}) / M(\text{Cu}) = (67.5\text{ g}) / (63.55\text{ g mol}^{-1}) = 1.06$ mol
This copper came from 100.0 g of ore, 89.0 g of which consists of CuS and Cu_2S (because there is an 11% impurity). 67.5 g of the 89.0 g come from copper atoms (the copper recovered). Therefore, 89.0 − 67.5 g = 21.5 g of the 89.0 g come from sulfur atoms. This corresponds to

amount of sulfur, $n = m(S) / M(S) = (21.5 \text{ g}) / (32.06 \text{ g mol}^{-1}) = 0.671$ mol
If we let the amount of CuS be a and the amount of Cu$_2$S be b, then we have

Amount of Cu $= a + 2b = 1.06$ mol   (1)
Amount of S $= a + b = 0.671$ mol   (2)

Therefore – subtract (2) from (1),

$b = 1.06 - 0.671$ mol $= 0.389$ mol

and – substitute back into (2)

$a = 0.671 - 0.389$ mol $= 0.282$ mol

Mass of CuS in ore $= a \times M(\text{CuS}) = 0.282$ mol $\times 95.61$ g mol^{-1} $= 27.0$ g
Mass of Cu$_2$S in ore $= b \times M(\text{Cu}_2\text{S}) = 0.389$ mol $\times 159.16$ g mol^{-1} $= 61.9$ g
Mass percent CuS in ore $= 100\% \times 27.0 / 89.0 = 30.3\% \approx 30\%$
Mass percent Cu$_2$S in ore $= 100\% \times 61.9 / 89.0 = 69.6\% \approx 70\%$

Section 5.10: Atom Economy, Atom Efficiency

5.59

(a)

The atom efficiency for all atoms—C, H, and Br—is clearly 100%. There are no waste products. Overall atom efficiency is also 100% and the E-factor = 0.

(b)

Atom efficiency is 100% for carbon and bromine. We see this by noting that C and Br do not appear among the waste products—H$_2$O is the only waste product. For oxygen and hydrogen, we have
O atom efficiency = 100% × (O atoms in desired product)/(O atoms in all reactants) = 100% × 0/1 = 0%

O atoms are only waste atoms

H atom efficiency = 100% × (H atoms in desired product)/(H atoms in all reactants) = 100% × 9/11 = 81.8%
The overall atom efficiency = 100% × M(1-bromobutane) / (M(1-bromobutane) + M(H$_2$O))
= 100% × (137.01 g mol^{-1}) / (137.01 g mol^{-1} + 18.02 g mol^{-1})
= 88.4%

$$\text{E-factor} = \frac{18.02 \text{ g mol}^{-1}}{137.01 \text{ g mol}^{-1}} = 0.131$$

The E-factor is also given by (100 − 88.4)/ 88.4 = 0.131

(c)

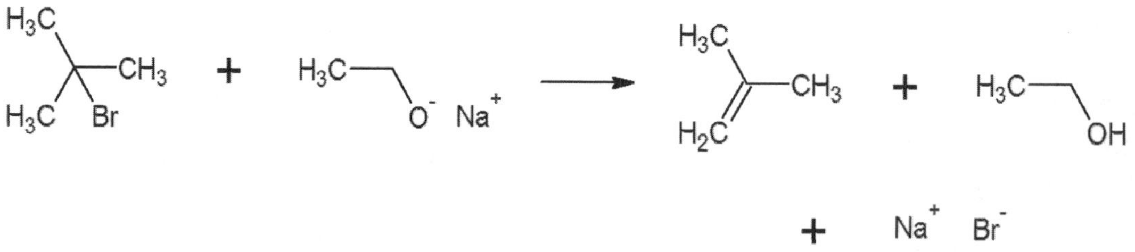

Viewing 2-methylpropene as the desired product, we compute atom efficiencies as follows:
Na, O, and Br atom efficiencies = 0%
Na, O, and Br atoms are only waste atoms
C atom efficiency = 100% × (C atoms in desired product)/(C atoms in all reactants) = 100% × 4/6 = 66.7%
H atom efficiency = 100% × (H atoms in desired product)/(H atoms in all reactants) = 100% × 8/14 = 57.1%
The overall atom efficiency = 100% × M(2-methylpropene) / ((M(2-methylpropene) M(ethanol) + M(sodium bromide))
= 100% × (56.10 g mol^{-1}) / (56.10 g mol$^-$1 + 46.07 mol^{-1} + 102.89 g mol^{-1})
= 27.4%

$$E\text{-factor} = \frac{(46.07 + 102.89) \text{ g mol}^{-1}}{56.10 \text{ g mol}^{-1}} = 2.66$$

The E-factor is also given by (100 − 27.4)/ 27.4 = 2.65

(d)

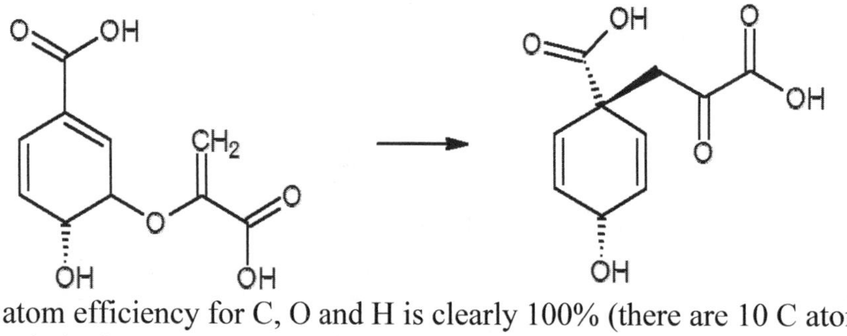

The atom efficiency for C, O and H is clearly 100% (there are 10 C atoms, 6 O atoms and 10 H atoms in both reactant and product). The overall atom efficiency is clearly also 100%.
The E-factor = (100 − 100)/100 = 0

SUMMARY AND CONCEPTUAL QUESTIONS

5.61

Given the balanced equation,
$I_2(g) + 3 Cl_2(g) \longrightarrow 2 ICl_3(g)$,
starting with

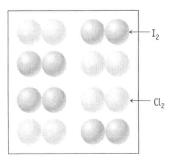

we end up with

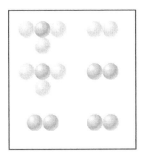

Cl_2 is the limiting reactant because some I_2 molecules remain unreacted.

5.63

(a) $2\,Fe(s) + 3\,Cl_2(g) \longrightarrow 2\,FeCl_3(s)$

Mass of Fe, $m = 1.54$ g

Amount of Fe, $n = m(Fe)\,/\,M(Fe) = (1.54\ g)\,/\,(55.85\ g\ mol^{-1}) = 0.0276$ mol

Amount of Cl_2 required for complete reaction, $n = \dfrac{3}{2} \times$ amount of Fe $= 0.0414$ mol

Mass of Cl_2 required, $m = n(Cl_2) \times M(Cl_2) = (0.0414\ mol) \times (70.90\ g\ mol^{-1}) = 2.94$ g

Amount of $FeCl_3$ produced, $n =$ amount of Fe $= 0.0276$ mol

Mass of $FeCl_3$ produced, $m = n(FeCl_3) \times M(FeCl_3) = (0.0276\ mol) \times (162.20\ g\ mol^{-1}) = 4.48$ g

(b) $FeCl_3(aq) + 3\,NaOH(aq) \longrightarrow Fe(OH)_3(s) + 3\,NaCl(aq)$

Mass of iron(III) chloride, $m = 2.0$ g

Amount of iron(III) chloride, $n = m(FeCl_3)\,/\,M(FeCl_3) = (2.0\ g)\,/\,(162.20\ g\ mol^{-1}) = 0.012$ mol

Mass of NaOH, $m = 4.0$ g

Amount of NaOH, $n = m(NaOH)\,/\,M(NaOH) = (4.0\ g)\,/\,(40.00\ g\ mol^{-1}) = 0.10$ mol

Since it takes amount of NaOH, $n = 3 \times$ amount of $FeCl_3 = 0.036$ mol to react all the iron (III) chloride with NaOH, there is clearly enough NaOH – so $FeCl_3$ is the limiting reagent.

Amount of $Fe(OH)_3$ produced, $n =$ initial amount of $FeCl_3 = 0.012$ mol

Mass of $Fe(OH)_3$ produced, $m = n(Fe(OH)_3) \times M(Fe(OH)_3) = (0.012\ mol) \times (106.87\ g\ mol^{-1}) = 1.3$ g

5.65

Only bromine, the brown liquid, remains in the beaker. Near the bottom of the beaker, it appears that unreacted Al foil remains. However, a closer look reveals that it is just variations in the thickness of the white powder residue. Aluminum, Al(s), is therefore the limiting reactant.

CHAPTER 6
Chemistry of Water, Chemistry in Water

IN-CHAPTER EXERCISES

Exercise 6.1—Temperature Dependence of Water Density
According to Figure 6.3 (from the text), the densities of ice and liquid water (presumably at 1 bar) at the temperatures indicated are tabulated as follows:

Temperature / °C	Density of Ice / g mL^{-1}	Density of Liquid Water / g mL^{-1}
−5	0.9175	
0	0.9169	0.99987
4		1.00000
10		0.99974

The volume of 100.0 g of ice at −5°C is
    100.0 g / 0.9175 g mL^{-1} = 109.0 mL
The other volumes are similarly computed—results are tabulated as follows:

Temperature / °C	Volume of 100.0 g of Ice / mL	Volume of 100.0 g of Liquid Water / mL
−5	108.99	
0	109.06	100.01
4		100.00
10		100.03

These data are displayed graphically as follows:

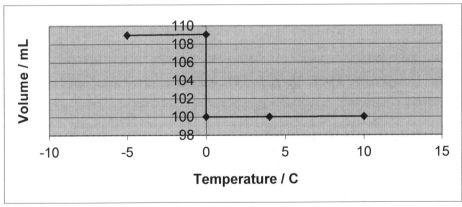

Exercise 6.3—Enthalpy Change of Vaporization

(a) Amount of methanol, $n(CH_3OH) = \dfrac{m}{M} = \dfrac{1.00 \times 10^3 \, \cancel{g}}{32.04 \, \cancel{g} \, mol^{-1}} = 31.2 \, mol$

Energy absorbed $= n \times \Delta_{vap}H = 31.2 \, \cancel{mol} \times 35.2 \, kJ \, \cancel{mol^{-1}} = 1.10 \times 10^3 \, kJ$

(b) To vaporize the same amount of ammonia requires

Energy absorbed $= n \times \Delta_{vap}H = 31.2 \, \cancel{mol} \times 23.3 \, kJ \, \cancel{mol^{-1}} = 727 \, kJ$

Exercise 6.5—Boiling point and Molar Enthalpy Change of Vaporization

The order of normal boiling points is the same as the order of enthalpy changes of vaporization. This is what we expect because both boiling point and enthalpy change of vaporization increase with increasing strength of intermolecular forces.

Exercise 6.7—Intermolecular Forces and Intramolecular Forces

(a) Water molecules in the cold morning air settle on even colder objects on the ground such as blades of grass. Here, the intermolecular force of attraction between water molecules and other water molecules or the molecules on the surface of the blades of grass is sufficient to allow water molecules to stick together on the blades of grass.
(b) When a piece of paper burns, carbon-carbon and carbon-hydrogen bonds are broken while new bonds with oxygen are formed. Intramolecular forces are the losing (in the breaking of reactant bonds) and winning forces (in the making of bonds) at work when paper burns.
(c) Intermolecular forces holding water molecules together, and next to clothes fibre molecules, are overcome when wet clothes are hung out to dry.
(d) Intermolecular forces cause water molecules to arrange themselves into a lattice of lowest potential energy. The freezer takes away the kinetic energy thereby liberated, leaving water in its crystalline, solid form—ice.

Exercise 6.9—Bond Polarities

(a) $\chi(C) = 2.5$, $\chi(H) = 2.2$, and $\chi(O) = 3.5$
The C=O and C–H bonds are polar, while C–C and C=C are non-polar.
(b) The C=O bond is the most polar, with O at the negative end.

Exercise 6.11—Bond Polarity and Molecular Polarity

CS_2 and CO_2 are non-polar because of their linear shape. The two bond dipoles in these molecules exactly cancel out, leaving NO net dipole moment.
SO_2 and H_2O are polar because of their bent shape. The two bond dipole vectors only partially cancel out, leaving a NET dipole moment.

Exercise 6.15—Dipole–Dipole Forces

(a) Bent SO_2 has a net dipole. So, there are dipole–dipole forces between neighbouring molecules.

(b) Linear CO_2 has no net dipole. There are no dipole–dipole forces between these molecules— only dispersion forces.

(c) HCl has just one polar bond, and so has a dipole moment. There are dipole–dipole forces between neighbouring molecules.

Exercise 6.17—Molecular Structure and Hydrogen Bonding

$$CH_3-O-H \cdots\cdots O-H$$
$$|$$
$$CH_3$$

shows the hydrogen bonding between neighbouring methanol molecules as they move past each other.

Exercise 6.19—Dispersion Forces

(a) Fluorine has lower melting and boiling points than bromine. Clearly, bromine has the stronger dispersion forces. This is because bromine atoms are bigger with more electrons. The bromine molecule charge distribution has larger fluctuations, and its neighbours are more easily polarized.

(b) Ethane has lower melting and boiling points than butane. Butane has the stronger dispersion forces. It is simply a bigger molecule—more C and H atoms—with more electrons. The butane molecule charge distribution has larger fluctuations, and its neighbours are more easily polarized.

Exercise 6.23—Intermolecular Forces and Properties

(a) In hexane, dispersion forces are the only intermolecular forces affecting vapour pressure.

(b) In water, hydrogen bonds (special dipole–dipole bonds) and dispersion forces are at work.

(c) In carbon tetrachloride, dispersion forces are the only intermolecular forces at work.

Exercise 6.27—Aquated Ions

$Mg^{2+}(aq)$ and $Br^-(aq)$ are the main species present in solution when some magnesium bromide, $MgBr_2(s)$, is dissolved in water. These species can be represented in a similar way to the unnamed aquated cation and anion in Figure 6.26 of the textbook. A more schematic representation is provided as follows:

Exercise 6.29—Solubilities of Molecular Substances

(a) Ammonia, $NH_3(g)$, is soluble in water because water molecules can make strong hydrogen bonds with ammonia molecules The –O–H groups in water interact with the and N–H groups in ammonia via hydrogen bonding.

(b) $HCl(g)$ dissolves in water as it dissociates into H^+ ion and Cl^- ions. The resulting ions are aquated (i.e., surrounded by a number of water molecules). The interaction of the ions with polar water molecules result in ion–dipole intermolecular force of attraction, which is favourable.

(c) Iodine, $I_2(s)$, is non-polar and insoluble in water. Only dispersion-like forces can operate between water and I_2. This interaction is weaker than the dispersion force between I_2 molecules and the hydrogen bonds in water.

(d) Octane, $C_8H_{18}(\ell)$, is non-polar and insoluble in water. Only dispersion-like forces can be made between water and octane. This interaction is weaker than the dispersion force between octane molecules and the hydrogen bonds between water molecules.

Exercise 6.31—Polar and Non-polar Parts of Molecules

Butan-1-ol, $CH_3CH_2CH_2CH_2OH(\ell)$, should be more soluble in hexane than butane -2,4-diol, $HOCH_2CH_2CH_2CH_2OH(\ell)$. The latter compound has two hydroxyl (–OH) groups vs. only one for the former. Like the –OH of water, these hydroxyls do not spontaneously mix with hexane— that is, not without the help of the favourable dispersion interactions of the alkyl chains. The energy of interaction between butan-1-ol molecules is too great to allow interpenetration by hexane molecules (in the absence of strong forces of attraction between butan-1-ol molecules and hexane molecules).

Exercise 6.37—Oxidation-Reduction Reactions

$Fe^{2+}(aq)$ ions are oxidized to $Fe^{3+}(aq)$ ions. The oxidizing agent is $MnO_4^-(aq)$ ions. $MnO_4^-(aq)$ ions are reduced to $Mn^{2+}(aq)$ ions. $Fe^{2+}(aq)$ ions are the reducing agent.

Exercise 6.39—Acids and Bases

(a) $H_3O^+(aq) + OH^-(aq) \longrightarrow 2\ H_2O(\ell)$

$Li^+(aq)$ and $Br^-(aq)$ ions are spectator ions—they do not appear in the net reaction.

(b) $H_2CO_3(aq) + OH^-(aq) \longrightarrow HCO_3^-(aq) + H_2O(\ell)$

and $HCO_3^-(aq) + OH^-(aq) \longrightarrow CO_3^{2-}(aq) + H_2O(\ell)$

$K^+(aq)$ ion is the spectator ion.

(c) $H_3O^+(aq) + CH_3NH_2(aq) \longrightarrow H_2O(\ell) + CH_3NH_3^+(aq)$

$NO_3^-(aq)$ is the spectator ion.

(d) $H_3Citrate(aq) + OH^-(aq) \longrightarrow H_2Citrate^-(aq) + H_2O(\ell)$

$H_2Citrate^-(aq) + OH^-(aq) \longrightarrow HCitrate^{2-}(aq) + H_2O(\ell)$

and $HCitrate^{2-}(aq) + OH^-(aq) \longrightarrow Citrate^{3-}(aq) + H_2O(\ell)$

$Na^+(aq)$ ion is the spectator ion.

Exercise 6.41—Aquation as Complexation

(a) $[Fe(OH_2)_6]^{2+}$
(b) $[Mn(OH_2)_6]^{2+}$
(c) $[Al(OH_2)_6]^{3+}$
(d) $[Cr(OH_2)_6]^{3+}$

These complex ions can be pictorially represented as follows.

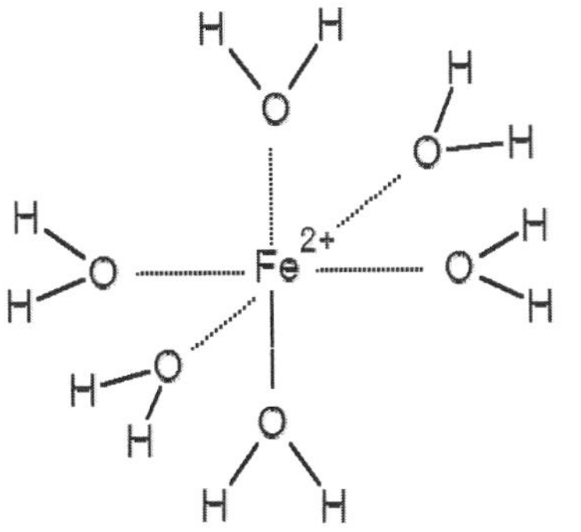

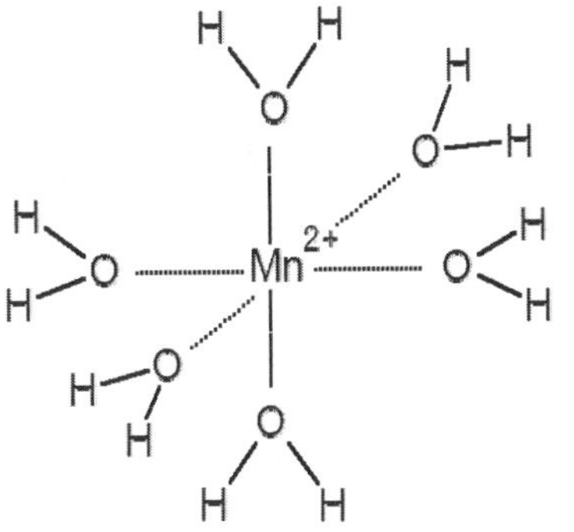

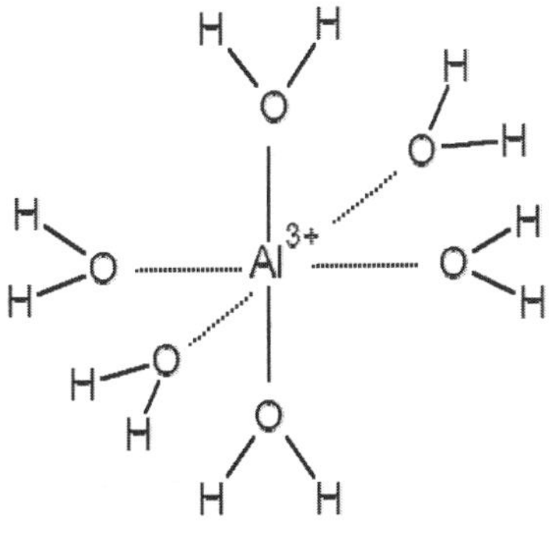

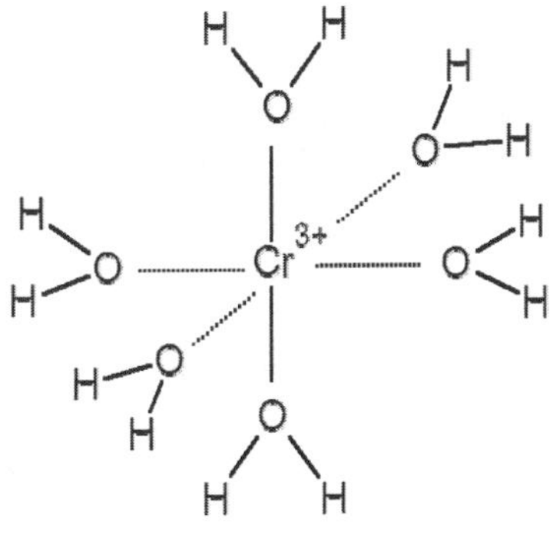

Exercise 6.43—Solution Concentration

Amount of Na_2CO_3, $n = \dfrac{m}{M} = \dfrac{25.3 \text{ g}}{105.99 \text{ g mol}^{-1}} = 0.239$ mol

Molarity of sodium carbonate solution

$$= c = \frac{\text{amount of solute (mol)}}{\text{volume (L)}} = \frac{0.239\,\text{mol}}{0.500\,\text{L}} = 0.478\,\text{mol}\,\text{L}^{-1}$$

REVIEW QUESTIONS

Section 6.1: Arsenic Ain't Arsenic

6.51

Discussions of arsenic toxicity do not refer to the element itself, rather arsenic-containing species. Arsenic compounds can be characterized into two categories: inorganic and organic.

6.53

Chemical species that are "bioavailable" or compounds that can readily be taken up into plants cells make them harmful to plants.
(a) If the soil contains significant concentrations of citrate/tannate (aq) anions as well as relatively high concentrations of "copper" (Cu^{2+}(aq) ions), then they will bind together and become complex ions that are no longer bioavailable and plants will not absorb them.
(b) If the soil contains relatively high concentrations of "copper" but very high concentrations of citrate/tannate (aq) anions, there will not be enough Cu^{2+}(aq) ions to quench, and the residual citrate/tannate (aq) anions become bioavailable.

6.55

Biological "species" is the basic form of classification for a group of organisms. Chemical "species" refers to an aggregate of atoms, charged or uncharged.

Section 6.2: The Remarkable Properties of Water

6.57

Water has a lower vapour pressure than hexane, 3.17 vs. 20.2 kPa at 25°C. It takes a higher temperature for the vapour pressure of water to reach 1 atm, than it does for hexane—i.e., water has a higher boiling point.

Section 6.3: Intermolecular Forces

6.59

(a) Bonds are broken and formed when iron rusts—these are intramolecular forces.
(b) Long-stranded polymer molecules are pulled together when a rubber band is stretched—chemical bonds are not made or broken. Intermolecular forces are at play here.
(c) Ultraviolet light, in the stratosphere, breaks O–O bonds to form O atoms. Intramolecular forces are overcome.

(d) Mothballs gradually "disappear" by sublimation wherein a solid material passes directly into the gas phase. Here, intermolecular forces holding the molecules in the solid phase are overcome. There is a steady outflux of naphthalene—the mothball material—until the mothball has disappeared.

6.61

(a) $C\rightarrow O > C\rightarrow N$
  i.e., $C\rightarrow O$ is more polar than $C\rightarrow N$
(b) $P\rightarrow Cl > P\rightarrow Br$
(c) $B\rightarrow O > B\rightarrow S$
(d) $B\rightarrow F > B\rightarrow I$

6.63

The dipole moments are ordered as HF > HCl > HBr > HI. In each case, the dipole points toward the halogen. While the dipole moment depends on the charge separation which is indicative of difference in electronegativity, it also depends on the bond distance. Note that even though the bond distance increases in this series, the dipole moments decrease. This indicates a decreasing sequence of differences in electronegativity which corresponds to a decreasing sequence of halogen electronegativities—F > Cl > Br > I.

6.65

(a) Linear $BeCl_2$ is non-polar because of symmetry—the bond dipoles cancel.
(b) Trigonal planar HBF_2 is polar because one of the three bonds is different—the bond dipoles do not cancel exactly. The negative end is at the two F atoms ($\chi = 4.0$).
(c) Tetrahedral CH_3Cl, is polar because one of the four bonds is different—the bond dipoles do not cancel. The negative end is at the Cl atom. Note that in this case the three H–C bond dipoles reinforce the C–Cl bond dipole.
(d) Trigonal pyramidal SO_3 is non-polar because of symmetry—the bond dipoles cancel.

6.67

(a) There are NO dipole–dipole forces between hexane molecules.
(b) There are dipole–dipole forces between H_2S molecules.
(c) There are NO dipole–dipole forces between methane molecules.

6.69

(b)
i.e., hydrogen bonding operates in only formic acid, HCOOH. In H_2Se, HI, CH_3COCH_3, hydrogen is bonded to a more electronegative atom, but not one of N, O, or F. Hydrogen bonds form only between these small (second period), electronegative atoms—the concentration of charge is important for the effect.

6.71

Kr atoms are more polarizable than Ne atoms because Kr atoms are bigger, with more electrons.

6.73

(a) When O_2 vaporizes, dispersion forces are overcome.
(b) When mercury vaporizes, metallic bonds are overcome.
(c) When CH_3I vaporizes, dipole–dipole and dispersion forces are overcome.
(d) When CH_3CH_2OH vaporizes, H-bonds (special dipole–dipole interactions) and dispersion forces are overcome.

6.75

In order of increasing force of attraction, we have
(a) < (b) < (c) < (d)
i.e., Ne < CH_4 < CO < CCl_4
The series simply reflects the size of dispersion forces—they increase with increasing number of electrons. The dipole-dipole forces in CO are not as big as the dispersion forces in CCl_4.

6.77

$\chi(O) > \chi(C) > \chi(H)$ (3.5, 2.5, 2.2, respectively).
For C–O bonds, $\Delta \chi = 3.5–2.5 = 1$. For C–H bonds, $\Delta \chi = 2.5–2.2 = 0.3$. For O–H bonds, $\Delta \chi = 3.5–2.2 = 1.3$. The greater the difference in electronegativities, the bigger the charge separation, the more polar the compound is.

6.79

"Dispersion" is the act of being dispersed, or distribution over a wide area. In dispersion forces, the weak intermolecular attraction forces are spread or distributed over the entirety of the non-polar/polar substances.

6.81

(a) There are 8 covalent bonds + 1 coordinate covalent bond (H^+ in hydronium ion) and **(b)** 3 hydrogen bonds.

Section 6.4: Explaining the Properties of Water

6.83

(a) The stronger are the intermolecular forces in a liquid, its normal boiling point is *higher*.
(b) The weaker are the intermolecular forces in a liquid, its equilibrium vapour pressure at a specified temperature is *higher*.
(c) The smaller the volume of water in a sealed flask, its equilibrium vapour pressure is *unchanged*.

(d) The higher the temperature of a liquid, its equilibrium vapour pressure is *higher*.

(e) The more volatile a liquid, its boiling point is *smaller*.

(f) The higher the boiling point of a liquid, its enthalpy change of vaporization is *higher*.

6.85

Propan-1-ol ($CH_3CH_2CH_2OH$) has a higher boiling point than methyl ethyl ether ($CH_3CH_2OCH_3$), a compound with the same empirical formula, because propan-1-ol molecules can form hydrogen bonds, whereas methyl ethyl ether cannot. Both molecules have dipole–dipole forces, but they are the especially strong hydrogen bonds in the case of propan-1-ol.

6.87

With respect to boiling point, we have
(a) $O_2 > N_2$ because the oxygen molecules are bigger—bigger dispersion forces
(b) $SO_2 > CO_2$ because SO_2 is bent—it has a net dipole moment, and dipole–dipole intermolecular forces
(c) HF > HI because of hydrogen bonding in the case of HF
(d) $SiH_4 < GeH_4$ because GeH_4 is bigger

6.89

The physical properties likely to be affected by hydrogen bonding are boiling/freezing point, vapour pressure, solubility, and viscosity.

6.91

(a) Water has a higher viscosity than hexane, in spite of its smaller dispersion forces, because of its strong hydrogen bonds.

(b) Glycerol (propane-1,2,3-triol, $HOCH_2CHOHCH_2OH$) is even more viscous than water because it has three O–H groups, and can form more hydrogen bonds than water.

6.93

The melting point of fumaric acid (287°C) is much higher than that of maleic acid (131°C) even though these substances are just *cis* and *trans* isomers.

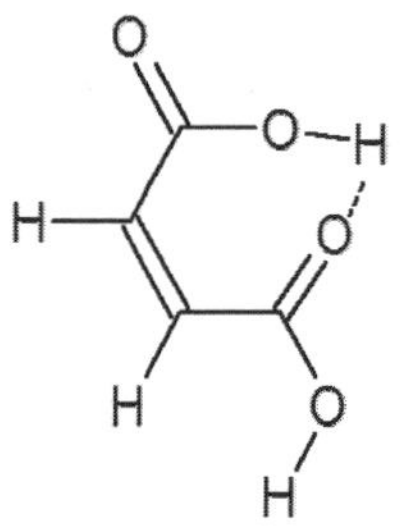

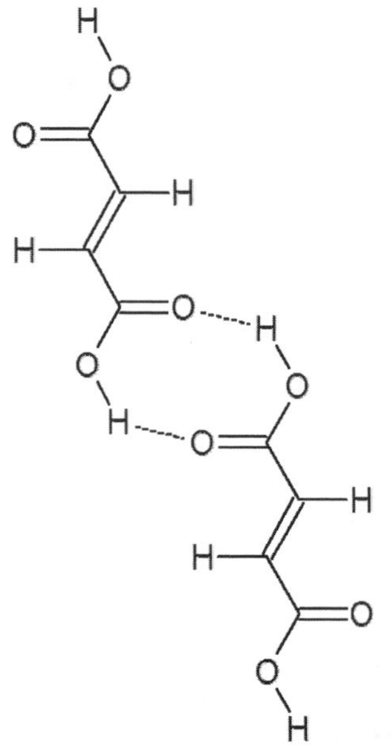

Maleic acid, on the left, makes a strong intramolecular hydrogen bond—this reduces opportunities for intermolecular hydrogen bonds, as an O and H are already hydrogen bonding. Strong intermolecular pairs of hydrogen bonds are formed between adjacent fumaric acid molecules.

Section 6.5: Water as a Solvent

6.95

"Aqua-" refers to anything that is water related. In inorganic chemistry, H_2O is referred to as an "aqua" ligand when it surrounds a metal. Therefore, anytime water surrounds an ion, it is termed aquation.

6.97

Solvated (aquated) K^+(aq) ions and SO_4^{2-}(aq) ions are the main species present when K_2SO_4(s) is dissolved in water. Here we see six water molecules solvating each of the ions.

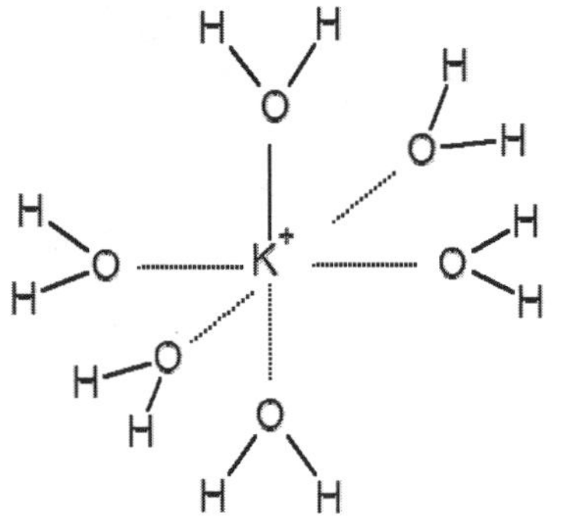

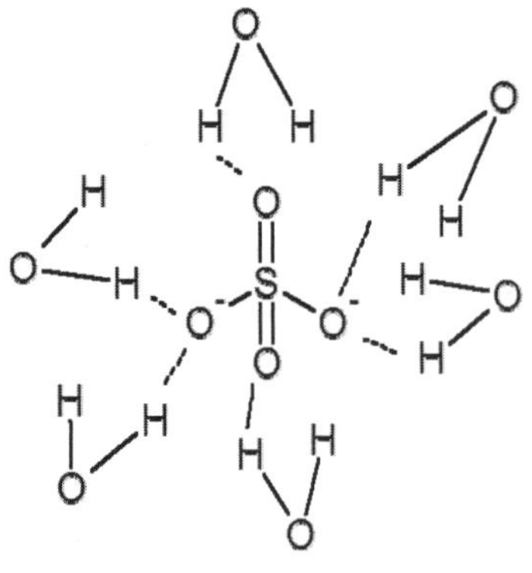

6.99

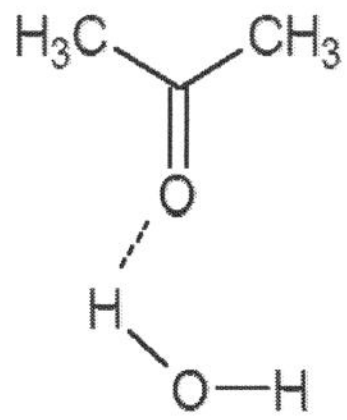

Acetone can hydrogen bond with water via its O atom—which interacts with an H on a water molecule. Otherwise, the interactions between acetone molecules are dipole–dipole intermolecular forces, in addition to dispersion forces, which always exist in between atoms, ions, and molecules.

6.101

(a) Because CCl_4 is non-polar, it is immiscible in water. Consequently, adding equal volumes of water and CCl_4 to a test tube would produce two colourless layers of liquid. CCl_4, the denser liquid, would be on the bottom.

(b) This would become evident when hexane is added. It will dissolve in the bottom, non-polar layer, increasing the volume of the layer.

Section 6.6: Self-Ionization of Water

6.103

The self-ionization reaction of ammonia,
$$2\,NH_3(\ell) \rightarrow NH_4^+(aq) + NH_2^-(aq)$$
produces equal concentrations of aquated ammonium ions and amide ions.

Section 6.7: Categories of Chemical Reaction in Water

6.105

(a) $Cd^{2+}(aq) + 2\,OH^-(aq) \rightarrow Cd(OH)_2(s)$
$Na^+(aq)$ ions and $Cl^-(aq)$ ions are spectator ions.

(b) $Ni^{2+}(aq) + CO_3^{2-}(aq) \rightarrow NiCO_3(s)$
$K^+(aq)$ ions and $NO_3^-(aq)$ ions are spectator ions.

(c) $Cu^{2+}(aq) + S^{2-}(aq) \rightarrow CuS(s)$
$NH_4^+(aq)$ ions and $SO_4^{2-}(aq)$ ions are spectator ions.

(d) $Ca^{2+}(aq) + C_2O_4^{2-}(aq) \rightarrow CaC_2O_4(s)$
$K^+(aq)$ ions and $NO_3^-(aq)$ ions are spectator ions.

6.107

$$2\,Na(\ell) \longrightarrow 2\,Na^+ + 2\,e^-$$
$$Cl_2(g) + 2\,e^- \longrightarrow 2\,Cl^-$$

Here, liquid sodium is oxidized to sodium cations. Chlorine is the oxidizing agent. Cl_2 is reduced to Cl^-. $Na(\ell)$ is the reducing agent.

6.109

(a)

    (i) $H^+(aq) + OH^-(aq) \rightarrow H_2O(\ell)$
    (ii) $H^+(aq) + OH^-(aq) \rightarrow H_2O(\ell)$

(b) In both cases, the products are water and a salt solution. These are really the same reaction—the reaction of hydronium and hydroxide to produce water molecules. The only difference is the spectator ions.

(c)

    (i) $K^+(aq)$ ions and $NO_3^-(aq)$ ions are spectator ions.
    (ii) $Na^+(aq)$ ions and $Cl^-(aq)$ ions are spectator ions.

(d)

    (i) Potassium nitrate, $KNO_3(s)$.
    (ii) Sodium chloride, $NaCl(s)$.

6.111

(a) Water, (b) chloride, (e) methylamine, and (f) cyanide can be Lewis bases—they have lone pairs of electron available for donation. (c) Methane and (d) ammonium have no lone pairs of electrons—they cannot be Lewis bases.

Section 6.8: Solution Concentration

6.113

Volume = 250 mL = 0.250 L
Amount of $AgNO_3$ required = volume × (concentration of $AgNO_3$) = (0.250 L) × (0.0200 mol L^{-1}) = 0.00500 mol
Mass of $AgNO_3$ required = $n(AgNO_3) \times M(AgNO_3)$ = (0.00500 mol) × (169.91 g mol^{-1}) = 0.850 g
To make the desired solution, carefully weigh 0.850 g of $AgNO_3(s)$, add it to the volumetric flask, add ~150 mL of de-ionized water, stopper the flask and shake to dissolve the $AgNO_3$ and ensure a homogenenous solution. After the $AgNO_3$ has dissolved, top up the volume to the 250 mL mark with de-ionized water—add the additional de-ionized water in steps, swirling between each step to ensure a homogeneous solution.

6.115

Amount of NaCl, n = (volume of NaCl solution) × (concentration of NaCl) = (1.0 L) × (0.1 mol L^{-1}) = 0.1 mol
Mass of NaCl, m = n(NaCl) × M(NaCl) = (0.1 mol) × (58.44 g mol^{-1}) = 6 g
Volume of Na_2CO_3 solution = 1250 mL = 1.250 L
Amount of Na_2CO_3, n = (volume of Na_2CO_3 solution) × (concentration of Na_2CO_3) = (1.250 L) × (0.060 mol L^{-1}) = 0.075 mol
Mass of Na_2CO_3, m = n(Na_2CO_3) × M(Na_2CO_3) = (0.075 mol) × (105.99 g mol^{-1}) = 7.9 g > 6 g
1250 mL of 0.060 mol L^{-1} Na_2CO_3 solution has the greater mass of solute.

6.117

Volume of solution = 250 mL = 0.250 L
Amount of $KMnO_4$, n = (volume of solution) × (concentration of $KMnO_4$) = (0.250 L) × (0.0125 mol L^{-1})
$\qquad$ = 0.00312 mol
Mass of $KMnO_4$, m = n($KMnO_4$) × M($KMnO_4$) = (0.00312 mol) × (158.04 g mol^{-1}) = 0.493 g

6.119

Amount of NaOH, n = m(NaOH) / M(NaOH) = (25.0 g) / (40.00 g mol^{-1}) = 0.625 mol
Volume of solution = (amount of NaOH) / (concentration of NaOH) = (0.625 mol) / (0.123 mol L^{-1}) = 5.08 L

6.121

Volume of solution = 250 mL = 0.250 L
Amount of $H_2C_2O_4$, n = Volume × concentration of $H_2C_2O_4$ = (0.250 L) × (0.15 mol L^{-1}) = 0.0375 mol
Mass of $H_2C_2O_4$, m = n($H_2C_2O_4$) × M($H_2C_2O_4$) = (0.0375 mol) × (90.04 g mol^{-1}) = 3.38 g

6.123

Amount of Na_2CO_3, n = m(Na_2CO_3) / M(Na_2CO_3) = (6.73 g) / (105.99 g mol^{-1}) = 0.0635 mol
Volume of solution = 250 mL = 0.250 L
Concentration of Na_2CO_3, c = n(Na_2CO_3) / Volume = (0.0635 mol) / (0.250 L) = 0.254 mol L^{-1}
$[Na^+]$ = 2 × c(Na_2CO_3) = 2 × 0.254 mol L^{-1} = 0.508 mol L^{-1}
$[CO_3^{2-}]$ = c(Na_2CO_3) = 0.254 mol L^{-1}

SUMMARY AND CONCEPTUAL QUESTIONS

6.125

The density remains 1.0 g mL^{-1} at 4°C regardless of its mass.

6.127

The addition of the word *change* is preferable because the enthalpy of vaporization alone is a function of state. When a substance is exposed to a specific set of conditions, this energy (ΔH) will elicit a change.

6.129

Water and octane are immiscible liquids. Entropy is the driving force toward a more probably distribution of molecular substances. Therefore, if we add water to octane and shake the mixture, the molecular substances will randomly distribute themselves and entropy maximization will be achieved. That is until mixing ceases and the two liquids return to their ordered state in separate layers.

6.131 	The figure on the left is a frame from the animation in e6.2 Molecular-level modelling activity depicting the evaporation of water on the surface. **(a)** The lone water molecule on the left is being pulled back into the liquid phase by **(b)** hydrogen bonds to adjacent water molecules.
6.133 	The figure on the left is a frame from the animation in e6.15 molecular-level modelling activity depicting a chloride ion being hydrated and leaving the surface of the NaCl lattice. **(a)** The water molecules have their partially positively charged hydrogen atoms pointed toward the negative chloride ion. **(b)** To dissolve the chloride ion, electrostatic attraction of the ion to sodium ions in the crystal lattice must be overcome by the attraction to dipolar water molecules. Also favouring solvation is the entropic effect that corresponds here to ions being whisked away once they become solvated. Dissolved ions thus have little chance to re-crystallize, whereas re-crystallized ions on the surface are always continually exposed to solvating water molecules.

6.135	The figure on the left is a frame from the animation in e6.21 molecular-level modelling activity depicting a hydroxide ion accepting a proton from another water molecule.
	**(a)** Hydroxide ions appear to diffuse very quickly through an aqueous solution just like hydronium ions. The transfer of hydroxide across a cluster of water molecules occurs via a series of proton transfers in the opposite direction. **(b)** In this case, the proton transfers are from water molecules to hydroxides.
6.137	The figures on the left are frames from the animation in e6.23 molecular-level modelling activity depicting the reduction of the silver ion, and the oxidation of a copper atom.
	**(a)** The silver ion approaches the copper atoms at the surface of the metal—it is attracted to the electrons of the metal. An electron is extracted from the surface and the silver atom adheres to the metal's surface. Every second electron extracted in this way results in a Cu^{2+} ion dislodging from the surface and becoming solvated by water molecules. **(b)** The metal is ionized—the attraction of an electron to a singly charged metal surface is overcome—and a silver ion accepts an electron (the reverse of ionization). Electron binding energies are involved in both these processes.

6.139	The figure on the left is a frame from the animation in e6.25 molecular-level modelling activity depicting proton transfer from a water molecule to an ammonia molecule.
	**(a)** Here a lone pair of electrons on the N atom of ammonia accepts a proton from a water molecule—the formerly bonding pair of electrons becomes a lone pair on the O atom of the hydroxide ion. **(b)** A covalent bond is broken and a new one made.

6.141

(a) $Ba^{2+}(aq) + CO_3^{2-}(aq) \longrightarrow BaCO_3(s)$

is a precipitation reaction, (i).

(b) $[Zn(OH_2)_6]^{2+} + 4CN^-(aq) \longrightarrow [Zn(CN)_4]^{2-} + 6H_2O(\ell)$

is a complexation reaction, (iv). The cyanide ions displace the water ligands bound to Zn^{2+} ions.

(c) $Zn(s) + Fe^{2+}(aq) \longrightarrow Zn^{2+}(aq) + Fe(s)$

is an oxidation-reduction reaction, (ii).

(d) $OH^-(aq) + CH_3CH_2COOH(aq) \longrightarrow H_2O(\ell) + CH_3CH_2COO^-(aq)$

is an acid-base reaction, (iii).

6.143

The bond between a Lewis acid and a Lewis base is a covalent bond formed with both of the electrons coming from the Lewis base.

CHAPTER 7
Chemical Reactions and Energy Flows

IN-CHAPTER EXERCISES

Exercise 7.1—Exothermic and Endothermic Processes

(a) Since $H^+(aq) + OH^-(aq) \longrightarrow H_2O(\ell)$ is an exothermic process (the change in energy is negative), 1 mol each of $H^+(aq)$ ions and $OH^-(aq)$ ions, that have not reacted, have more energy than 1 mol of water. The difference is released as heat.

(b) If this reaction takes place in a test tube, where you are holding your fingers would feel hot. This is because the decrease of energy of the system requires removal of excess energy—it leaves in the form of heat that your fingers sense.

Exercise 7.3—Energy Conversion

The chemical potential energy stored in a battery can be converted to the mechanical energy of sound waves—from your mp3 player—the electrical and magnetic energy of an image taken by your digital camera, or the light energy emitted by a flashlight—to name a few possibilities.

Exercise 7.5—System and Surroundings

What constitutes the system can be defined at will. The following are sensible suggestions:

(a) The system is the contents of the combustion chamber of the gas furnace—a mixture of air and methane. The surroundings are the furnace and everything around it.

(b) The system is the water drops plus the air around you. The surroundings consist of your body and the sun—they provide the heat that evaporates the water drops.

(c) The water, initially at 25°C, is the system. The container and the rest of the freezer contents—including the air—are the surroundings.

(d) The aluminum and $Fe_2O_3(s)$ mixture is the system (initially—later it consists of $Al_2O_3(s)$ and iron). The flask and the laboratory bench are the surroundings.

Exercise 7.7—Enthalpy Change and Change of State

When 1.0 L of water at 0°C freezes to form ice at 0°C,

heat evolved $= -$ (change in enthalpy)

$\qquad\qquad = $ (amount of water – in mol) $\times$ (molar enthalpy of fusion – in kJ mol^{-1})

mass of water $=$ volume $\times$ density $= (1.0 \text{ L}) \times (1.0 \text{ g mL}^{-1})$

$= (1.0 \times 10^3 \text{ mL}) \times (1.0 \text{ g mL}^{-1}) = 1.0 \times 10^3 \text{ g}$

Heat evolved, $q = -6.00 \text{ kJ } \cancel{mol^{-1}} \times \dfrac{1.0 \times 10^3 \cancel{g}}{18.02 \cancel{g} \cancel{mol^{-1}}} = -333 \text{ kJ}$

Exercise 7.9—Enthalpy Change and Change of State

Mass of mercury = volume × density = $(1.00 \text{ mL}) \times (13.6 \text{ g mL}^{-1}) = 13.6$ g
Heat released upon freezing = (molar enthalpy change of fusion of mercury) × (amount of mercury)

$$= q = -2.3 \text{ kJ mol}^{-1} \times \frac{13.6 \text{ g}}{200.6 \text{ g mol}^{-1}} = -0.16 \text{ kJ}$$

Exercise 7.13—Using a Calorimeter to Measure $\Delta_r H$

First, we assume that all of the heat evolved stays within the calorimeter. This is the purpose of the styrofoam insulation—the coffee cup. The specific heat capacity of the reaction solution is $4.20 \text{ J K}^{-1} \text{ g}^{-1}$, while the density of the solution is 1.00 g mL^{-1}.
Mass inside calorimeter = (total volume) × (density of solution) = $(200 + 200 \text{ mL}) \times (1.00$ g $\text{mL}^{-1}) = 400$ g
The heat evolved when 200 mL of 0.400 mol L^{-1} HCl solution is mixed with 200 mL of 0.400 mol L^{-1} NaOH solution in a coffee-cup calorimeter is given by
        Heat evolved = (specific heat capacity) × (mass inside calorimeter) × (temperature change)
$$= (4.20 \text{ J K}^{-1} \text{ g}^{-1}) \times (400 \text{ g}) \times (27.78 - 25.10 \text{ K}) = 4502.4 \text{ J}$$
The reaction in the calorimeter can be written
        $H_3O^+(aq) + OH^-(aq) \rightarrow 2 H_2O(\ell)$
so the $H_3O^+(aq)$ and $OH^-(aq)$ ions react in a 1:1 ratio of amounts. HCl and NaOH are strong electrolytes in water, so:
        In 200 mL of 0.400 mol L^{-1} HCl solution, $n(H^+) = (0.200 \text{ L}) \times (0.400 \text{ mol L}^{-1}) = 0.0800$ mol
        In 200 mL of 0.400 mol L^{-1} NaOH solution, $n(OH^-) = (0.200 \text{ L}) \times (0.400 \text{ mol L}^{-1}) = 0.0800$ mol
The reacting species are present in stoichiometric amounts: neither is a limiting reagent, so 0.0800 mol of each ion reacts.
Molar enthalpy change of neutralization of $H^+(aq)$ ions = heat liberated when 1 mol of $H^+(aq)$ ions react
        $= -$ (heat evolved when 0.0800 mol of $H^+(aq)$ ions react) / (0.0800 mol) $= -$ (4502.4 J) / (0.0800 mol) $= -56300$ J mol^{-1}
        $= -56.3$ kJ mol^{-1}
The minus sign denotes an exothermic reaction. Note that the reaction has a molar enthalpy change of -56.3 kJ mol^{-1} at 25.10 °C. In the experiment the temperature increased to 27.78°C. Imagine removing the heat evolved from the calorimeter, returning its temperature to 25.10°C.

Exercise 7.17—Standard Enthalpy Change of Reaction

(a) $CO(g) + \frac{1}{2} O_2(g) \longrightarrow CO_2(g)$

The standard enthalpy change of this reaction, $\Delta_r H°$, is the heat absorbed (hence a negative number when is evolved) per mole of CO(g) that reacts at a constant temperature of 25°C, in vessel in which each of CO(g), O_2(g) and CO_2(g) have a partial pressure of 1 bar.

(b) $Mg(s) + 2 H^+(aq) \longrightarrow Mg^{2+}(aq) + H_2(g)$

The standard enthalpy change of this reaction, $\Delta_r H°$, is the heat absorbed per mole of Mg(s) that reacts in a reaction mixture at 25 °C in which simultaneously there is pure Mg(s), in solution $[H^+] = [Mg^{2+}] = 1$ mol L^{-1}, and $H_2(g)$ is present at 1 bar pressure.

(c) $H^+(aq) + OH^-(aq) \longrightarrow H_2O(\ell)$

The standard enthalpy change of this reaction, $\Delta_r H°$, is the heat absorbed per mole of $H^+(aq)$ ions that react, under constant pressure conditions of 1 bar, and constant temperature at 25°C, if in solution $[H^+] = [OH^-] = 1$ mol L^{-1}.

Exercise 7.19—Using Hess's Law

$$
\begin{array}{ll}
C(s) + O_2(g) \longrightarrow CO_2(g) & \Delta_r H°_1 = -393.5 \text{ kJ mol}^{-1} \\
2\,S(s) + 2\,O_2(g) \longrightarrow 2\,SO_2(g) & \Delta_r H°_2 = 2 \times (-296.8 \text{ kJ mol}^{-1}) \\
CO_2(g) + 2\,SO_2(g) \longrightarrow CS_2(\ell) + 3\,O_2(g) & \Delta_r H°_3 = -(-1103.9 \text{ kJ mol}^{-1}) \\
\hline
C(s) + 2\,S(s) \longrightarrow CS_2 & \Delta_r H° = 116.8 \text{ kJ mol}^{-1}
\end{array}
$$

Exercise 7.21—Standard Molar Enthalpy Changes of Formation

(a) The standard molar enthalpy of formation of bromine, $Br_2(\ell)$, is the standard enthalpy change of the reaction:
$$Br_2(\ell) \rightarrow Br_2(\ell) \text{ at } 25°C$$

(b) The standard molar enthalpy of formation of solid iron(III) chloride, $FeCl_3(s)$, is the standard enthalpy change of the reaction:
$$Fe(s) + Cl_2(g) \rightarrow FeCl_3(s) \text{ at } 25°C$$

(c) The standard molar enthalpy of formation of solid sucrose, $C_{12}H_{22}O_{11}(s)$, is the standard enthalpy change of the reaction:
$$12\,C(\text{graphite}) + 11\,H_2(g) + \frac{11}{2}O_2(g) \rightarrow C_{12}H_{22}O_{11}(s) \text{ at } 25°C$$

Exercise 7.27—Using Bond Energies to Estimate Enthalpy Change of Reaction

Acetone   **Isopropanol**

$$
\begin{aligned}
\Delta_r H &= \sum D(\text{bonds broken}) - \sum D(\text{bonds formed}) \\
&= D(\text{C=O}) + D(\text{H-H}) - (D(\text{C-O}) + D(\text{C-H}) + D(\text{O-H})) \\
&= 745 \text{ kJ mol}^{-1} + 436 \text{ kJ mol}^{-1} - (358 \text{ kJ mol}^{-1} + 413 \text{ kJ mol}^{-1} + 463 \text{ kJ mol}^{-1}) \\
&= -53 \text{ kJ mol}^{-1}
\end{aligned}
$$

Exercise 7.31—Energy from Food

(a)   Breaking the P–O bond in ATP is an exothermic process. FALSE
Breaking bonds is always endothermic—it requires energy input.

(b) Making a new bond between the phosphorus atom in the phosphate group being cleaved off ATP and the OH group of water is an exothermic process. TRUE
Making a bond (in the absence of other concurrent processes) is always exothermic—it releases energy.

(c)   Breaking bonds is an endothermic process. TRUE

(d) The energy released in the hydrolysis of ATP may be used to run endothermic reactions in a cell. TRUE.
The cell uses ATP as an energy source—it couples other reactions to the hydrolysis of ATP to harness the chemical energy released in the latter process. Processes achieved this way are mostly endothermic—they need the energy input.

REVIEW QUESTIONS

Section 7.1: Powering Our Planet with Hydrogen?

7.33

We are using energy to obtain hydrogen in order to provide us with energy because we are running out of non-renewable energy sources and need to find alternatives. Hydrogen is an attractive option since its combustion does not produce CO_2, a greenhouse gas. Its high energy density stands to it to release 143 kJ per gram of H_2 (better than methane, gasoline and coal). A parallel to this would be using energy to obtain oil.

7.35

	Coal	**Natural Gas**	**Gasoline**	**H$_2$**
Energy density weight/volume	29–37 kJ g^{-1} (~pure C)	50 kJ g^{-1} (25% H w/w)	47 kJ g^{-1} (~17% H w/w)	121 kJ g^{-1} (high w/w)
Renewability/ Sustainability	No (150–? yrs)	No (80–200 yrs)	No (30–80 yrs)	Yes
Portability	Yes	Dangerous	Yes but dangerous (oil spills)	Not yet (SAFE storage/delivery an issue)
Suitability for use in transport	Yes (trains)	Yes	Yes	Not yet (Yes in rockets)
Environmental Advantages	Abundant, affordable, high energy	Burns clean, can be renewable (landfills)	Economical, smaller footprint than coal	Clean, no harmful emissions, abundant
Disadvantages	Burning = harmful by-products (acid rain/ash), CO$_2$, mining	Highly volatile, must be treated with sulphur (smell), greenhouse gas emissions	NO/CO$_2$ emission, resources depleting quickly, wars over oil, price fluctuation	Expensive, storage, flammable, depends on fossil fuels to separate from oxygen

Other: solar, wind, biomass, water, geothermal
Nuclear energy: non-renewable in principle

7.37

There are many possible answers, depending upon the Internet query. These are some examples for photovoltaic cells. Four scarce elements that might limit the implementation of photovoltaic cells are tellurium (Te), selenium (Se), cadmium (Cd), and indium (In).

Section 7.2: Chemical Changes and Energy Redistribution

7.39

(a) $H_2O(\ell) \longrightarrow H_2O(s)$

is an exothermic process—intermolecular bonds between water molecules are formed during freezing.

(b) $2\,H_2(g) + O_2(g) \longrightarrow 2\,H_2O(g)$

is an exothermic process. Though bonds are broken and made in this process, we recognize that it is the combustion of hydrogen—a fuel. The reaction releases energy that is harnessed by hydrogen cars. Perhaps you have seen a hydrogen balloon ignited? It is clearly an exothermic process.

(c) $H_2O(\ell, 25°C) \longrightarrow H_2O(\ell, 15°C)$

is an exothermic process. Cooling of liquid water (or any liquid) requires taking energy away—enthalpy of the system goes down.

(d) $H_2O(\ell) \longrightarrow H_2O(g)$

is an endothermic process. Energy must be put into the liquid water to break its intermolecular bonds and form water vapour.

7.41

The system could be the combustion chamber within the stove. The surroundings would define anything that the system can exchange energy with. Therefore, you would be part of the surroundings. Other things in the surroundings could be the stove itself and anything around it (e.g., pots, pans).

Section 7.3: Energy: Its Forms and Transformations

7.43

Many answers are possible; here are some examples. When a car is driven uphill, the gravitational potential energy increases. The car exhibits kinetic energy while in motion. The gasoline in the combustion chamber has chemical potential energy that is then transformed into heat and thrust for the car.

7.45

(a) Your energy has decreased as the kinetic energy (due to motion) has decreased. You (system) have transferred energy to the surroundings (stadium)
(b) The energy within the stadium has increased. This is due to the law of conservation of energy that energy in the universe is constant, but can be transformed from one form to the next.

Section 7.4: Energy Flows between System and Surroundings

7.47

An object that feels hot is the result of an exothermic process. An object that feels cold is the result of an endothermic process.

7.49

(a) The energy of the surroundings increases by 385 J.
(b) The temperature of the air must be cooler than that of the copper block, as heat flows from an object of higher temperature to lower temperature.
(c) The specific heat (c) for copper is 0.385 J g^{-1} °C^{-1}. Therefore, the change in air temperature can be calculated via the following equation:

$$\Delta T = \frac{q}{m \times c} = \frac{385 \, \cancel{J}}{(1000 \, \cancel{g} \times 0.385 \, \cancel{J} \cancel{g^{-1}} \, {}^{\circ}C^{-1})} = 1^{\circ}C$$

Therefore, the temperature increased by 1°C.

Section 7.5: Enthalpy Changes Accompanying Changes of State

7.51

Amount of benzene, n = (125 g) / (78.11 g mol^{-1}) = 1.60 mol
Heat required to vaporize 125 g of benzene, q = (1.60 mol) × (30.8 kJ mol^{-1}) = 49.3 kJ

7.53

Heat required to convert 60.1 g of ice to liquid water at 0.0°C
    = (333 J g^{-1}) × (60.1 g) = 2.00×10^4 J = 20.0 kJ
Heat required to raise temperature of liquid water from 0.0°C to 100.0°C
    = (4.18 J K^{-1} g^{-1}) × (60.1 g) × (100 K) = 2.51×10^4 J = 25.1 kJ
Heat required to convert liquid water to vapour at 100.0 °C
    = (2260 J g^{-1}) × (60.1 g) = 1.36×10^5 J = 136 kJ
Total heat required for the three steps = (20.0 + 25.1 + 136) kJ = 181 kJ

Section 7.6: Enthalpy Change of Reaction, $\Delta_r H$

7.55

$$H_2(g) + \frac{1}{2}O_2(g) \longrightarrow H_2O(\ell) \qquad \Delta_r H = -285.8 \, kJ$$

Decomposition of liquid water to hydrogen and oxygen gases is the reverse of this reaction. Formation of water is exothermic, so its decomposition is endothermic (heat is absorbed)

Heat absorbed, q = (285.8 kJ $\cancel{mol^{-1}}$) × {(12.6 $\cancel{g}$) / (18.02 $\cancel{g}$ $\cancel{mol^{-1}}$)} = 200 kJ

7.57

$$CH_3OH(\ell) + CO(g) \longrightarrow CH_3COOH(\ell) \qquad \Delta_r H = -355.9 \, kJ$$

The reaction is exothermic. Producing 1.00 L of acetic acid in this way evolves

q = (−355.9 kJ $\cancel{mol^{-1}}$) × {(1.00×10^3 $\cancel{mL}$) × (1.044 $\cancel{g}$ $\cancel{mL^{-1}}$) / (60.05 $\cancel{g}$ $\cancel{mol^{-1}}$)}
    = −6190 kJ

7.59

Heat absorbed by dissolution of 5.44 g of NH₄NO₃(s)
    = (4.20 J K^{-1} g^{-1}) × (155.4 g) × (18.6 − 16.2 K) = 1.57×10^3 J = 1.57 kJ
Note the reversal of initial and final temperatures because we compute heat "absorbed"—rather than heat evolved.
Amount of NH₄NO₃(s) = (5.44 g) / (80.05 g mol^{-1}) = 0.0680 mol

Per mole of $NH_4NO_3(s)$ that dissolves,
heat absorbed $= (1.57\text{ kJ}) / (0.0680\text{ mol}) = 23.1\text{ kJ}$

7.61

(a) Pure $H_2O(\ell)$ is the standard state of water at 25°C.
(b) Pure $NaCl(s)$ is the standard state of sodium chloride at 25°C.
(c) Pure $Hg(\ell)$ is the standard state of mercury at 25°C.
(d) $CH_4(g)$ at 1 bar pressure is the standard state of methane at 25°C.
(e) The standard state for $Na^+(aq)$ ions in solution, regardless of their origin, is $[Na^+] = 1.0$ mol L^{-1}.

Section 7.7: Hess's Law

7.63

$$\frac{1}{2}\{N_2(g) + 3H_2(g) \longrightarrow 2NH_3(g)\} \qquad \Delta_r H^\circ = \frac{1}{2}\times(-91.8\,\text{kJ mol}^{-1})$$

$$\frac{1}{4}\{4NH_3(g) + 5O_2(g) \longrightarrow 4NO(g) + 6H_2O(g)\} \qquad \Delta_r H^\circ = \frac{1}{4}\times(-906.2\,\text{kJ mol}^{-1})$$

$$\frac{3}{2}\{H_2O(g) \longrightarrow H_2(g) + \frac{1}{2}O_2(g)\} \qquad \Delta_r H^\circ = -\frac{3}{2}\times(-241.8\,\text{kJ mol}^{-1})$$

$$\frac{1}{2}N_2(g) + \frac{1}{2}O_2(g) \longrightarrow NO(g) \qquad \Delta_r H^\circ = 90.3\,\text{kJ mol}^{-1}$$

Section 7.8: Standard Molar Enthalpy Change of Formation

7.67

$$4NH_3(g) + 5O_2(g) \longrightarrow 4NO(g) + 6H_2O(g)$$

$\Delta_r H^\circ = 4\times\Delta_r H^\circ[NO(g)] + 6\times\Delta_r H^\circ[H_2O(g)] - 4\times\Delta_r H^\circ[NH_3(g)] - 5\times \Delta_r H^\circ[O_2(g)]$
$= (4 \times 90.29\text{ kJ mol}^{-1}) + (6 \times -241.83\text{ kJ mol}^{-1}) - (4 \times -45.90\text{ kJ mol}^{-1}) - (5 \times 0)$
$= -906.22\text{ kJ mol}^{-1}$

$$\underline{2N_2(g) + 6H_2(g) + 5O_2(g)}$$

$\uparrow -4\,\Delta_f H^\circ[NH_3(g)] \qquad\qquad |$

$\underline{4NH_3(g) + 5O_2(g)} \qquad | \qquad\qquad\qquad |$

$\qquad\qquad\qquad\qquad | \qquad\qquad | \; 4\times\Delta_r H^\circ[NO(g)] + 6\times\Delta_r H^\circ[H_2O(g)]$

$4\times\Delta_r H^\circ[NO(g)] + 6\times\Delta_r H^\circ[H_2O(g)] \qquad | \qquad\qquad |$

$\qquad - 4\times\Delta_r H^\circ[NH_3(g)] \qquad\qquad | \qquad\qquad |$

$\qquad\qquad\qquad\qquad\qquad\qquad \downarrow \qquad\qquad \downarrow \; \underline{4NO(g) + 6H_2O(g)}$

7.69

$$C_{10}H_8(s) + 12\,O_2(g) \longrightarrow 10\,CO_2(g) + 4\,H_2O(l) \qquad \Delta_rH° = -5156.1\ \text{kJ mol}^{-1}$$

$\Delta_rH° = 10\ \Delta_fH°[CO_2(g)] + 4\ \Delta_fH°[H_2O(\ell)] - \Delta_fH°[C_{10}H_8(s)] - 12\ \Delta_fH°[O_2(g)]$

$\qquad = -5156.1\ \text{kJ mol}^{-1}$

Note that $\Delta_fH°[O_2(g)] = 0$

We can solve the above equation for the unknown, $\Delta_fH°[C_{10}H_8(s)]$.

$\Delta_fH°[C_{10}H_8(s)] = 5156.1\ \text{kJ} + 10\ \Delta_fH°[CO_2(g)] + 4\ \Delta_fH°[H_2O(\ell)]$

$\qquad\qquad = 5156.1\ \text{kJ mol}^{-1} + (10 \times (-393.509) + 4 \times (-285.83))\ \text{kJ mol}^{-1} = 7.69\ \text{kJ mol}^{-1}$

7.71

$$Mg(s) + 2\,H_2O(\ell) \longrightarrow Mg(OH)_2(s) + H_2(g)$$

$\Delta_rH° = \Delta_fH°[Mg(OH)_2(s)] + \Delta_fH°[(H)_2(g)] - \Delta_fH°[Mg(s)] - 2 \times \Delta_fH°[H_2O(\ell)] = \{-924.54 + 0 - 0 - 2 \times (-285.83)\}\ \text{kJ} = -352.9\ \text{kJ mol}^{-1}$

That is, when 1.000 mol of Mg(s) reacts 352.9 kJ of energy is released.

To warm 25 mL of water from 25°C to 85°C requires $(4.20\ \text{J g}^{-1}\ \text{K}^{-1}) \times ((25\ \text{mL}) \times (1.00\ \text{g mL}^{-1})) \times (85 - 25)\ \text{K} = 6300\ \text{J}$

Amount of Mg(s) that reacts to evolve this much heat, $n = (6300\ \text{J}) / (352.9\ \text{kJ mol}^{-1})$

$= (6300\ \text{J}) / (352.9 \times 10^3\ \text{J mol}^{-1}) = 1.79 \times 10^{-2}\ \text{mol}$

Mass of Mg(s) required, $m = (1.79 \times 10^{-2}\ \text{mol}) \times (24.31\ \text{g mol}^{-1}) = 7.36 \times 10^{-4}\ \text{g} = 0.74\ \text{mg}$

7.73

$$N_2H_4(\ell) + O_2(g) \longrightarrow N_2(g) + 2\,H_2O(g)$$
hydrazine

$$N_2H_2(CH_3)_2(\ell) + 4\,O_2(g) \longrightarrow 2\,CO_2(g) + 4\,H_2O(g) + N_2(g)$$
1,1-dimethylhydrazine

For hydrazine combustion,

$\Delta_rH° = 2\ \Delta_fH°[H_2O(g)] + \Delta_fH°[N_2(g)] - \Delta_fH°[N_2H_4(\ell)] - \Delta_fH°[O_2(g)] = \{2 \times (-241.83) + 0 - (50.6) - 0)\}\ \text{kJ mol}^{-1}$

$\qquad\qquad = -534.3\ \text{kJ mol}^{-1}$ (per mol of hydrazine)

Enthalpy change per g of hydrazine $= (-534.3\ \text{kJ mol}^{-1}) / (32.05\ \text{g mol}^{-1}) = -16.67\ \text{kJ g}^{-1}$

For 1,1-dimethylhydrazine combustion,

$\Delta_rH° = 2\ \Delta_fH°[CO_2(g)] + 4\ \Delta_fH°[H_2O(g)] + \Delta_fH°[N_2(g)] - \Delta_fH°[N_2H_2(CH_3)_2(\ell)] - 4\ \Delta_fH°[O_2(g)] = \{2 \times (-393.509) + 4 \times (-241.83) + 0 - (48.9) - 4 \times (0)\}\ \text{kJ mol}^{-1}$

$\qquad\qquad = -1803.2\ \text{kJ mol}^{-1}$ (per mol of 1,1-dimethylhydrazine)

Enthalpy change per g of 1,1-dimethylhydrazine $= (-1803.2\ \text{kJ mol}^{-1}) / (60.10\ \text{g mol}^{-1}) = -30.00\ \text{kJ g}^{-1}$

1,1-dimethylhydrazine evolves more heat on a per gram basis.

Section 7.9: Enthalpy Change of Reaction from Bond Energies

7.75

$O_3 + O \longrightarrow 2\,O_2$ $\Delta_r H° = $ -394 kJ mol^{-1}

$\Delta_r H° = 2\,D[\text{O–O in O}_3] - 2\,D[\text{O=O in O}_2] = -394$ kJ mol^{-1}

So, $D[\text{O–O in O}_3] = \frac{1}{2} \times (-394$ kJ mol$^{-1} + 2\,D[\text{O=O in O}_2]\,) = \frac{1}{2} \times (-394 + 2 \times 498\,)$ kJ mol^{-1}

$= 301$ kJ mol^{-1}

This value is between the oxygen single and double bond energies, 146 and 498 kJ mol^{-1}, respectively.

7.77

$2\,CH_3OH(g) + 3\,O_2(g) \longrightarrow 2\,CO_2(g) + 4\,H_2O(g)$

(a) From bond energies,

$\Delta_r H° = 2 \times (\,3\,D[\text{C–H}] + D[\text{C–O}]) + 3\,D[\text{O=O}] - (\,2 \times 2\,D[\text{C=O}] + 4 \times 2\,D[\text{O–H}]\,)$

$= 6\,D[\text{C–H}] + 2\,D[\text{C–O}] + 3\,D[\text{O=O}] - (\,4\,D[\text{C=O}] + 6\,D[\text{O–H}]\,)$

$= \{6 \times 413 + 2 \times 358 + 3 \times 498 - (\,4 \times 745 + 6 \times 463\,)\}$ kJ mol$^{-1} = -1070$ kJ mol^{-1}

(b) From enthalpies of formation,

$\Delta_r H° = 2\,\Delta_f H°[\text{CO}_2(g)] + 4\,\Delta_f H°[\text{H}_2\text{O}(g)] - 2\,\Delta_f H°[\text{CH}_3\text{OH}(g)] - 3\,\Delta_f H°[\text{O}_2(g)]$

$= \{2 \times (-393.509) + 4 \times (-241.83) - 2 \times (-201.0) - 3 \times (0)\}$ kJ mol^{-1}

$= -1352.3$ kJ mol^{-1}

This is the more correct enthalpy change for this reaction. The value computed in part (a) from average bond energies is only an estimate—about 20% off in this case.

Section 7.10: Energy from Food

7.79

$C_6H_{12}O_6(s) + 6\,O_2(g) \longrightarrow 6\,CO_2(g) + 6\,H_2O(\ell)$ $\Delta_r H° = -2803$ kJ

$\Delta_r H° = 6\,\Delta_f H°[\text{CO}_2(g)] + 6\,\Delta_f H°[\text{H}_2\text{O}(\ell)] - \Delta_f H°[\text{C}_6\text{H}_{12}\text{O}_6(s)] - 6\,\Delta_f H°[\text{O}_2(g)] = -2803$ kJ

$= \{6 \times (-393.509) + 6 \times (-285.83) - \Delta_f H°[\text{C}_6\text{H}_{12}\text{O}_6(s)] - 6 \times (0)\}$ kJ mol^{-1}

Therefore,

$\Delta_r H°[\text{C}_6\text{H}_{12}\text{O}_6(s)] = \{6 \times (-393.509) + 6 \times (-285.83) + 2803\}$ kJ mol$^{-1} = -1273$ kJ mol^{-1}

SUMMARY AND CONCEPTUAL QUESTIONS

7.81

(a) An exothermic reaction releases energy which must be removed to return the system to its original temperature—heat leaves the system. An endothermic reaction absorbs energy which must be supplied for the system to stay at its original temperature—heat enters the system.

(b) The system is the set of all substances of interest—e.g., the reactants and products of a reaction. The surroundings consist of everything else.

(c) The specific heat capacity of substance is the amount of heat (usually expressed in J) required to raise the temperature of exactly 1 g of the substance by 1°C—assuming no phase transitions occur during heating.

(d) A state function is anything that depends only on the "state" of a system. A state function is any property of the system, such as enthalpy or entropy, which has a particular value at a specified set of conditions, regardless of its history (how those conditions were obtained).

(e) The standard state of a substance is the stable form of the substance at 1 bar pressure and—unless specified otherwise—25°C.

(f) The enthalpy change of reaction, $\Delta_r H$, is the change in enthalpy when the extent of reaction is 1 mol (i.e., reactants form products with numbers of moles given by the stoichiometric coefficients), and the temperature of the products is returned to the initial temperature of reactants. It is the difference between the total enthalpy of all the products and the total enthalpy of all the reactants.

(g) The standard enthalpy change of reaction, $\Delta_r H°$, is the enthalpy change of reaction under standard conditions—all reactants and products are in their standard states at the specified temperature.

(h) The standard molar enthalpy change of formation, $\Delta_f H°$, is the standard enthalpy change of a formation reaction wherein 1 mol of a substance is formed from its elemental substances in their standard states.

7.83

A perpetual motion machine is impossible as soon as there is friction or other forms of energy dissipation. Because energy is constantly lost to friction, there must be a constant supply of incoming useful energy—here we invoke conservation of energy. For the machine to run forever, it must have an infinite supply of energy—impossible in a finite machine.

7.85

We can combust sulfur to $SO_3(g)$ and $Ca(s)$ to $CaO(s)$, separately in calorimeters, measuring the enthalpy change for both processes. We can also combine $SO_3(g)$ and $CaO(s)$ in a calorimeter, and measure the enthalpy change for this reaction. The enthalpy change for the desired reaction is the sum of the three enthalpy changes just described—just as the desired reaction itself is the sum of the three reactions.

$$\frac{1}{8}S_8(s) + \frac{3}{2}O_2(g) \longrightarrow SO_3(g)$$
$$Ca(s) + \frac{1}{2}O_2(g) \longrightarrow CaO(s)$$
$$CaO(s) + SO_3(g) \longrightarrow CaSO_4(s)$$
$$\overline{Ca(s) + \frac{1}{8}S_8(s) + 2O_2(g) \longrightarrow CaSO_4(s)}$$

Calorimetric measurements involve reacting known amounts of reactants in a calorimeter and measuring the temperature change.

$$\Delta_r H° = \Delta_r H°_1 + \Delta_r H°_2 + \Delta_r H°_3 = \{\Delta_f H°[SO_3(g)] + \Delta_f H°[CaO(s)] + (-402.7)\} \text{ kJ mol}^{-1}$$
$$= \{(-395.77) + (-635.09) + (-402.7)\} \text{ kJ mol}^{-1} = -1433.6 \text{ kJ mol}^{-1}$$

7.87

A bond dissociation energy is the enthalpy change associated with breaking a bond. All bond breaking processes are endothermic—the enthalpy change is positive. It always takes energy input to break a bond.

7.89

Volume of rain water $=$ thickness $\times$ area $=$ (25 mm) $\times$ (1 km^2) $=$ (2.5 cm) $\times$ (10^5 cm)^2 $=$ 2.5 $\times 10^{10}$ cm^3

Mass of rain water $=$ volume $\times$ density $=$ (2.5 $\times 10^{10}$ cm^3) $\times$ (1.0 g cm^{-3}) $=$ 2.5 $\times 10^{10}$ g

Amount of rain water $=$ (2.5 $\times 10^{10}$ g) / (18.02 g mol^{-1}) $=$ 1.4 $\times 10^9$ mol

Heat released upon condensation of this much water vapour $=$ (44.0 kJ mol^{-1}) $\times$ (1.4 $\times 10^9$ mol) $=$ 6.2 $\times 10^{10}$ kJ

This is more than 10 000 times the energy released when a tonne of dynamite explodes.

CHAPTER 8
Modelling Atoms and Their Electrons

IN-CHAPTER EXERCISES

Exercise 8.1—Chemical Periodicity

(a) Element with atomic number 8 greater than F—i.e., atomic number = 17—is Cl.
Element with atomic number 18 greater than Cl—i.e., atomic number = 35—is Br.
Element with atomic number 18 greater than Br—i.e., atomic number = 53—is I.
Element with atomic number 32 greater than I—i.e., atomic number = 85—is At.
These are all halogens.

(b) The atomic numbers of the group 15 elements, N, P, As, Sb & Bi, are
7, 15, 33, 51, and 83.
Differences between successive atomic numbers = 8, 18, 18, and 32.

Exercise 8.3—Periodic Trends of Oxidizing and Reducing Abilities

(a) Bromine, $Br_2(g)$, to the right of arsenic, As(s), is the more powerful oxidizing agent.
(b) Sodium, Na(s), to the left of silicon, Si(s), is the more powerful reducing agent.
(c) Chlorine, $Cl_2(g)$, well to the right of aluminium, Al(s), is most likely to use to bring about oxidation of a substance that is difficult to oxidize.
(d) The elements with atomic numbers 34, 35, 36, 37, and 38 are Se, Br, Kr, and Rb.
Rubidium, Rb, is the lowest and furthest to the left of these. It is the most powerful reducing agent.

Exercise 8.9—Charges on Monatomic Ions of the Elements

(a) The ions of barium, a group 2 element, are normally Ba^{2+}.
(b) The ions of selenium, a group 16 element, are normally Se^{2-}.

Exercise 8.15—Periodic Trends of Properties

(a) In order of increasing atomic radius, we have
O < C < Si
Here, we move to the left and then down within the periodic table.
(b) The first ionization energy of O is larger than that of C and Si.
In order of increasing ionization energy, we have
Si < C < O
Here, we move up and then to the right within the periodic table.
(c) Which has the greatest electron affinity: O or C?
The electron affinity of O is larger than that of C and Si.
In order of increasing electron affinity, we have

Si < C < O

Here, we move up and then to the right within the periodic table.

Exercise 8.17—Energy of a Spectral Line of Atomic Hydrogen

The energy of the photon emitted when an H atom in the $n = 4$ state changes to the $n = 1$ state is

$$E_{photon} = E_4 - E_1 = \frac{-2.179 \times 10^{-18}\,J}{4^2} - \frac{-2.179 \times 10^{-18}\,J}{1^2}$$

$$= \left(-\frac{1}{16} + \frac{1}{1}\right)(2.179 \times 10^{-18}\,J) = 2.043 \times 10^{-18}\,J$$

The frequency and the wavelength of the third line of the Lyman series are given by

$$\nu = \frac{E_{photon}}{h} = \frac{2.043 \times 10^{-18}\,\cancel{J}}{6.626 \times 10^{-34}\,\cancel{J}\,s} = 3.083 \times 10^{15}\,s^{-1} = 3083\,THz$$

$$\lambda = \frac{c}{\nu} = \frac{(2.998 \times 10^8\,m\,\cancel{s^{-1}})}{3.083 \times 10^{15}\,\cancel{s^{-1}}} = 9.724 \times 10^{-8}\,m = 97.24\,nm$$

Exercise 8.19—Using de Broglie's Equation

We need the mass of the golf ball in kg
$1.0 \times 10^2\,g = 1.0 \times 10^{-1}\,kg$
Also, note $1\,J\,s = 1\,(kg\,m^2\,s^{-2}) \times s = 1\,kg\,m^2\,s^{-1}$

$$\lambda = \frac{h}{m\nu} = \frac{6.626 \times 10^{-34}\,\cancel{kg}\,m^2\,\cancel{s^{-1}}}{(1.0 \times 10^{-1}\,\cancel{kg}) \times (30\,\cancel{m}\,\cancel{s^{-1}})} = 2.2 \times 10^{-34}\,m$$

$$= 2.2 \times 10^{-25}\,nm$$

This de Broglie wavelength (of a macroscopic object—a golf ball) is even very small on the scale of an atom.

To have a wavelength of 5.6×10^{-3} nm, the ball must travel

$$\nu = \frac{h}{m\lambda} = \frac{6.626 \times 10^{-34}\,\cancel{kg}\,m\,\cancel{s^{-1}}}{(1.0 \times 10^{-1}\,\cancel{kg}) \times (5.6 \times 10^{-12}\,\cancel{m})} = 1.2 \times 10^{-21}\,m\,s^{-1}$$

Exercise 8.23—Standing Waveforms from the Wave Equation

(a) When $n = 2$, the value of l must be **0 or 1**.
(b) When $l = 1$, the value of m_l must be **−1, 0 or +1**, and the subshell is labelled the **p subshell**.
(c) When $l = 2$, the subshell is called a **d subshell**.
(d) When a subshell is labelled s, the value of l **is 0** and m_l **has the value 0**.
(e) When a subshell is labelled p, it has **3 orbitals**.
(f) When a subshell is labelled f, there are **7** values of m_l and it has **7 orbitals**.

Exercise 8.25—Quantum Numbers of Electrons in Atoms

(a) $n = 4, l = 2, m_l = 0, m_s = 0$ is not valid because m_s is always $\pm \frac{1}{2}$ because $s = \frac{1}{2}$ for an electron. m_s is never equal to zero.

(b) $n = 3$, $l = 1$, $m_l = -3$, $m_s = -\frac{1}{2}$ is not valid because $m_l = -3$ does not go with $l = 1$. $m_l = -1$, 0, or 1 are the "allowed" values.

(c) $n = 3$, $l = 3$, $m_l = -1$, $m_s = +\frac{1}{2}$ is not valid because l must be less than or equal to $n - 1$.

Exercise 8.27—Orbital Shapes

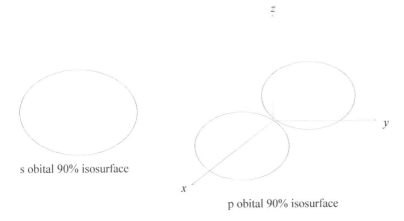

s obital 90% isosurface

p obital 90% isosurface

Exercise 8.33—Effective Nuclear Charge

$$Z^*\left(_{12}Mg\right) = +12 - [(1 \times 0.35) + (8 \times 0.85) + (2 \times 1.0)] = +12 - 9.15 = +2.85$$

$$Z^*\left(_{15}P\right) = +15 - [(4 \times 0.35) + (8 \times 0.85) + (2 \times 1.0)] = +15 - 10.2 = +4.8$$

$$Z^*\left(_{18}Ar\right) = +18 - [(7 \times 0.35) + (8 \times 0.85) + (2 \times 1.0)] = +18 - 11.25 = +6.75$$

These estimates show that valence electrons are held increasingly tightly by these elements in this order. This is consistent with the increasing first ionization energies of the elements.

Exercise 8.35—Effective Nuclear Charge and Atom Sizes

Although $_{84}Po$ has a lot more electrons than $_{11}Na$, the electrons of Po are held so much more tightly that it is smaller. The effective nuclear charges, +5.45 and +2.2, show that even the valence electrons of Po are held more tightly than the valence electrons of Na. The valence electrons determine the size of the atom.

Exercise 8.39—Effective Nuclear Charge and Charges on Simple Ions

(a) N^{3-} ion configuration: $1s^2\,2s^2\,2p^6$

$$Z^*\left(N^{3-}\right) = +7 - [(7 \times 0.35) + (2 \times 0.85)] = +7 - 4.15 = +2.85$$

Compare this with the N^{4-} ion: $1s^2\,2s^2\,2p^6\,3s^1$

$$Z^*\left(N^{4-}\right) = +7 - [(0 \times 0.35) + (8 \times 0.85) + (2 \times 1.0)] = +7 - 8.8 = -1.8$$

N can form a 3– ion because the valence electrons are still held by an effective nuclear charge of +2.85. However, a fourth additional electron, in the $3s$ orbital, would experience no attraction to the nuclear charge because of the degree of repulsion (shielding) by other electrons.

(b) S^{2-} ion configuration: $1s^2\,2s^2\,2p^6\,3s^2\,3p^6$

$$Z*(S^{2-}) = +16 - [(7 \times 0.35) + (8 \times 0.85) + (2 \times 1.0)] = +16 - 11.25 = +4.75$$

The S^{3-} ion has an electron in the next shell: $1s^2\ 2s^2\ 2p^6\ 3s^2\ 3p^6\ 4s^1$

$$Z*(S^{3-}) = +16 - [(0 \times 0.35) + (8 \times 0.85) + (10 \times 1.0)] = +16 - 16.8 = -0.8$$

S cannot form such an ion because its valence electron would be repelled by a net negative effective nuclear charge. The valence electron in the $4s$ orbital is well-shielded by the filled shells.

Exercise 8.41—Effective Nuclear Charge and Electronegativities

Electron configuration of $_{12}$Mg: $1s^2\ 2s^2\ 2p^6\ 3s^2$

$$Z*(Mg) = +12 - [(1 \times 0.35) + (8 \times 0.85) + (2 \times 1.0)] = +12 - 9.15 = +2.85$$

This value is a measure of the net force of attraction between the nucleus of a Mg atom and electrons in its valence shell. If a Mg atom were to attract another electron into its valence shell, we would have

$$Z*(Mg^-) = +12 - [(2 \times 0.35) + (8 \times 0.85) + (2 \times 1.0)] = +12 - 9.5 = +2.5$$

Electron configuration of $_{17}$Cl: $1s^2\ 2s^2\ 2p^6\ 3s^2\ 3p^5$

$$Z*(Cl) = +17 - [(6 \times 0.35) + (8 \times 0.85) + (2 \times 1.0)] = +17 - 10.9 = +6.1$$

This value illustrates that the force of attraction between the nucleus of a Cl atom and electrons in its valence shell is much greater than is the case for Mg atoms, and so the electronegativity of Cl is greater than that of Mg. Even if another electron is attracted into the valence shell of a Cl atom, it experiences strong attraction:

$$Z*(Cl^-) = +17 - [(7 \times 0.35) + (8 \times 0.85) + (2 \times 1.0)] = +17 - 11.25 = +5.75$$

REVIEW QUESTIONS

Section 8.1: Horseflies, Elephants and Electrons

8.43

There are many answers possible, here is an example: We walk at a speed of 2 km h^{-1}. When we wish to stop, the energy emitted is the difference (gap) between the action and at rest. However, not all energy gaps will be the same for each action. Other forms of energy usage include eating, metabolism/digestion, breathing, and temperature regulation.

8.45

(a) $\lambda = \dfrac{h}{mv} = \dfrac{6.626 \times 10^{-34}\ \text{kg m}^2\ \text{s}^{-1}}{(5000\ \text{kg}) \times (5\ \text{m s}^{-1})} = 2.65 \times 10^{-38}\ \text{m} = 2.65 \times 10^{-29}\ \text{nm} = 3 \times 10^{-29}$

This value for a macroscopic object, such as an elephant, is very small even when compared to an atom. Therefore, the elephant can be referred to as a particle rather than a wave.

(b) $\lambda = \dfrac{h}{mv} = \dfrac{6.626 \times 10^3 \text{ kg m}^2 \text{ s}^{-1}}{(5000 \text{ kg}) \times (5 \text{ m s}^{-1})} = 0.265 \text{ m} = 2.65 \times 10^8 \text{ nm} = 3 \times 10^8$

The elephant position may not be located with any degree of certainty until it is still.

Section 8.2: Periodic Variation of Properties of the Elements

8.47

(a) Calcium is a better electrical conductor than arsenic. It is a metal, while arsenic is a semi-metal.
(b) Sodium is a better thermal conductor than sulfur. It is a metal, while sulfur is a non-metal.
(c) Lead, the metal, is more malleable than non-metal carbon.

8.49

The next three elements, beyond fluorine, that occur as diatomic molecules and are powerful oxidizing agents are chlorine, bromine, and iodine. The associated atomic numbers are 17, 35, and 53. From fluorine to chlorine, the atomic number increases by 8. From chlorine to bromine, the atomic number increases by 18. From bromine to iodine, the atomic number increases by 18.

8.51

(a) Covalent bond distances are estimated as the sum of the atomic radii of the two bonded atoms. So, the H–O bond distance in H_2O is about 37 + 66 pm = 103 pm while the H–S bond distance is about 37 + 104 pm = 141 pm
(b) The covalent radius of Br is ½ the bond distance in Br_2,
½ × 228 pm = 114 pm
The estimated bond distance in BrCl is
114 + 99 pm = 213 pm

8.55

(a) $O^-(g) \longrightarrow O(g) + e^-(g)$
is endothermic. Work must be done to remove an electron from the ion, O^-. The electron affinity of O is positive (+141 kJ mol^{-1}, see Table 8.7).
(b) $O(g) + e^-(g) \longrightarrow O^-(g)$
is exothermic. As for atoms of all elements except those with negative electron affinity, energy is released when an electron attaches to the O atom to form an O^- ion.
(c) $Ne^-(g) \longrightarrow Ne(g) + e^-(g)$
is exothermic. Ne has a negative electron affinity. Energy is released when an electron is removed from a Ne$^-$ ion. Another way of seeing this is that work must be done to attach an electron to a Ne atom.

8.57

(a) Among B, Al, C and Si, Al has the most metallic character. It is lowest and furthest to the left within the periodic table.

(b) Al has the largest atomic radius. It is lowest and furthest to the left within the periodic table.

(c) C has the greatest electron affinity. It is highest and furthest to the right within the periodic table.

(d) In order of increasing first ionization energy, we have
Al < B < C
Here, we move up and to the right within the periodic table.

8.59

(a) Phosphorus, P, has the electron configuration, $1s^2 2s^2 2p^6 3s^2 3p^3$

(b) Beryllium, Be, is the alkaline earth (group 2) element with the smallest atomic radius.

(c) Nitrogen, N, is the group 15 element with the largest ionization energy.

(d) Technetium, Tc, is the element whose 2+ ion has the configuration $[\text{Kr}]4d^5$

(e) Fluorine, F, is the group 17 element with the greatest electron affinity.

(f) Zinc, Zn has the electron configuration, $[\text{Ar}]3d^{10}4s^2$

8.61

(a) Among S, Se, and Cl, Se has the largest atoms. Se is lowest and furthest to the left within the periodic table.

(b) Br^- ions are larger than Br atoms—one extra electron, same nuclear charge.

(c) Among Si, Na, P, or Mg, Na has the largest difference between the first and second ionization energy—a characteristic of the group 1 elements. The second ionization step requires removal of an electron from the filled second shell of Na^+. The second ionization step for the other atoms requires removal of a less tightly held third shell electron.

(d) Among the group 15 elements, N, P, or As, N has the largest ionization energy. N is the highest in the group.

(e) Among O^{2-}, N^{3-} and F^-, N^{3-} has the largest radius—same number of electrons, lowest nuclear charge.

Section 8.3: Experimental Evidence about Electrons in Atoms

8.63

(a) The line at 253.652 nm has the most energetic light.

(b)

$$v = \frac{c}{\lambda} = \frac{(2.998\times10^8 \text{ m s}^{-1})}{2.53652\times10^{-7} \text{ m}} = 1.18193\times10^{15} \text{ s}^{-1} = 1181.93\,\text{THz}$$

$$E_{\text{photon}} = hv = (6.626\times10^{-34}\,\text{J s}) \times (1.18193\times10^{15}\,\text{s}^{-1}) = 7.83147\times10^{-19}\,\text{J}$$

(c) The 404.656 nm and 435.833 nm lines are shown in the figure. They are blue.

8.65

The least energetic line in the Lyman series is associated with the $n_{initial} = 2$ to $n_{final} = 1$ transition of H atoms.

The frequency of this radiation is

$$\frac{E_{photon}}{h} = \frac{-2.179 \times 10^{-18}\ \cancel{J}}{6.626 \times 10^{-34}\ \cancel{J}\ s} \times \left(\frac{1}{n_{initial}^2} - \frac{1}{n_{final}^2} \right)$$

$$= -(3.289 \times 10^{15}\ s^{-1}) \times \left(\frac{1}{4} - \frac{1}{1} \right)$$

$$= (3.289 \times 10^{15}\ s^{-1}) \times (0.7500) = 2.466 \times 10^{15}\ s^{-1} = 2466\ \text{THz}$$

8.71

$$\lambda = \frac{h}{mv} = \frac{6.626 \times 10^{-34}\ \cancel{kg}\ m^2\ \cancel{s}^{-1}}{(9.109 \times 10^{-31}\ \cancel{kg}) \times (2.5 \times 10^6\ \cancel{m}\ \cancel{s}^{-1})} = 2.9 \times 10^{-10}\ m$$

$$= 0.29\ \text{nm}$$

Section 8.4: The Quantum Mechanical Model of Electrons in Atoms

8.73

(a) The allowed values of l are 0, 1, 2 and 3 (i.e., up to $n-1$), when $n = 4$.
(b) The allowed values of m_l are $-2, -1, 0, 1$ and 2 (i.e., from $-l$ to l), when $l = 2$.
(c) For a $4s$ orbital, the allowed values of n, l, and m_l are
$n = 4$, $l = 0$ and $m_l = 0$
(d) For a $4f$ orbital, the allowed values of n, l, and m_l are
$n = 4$, $l = 3$ and $m_l = -3, -2, -1, 0, 1, 2$ and 3

8.75

(a) The quantum numbers $n = 4$, $l = 2$, $m_l = -2$ give rise to a $4d$ orbital.
(b) The quantum numbers $n = 4$, $l = 2$, $m_l = 0$ lead to a solution for a $4d$ orbital.
(c) The quantum numbers $n = 3$, $l = 0$, $m_l = 0$ lead to a solution for a $3s$ orbital.
(d) The quantum numbers $n = 2$, $l = 1$, $m_l = -1$ lead to a solution for a $2p$ orbital.
(e) The quantum numbers $n = 5$, $l = 3$, $m_l = 2$ lead to a solution for a $5f$ orbital

8.77

(b) $2d$ and **(d)** $3f$ do NOT denote possible orbitals. $2d$ means $n = 2$ and $l = 2$ – but l cannot exceed 1 when $n = 2$.
(a) $2s$, **(c)** $3p$, **(e)** $4f$ and **(f)** $5s$ are possible orbitals.

8.81

(a) $n = 3$, $l = 0$, and $m_l = +1$
defines NO orbital
$m_l = 0$ when $l = 0$

(b) $n = 5$, $l = 1$
defines 3 orbitals
$m_l = -1$, 0 and 1

(c) $n = 7$, $l = 5$
defines 11 orbitals
$m_l = -5$, -4, ... , 4 and 5

(d) $n = 4$, $l = 2$, and $m_l = -2$
defines just one orbital

8.83

(a) $n = 2$, $l = 2$, $m_l = 0$, $m_s = +\frac{1}{2}$ is NOT a valid set of quantum numbers because l cannot exceed 1 when $n = 2$.

(b) $n = 2$, $l = 1$, $m_l = -1$, $m_s = 0$ is NOT a valid set of quantum numbers because m_s is never equal to 0.

(c) $n = 3$, $l = 1$, $m_l = +2$, $m_s = +\frac{1}{2}$ is NOT a valid set of quantum numbers because $|m_l|$ must be less than or equal to 1 when $l = 1$.

8.85

Orbital Type	Number of Orbitals in Given Sub-shell	Number of Nodal Surfaces
s	1	0
p	3	1
d	5	2
f	7	3

Section 8.5: Electron Configurations in Atoms

8.87

Gallium atoms have the following ground state electron configuration:

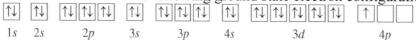
1s   2s   2p   3s   3p   4s   3d   4p

The highest energy electron is the lone 4p electron. It has quantum numbers, $n = 4$, $l = 1$ and $m_l = -1$, 0 or 1.

8.89

(a) Chlorine, Cl, has the ground-state configuration $1s^2 2s^2 2p^6 3s^2 3p^5$.

(b) Phosphorus atoms have the following ground state electron configuration: $1s^2 2s^2 2p^6 3s^2 3p^3$
or

1s   2s   2p   3s   3p

(c) There are 2 4s valence electrons in Ca. The quantum numbers for these electrons are $n = 4$, $l = 0$, $m_l = 0$ and $m_s = \pm \frac{1}{2}$

8.91

(a) Phosphorus atoms have the following ground state electron configuration: $1s^2 2s^2 2p^6 3s^2 3p^3$
or

↑↓	↑↓	↑↓ ↑↓ ↑↓	↑↓	↑ ↑ ↑

$1s$ $2s$ $2p$ $3s$ $3p$

(b) Chlorine atoms have the following ground state electron configuration: $1s^2 2s^2 2p^6 3s^2 3p^5$ or

↑↓	↑↓	↑↓ ↑↓ ↑↓	↑↓	↑↓ ↑↓ ↑

$1s$ $2s$ $2p$ $3s$ $3p$

Chlorine being to the right of phosphorus, in the same period, has more electrons.

Section 8.6: Shielding and Effective Nuclear Charge

8.95

(a) Elements with atomic numbers 17, 18, and 19—i.e., Cl, Ar, and K—have the following effective nuclear charges:

$$Z^*(\text{Cl}) = +17 - [(6 \times 0.35) + (8 \times 0.85) + (2 \times 1.0)] = +17 - 10.9 = +6.1$$

$$Z^*(\text{Ar}) = +18 - [(7 \times 0.35) + (8 \times 0.85) + (2 \times 1.0)] = +18 - 11.25 = +6.75$$

$$Z^*(\text{K}) = +19 - [(0 \times 0.35) + (8 \times 0.85) + (10 \times 1.0)] = +19 - 16.8 = +2.2$$

Here, we see that argon is unreactive because it holds its electrons very tightly. It is therefore hard to remove an electron. Adding an electron produces an effective nuclear charge of +1.2 (1 less than that of K)—a much lower value. This is a very unfavourable process. Chlorine similarly holds its electrons very tightly, but it can add an electron to yield an anion with an effective nuclear charge of +5.75 (1 less than Ar). This is a very stable anion, making its formation via reaction of chlorine very favourable—chlorine is highly reactive. K is similarly very reactive. In this case, it is because the valence electron is easily lost—K forms the +1 cation.

(b)

$$Z^*(\text{Ne}^-) = +10 - [(0 \times 0.35) + (8 \times 0.85) + (2 \times 1.0)] = +10 - 8.8 = +1.2$$

$$Z^*(\text{Ar}^-) = +18 - [(0 \times 0.35) + (8 \times 0.85) + (10 \times 1.0)] = +18 - 16.8 = +1.2$$

$$Z^*(\text{Kr}^-) = +36 - [(0 \times 0.35) + (18 \times 0.85) + (18 \times 1.0)] = +36 - 33.3 = +2.7$$

The valence electron in these anions is weakly held. Though the effective nuclear charge of Kr⁻ is larger, its radius is so large that the electron is still quite weakly held. The neutral atoms all have much larger effective nuclear charges. They are quite unreactive with respect to formation of the anion. Nevertheless, krypton compounds have been synthesized—not so for neon and argon. This latter observation is consistent with Kr⁻ having the larger effective nuclear charge—making it more stable.

(c) Because Z^* is so large for Ne, Ar and Kr atoms, the valence electrons are held very tightly. This reduces the polarizability of these atoms, and correspondingly reduces the dispersion forces between them. These elements occur mostly as monatomic gases. They have very low melting and boiling points.

Section 8.7: Rationalizing the Periodic Variation of Properties

8.97

(a) The element with successive ionization energies, 590, 1145, 4912, 6491, and 8153 kJ mol^{-1}, is in group 2.

(b) Since the element is in the fourth period, it must be Ca.

(c) We deduce this because the largest relative jump in ionization energy comes between the second and third. Group 2 elements have relatively low first and second ionization energies. The third ionization energy is much larger because the third electron is taken from the next highest shell which has a much higher effective nuclear charge.

Section 8.8: Modelling Atoms and Their Electrons: A Human Activity

8.99

No answers provided. They will come about through the exploration of this activity.

SUMMARY AND CONCEPTUAL QUESTIONS

8.101

The Bohr model treats electrons as point particles with definite position and velocity. However, experiments show that electrons have wave properties which make their position and velocity intrinsically uncertain. The Bohr model introduced the wavelength of the electron through an ad hoc quantization condition to give the correct Rydberg formula for the emission lines of hydrogen. It is now known that the wave character of electrons is of a more fundamental nature.

8.103

The square of the wave function, ψ, at each point gives the probability density (probability per unit volume) of finding the electron at a defined position in space around the nucleus.

8.105

Altogether, in the first three electron shells—in a universe with quantum numbers $N = 1, 2, 3, ..., \infty$, $L = N$ and $M = -1, 0, +1$, there are $3 + 3 + 3 = 9$ orbitals.

8.107

Among K^{2+}, Cs^+, Al^{4+}, F^{2-}, and Se^{2-}, Se^{2-} is the one most likely to be found in a chemical compound. Two is the normal valence of Se—the only negative valence normally exhibited by selenium. The other species are not normal valences. For K and Cs, it is +1; for Al, +3, and for F, −1.

Chapter 9
Molecular Structures, Shapes, and Stereochemistry—Our Evidence

IN-CHAPTER EXERCISES

Exercise 9.1—Differentiate Isomers by ^{13}C NMR Spectroscopy

$CH_3CH_2CH_2CH_2OH$ $CH_3CH_2OCH_2CH_3$
butan-1-ol   diethyl ether
Diethyl ether has only two distinct C atoms, because of symmetry. Butan-1-ol has four.

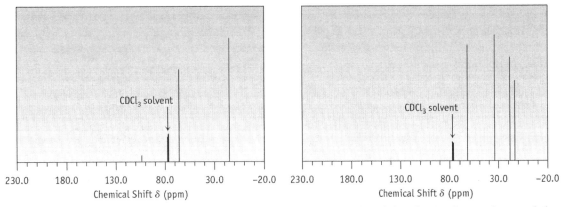

The spectrum on the left shows two peaks, while that on the right shows four—beyond the solvent and reference peaks. The spectra on the left and right are those of diethyl ether and butan-1-ol, respectively.

Exercise 9.3—Differentiate Positional Isomers by ^{13}C NMR Spectroscopy

$CH_3CH_2CH_2NH_2$ $CH_3CH(NH_2)CH_3$
propan-1-amine   propan-2-amine
Propan-2-amine has only two distinct C atoms, because of symmetry. Propan-1-amine has three.

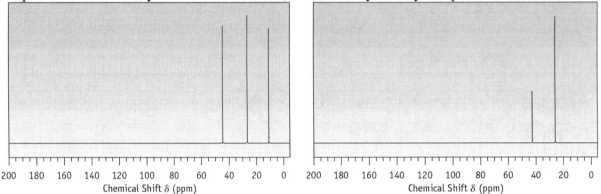

The spectrum on the right shows two peaks, while that on the left shows three. The spectra on the right and left are those of propan-2-amine and propan-1-amine, respectively.

Exercise 9.5—^{1}H NMR and ^{13}C NMR Spectroscopy to Determine Structure

(a) CH_4 methane
1 peak in both carbon-13 and proton spectra—the 4 H atoms are equivalent.

(b) CH_3CH_3 ethane
1 peak in both carbon-13 and proton spectra—the 2 C atoms are equivalent, as are the 6 H atoms.

(c) $CH_3CH_2CH_3$ propane
2 peaks in both carbon-13 and proton spectra—the 2 methyl C atoms are equivalent, while the 6 methyl H atoms and 2 methylene H atoms are equivalent.

(d) cyclohexane

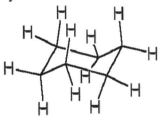

All C atoms are equivalent by symmetry—1 peak in carbon-13 spectrum.
There are two kinds of H atoms—axial and equatorial. However, at ordinary temperatures, the ring flips—chair to boat to chair (facing opposite direction)—back and forth so fast that the NMR experiment "sees" the average environment—i.e., we see one peak in the proton spectrum. If the sample were cooled sufficiently, it may be possible to see separate axial and equatorial peaks.

(e) CH_3OCH_3 dimethyl ether
1 peak in both carbon-13 and proton spectra

(f) benzene

1 peak in both carbon-13 and proton spectra

(g) $(CH_3)_3COH$ 1,1-dimethylethanol

2 peaks in both carbon-13 and proton spectra

(h) CH_3CH_2Cl chloroethane

2 peaks in both carbon-13 and proton spectra

(i) $(CH_3)_2C=C(CH_3)_2$ 2,3-dimethylbut-2-ene

2 peaks in carbon-13 spectrum and 1 peak in proton spectrum

Exercise 9.7—Newman Projections to Show Conformations

These are the Newman projections of the three staggered conformers. The conformer on the left has the lowest energy. The other two have the same energy.

These are the Newman projections of the three eclipsed conformations. The conformation on the left has the highest energy. The other two have the same energy.

Exercise 9.9—Skeletal Structural Representations

(a)	pyridine	C_5H_5N	one H atom is bonded to each C atom
(b)	cyclohexanone	$C_6H_{10}O$	two H atoms are bonded to each C atom, except the C=O carbon
(c)	indole	C_8H_6NH	one H atom is bonded to each C atom, except the two C atoms belonging to both rings

Exercise 9.15—Chair Conformers of Cyclohexane Molecules

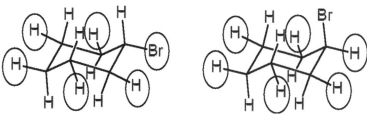

Here are the equatorial (on the left) bromine and axial bromine conformers of bromocyclohexane. The equatorial positions are circled.

Exercise 9.17—Optical Activity

Since $[\alpha]_D = -16°$ is negative for cocaine, cocaine is termed levorotatory.

Exercise 9.21—*R/S* Designation

From highest to lowest priority of substituents, we have
(a) —Br > —CH₂CH₂OH > —CH₂CH₃ > —H
(b) —OH > —CO₂CH₃ > —CO₂H > —CH₂OH
(c) —Br > —Cl > —CH₂Br > —CH₂Cl

Exercise 9.23—*R/S* Designation

At the top and bottom (respectively) carbon stereocentre, the configurations are
(a) *R* and *R*,
(b) *S* and *R*, and
(c) *R* and *S*

Exercise 9.25—Meso Stereoisomers

(a) 2,3-Dibromobutane is a meso compound—it is achiral. The molecule on the right is identical to its mirror image on the left

(b) 2,3-Dibromopentane: The compounds on the left and right are enantiomers

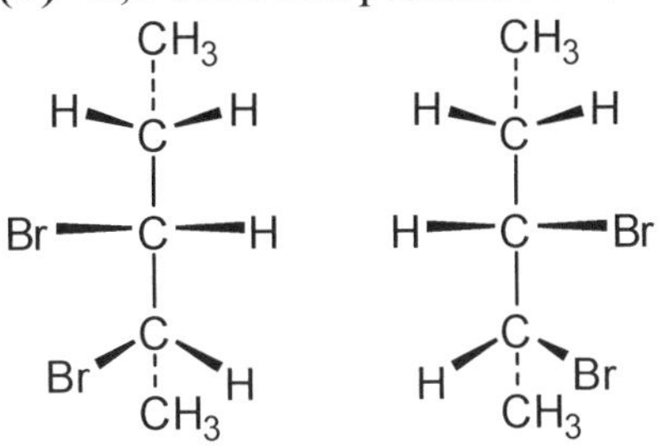

(c) 2,4-Dibromopentane is a meso compound—it is achiral. The molecule on the right is identical to its mirror image on the left

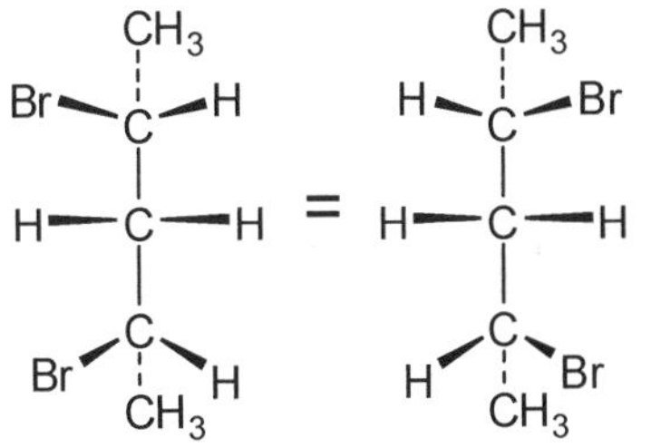

Exercise 9.27—Stereoisomers

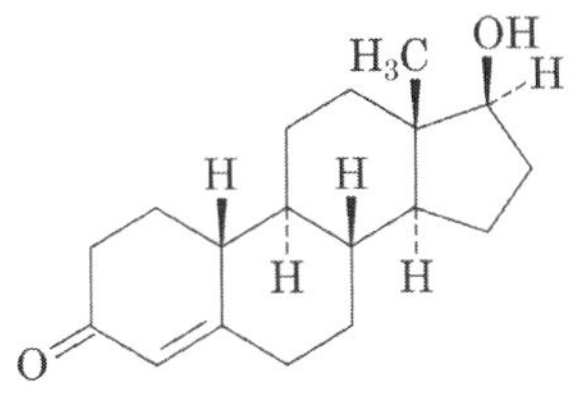

Nandrolone

Nandrolone has 6 stereocentres. There could in principle be as many as $2^6 = 64$ different stereoisomers.

REVIEW QUESTIONS

Section 9.1: Molecular Handshakes and Recognition

9.29

Intermolecular forces describe the forces of attraction between molecules. An example of this is hydrogen bonding in water.
Intramolecular forces describe the forces of attraction between atoms within the same molecule. An example of this is ionic bonding in NaCl.

9.31

The liquid deposits from geckos should contain both hydrophobic (fatty acid chains/lipids) and hydrophilic (phosphate group, glycerol) regions.

9.33

(a)+(b) The most important intermolecular force within and between β-sheets of β-keratins is hydrogen bonding. As stated in the answer for Question 9.32, this hydrogen bonding is rarely perfect, and can exist when strands are parallel or antiparallel to each other. These simple bonding modes can result in more complicated motifs.

9.35

Bacteria that are in close proximity can turn on genes that coordinate their behaviour (e.g., obtain nutrient sources, exchange DNA, create protective environment to resist antibiotics).

9.37

There is a large individual variation in the human senses. The varied reactions suggests that these people house molecular recognition receptors that could be either drastically or slightly different, which causes the brain to recognize these molecular handshakes as different odours.

9.39

Self-assembly is the ability for molecules to arrange/organize themselves in an ordered way as a result of molecular recognition. These molecules in close proximity must detect/recognize a complement group to create an interaction spontaneously.

Section 9.2: Experimental Tools for Molecular Structures and Shapes

9.41

Recall, the two different carvone molecules have the same molecular formula but different stereochemistry (they are chiral molecules).
(a) High-resolution mass spectrometry can only provide limited information about atom connectivity.
(b) Infrared spectroscopy determines which functional groups (C=O and C=C) are bound to the carbon skeleton and provides some information about connectivity.
(c) ^{13}C NMR spectroscopy will map carbon-hydrogen framework.
(d) X-ray crystallography shows spatial relationship of atoms, bond distances and angles, full structure connectivity (solid-state molecular structure).
(e) Polarimetry will determine configuration (stereochemistry).
While **(a)**–**(d)** will provide information about connectivity, only polarimetry can definitively tell the two forms of carvone apart.

Section 9.3: X-ray Crystallography

9.43

(a) $CH_3CH_2CH=CH=CH_2CH_2CH_3$
   is a hydrocarbon with seven peaks in its ^{13}C NMR spectrum.
(b) $(CH_3)_2CHCH_2CH_2CH_3$
   is a six-carbon compound with only five peaks in its ^{13}C NMR spectrum.
(c) $(CH_3)_2CHC(O)H$
   is a four-carbon compound with three peaks in its ^{13}C NMR spectrum and a carbonyl
   functional group.

9.45

NMR uses photons with lower energy than those used in IR spectroscopy. 400 MHz is a much
lower frequency than the 10–100 THz range of IR spectroscopy. NMR causes transitions
between spin energy levels that are very close. IR causes vibrational energy transitions that are
much higher in energy.

9.47

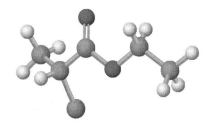

Reasonable values for the carbon atom chemical shifts are as follows:
• carbonyl carbon atom at 200 ppm
• two methyl carbon atoms at around 20 ppm—two distinct peaks
• C–O methylene carbon atom at 70 ppm
• C–Cl methine carbon atom at 50 ppm
The carbon-13 spectrum might look something like:

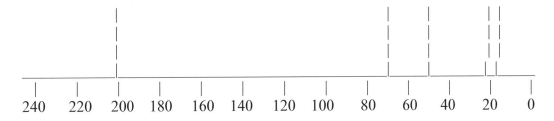

9.49

(a) Chloroform absorbs 77.23 ppm downfield from TMS.
(b) A chemical shift of δ 77.23 corresponds to 77.23 × 600 Hz = 46338 Hz downfield from TMS on a 600 MHz spectrometer.
(c) In δ units, the chemical shift of chloroform is δ 77.23 on a 600 MHz spectrometer.

9.51

In order of increasing ^{13}C chemical shift, the one peak exhibited by these compounds are expected to be ordered as follows:
CH_4 < CH_2Cl_2 < $HC\equiv CH$ < benzene < $HC(O)C(O)H$ (two carbonyl carbon atoms bonded together)

Section 9.5: Conformations of Alkanes—Rotation about Single Bonds

9.53

Possible skeletal structures:
(a) C_4H_8

(b) C_3H_6O

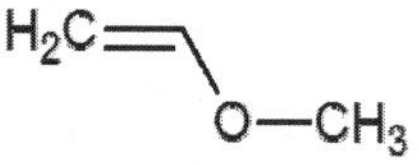

(c) C_4H_9Cl

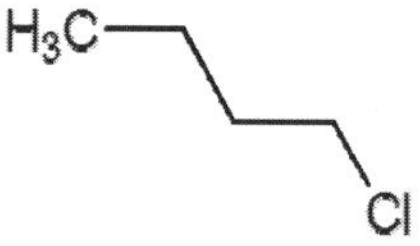

9.55

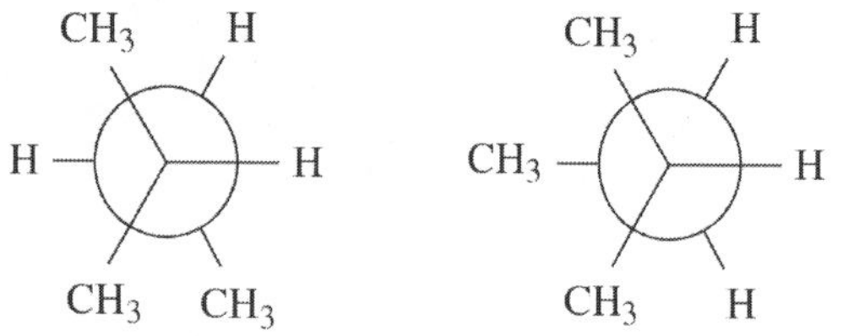

The staggered conformer on the left is expected to have the lower energy because the methyl substituent on the back carbon atom is close to both of the methyl substituents on the front carbon atom in the right conformer (two gauche interactions), but only one in the left conformer (one gauche interaction).

9.57

In bromoethane, there are three atom–atom interactions in the eclipsed conformation giving rise to the rotation barrier of 15.0 kJ mol^{-1} about the C–C bond—two H–H interactions and one H–Br interaction. The two H–H interactions account for 2×3.8 kJ mol^{-1} = 7.6 kJ mol^{-1} of the barrier. The H–Br interaction accounts for $15.0 - 7.6$ kJ mol^{-1} = 7.4 kJ mol^{-1}.

Section 9.6: Restricted Rotation about Bonds

9.59

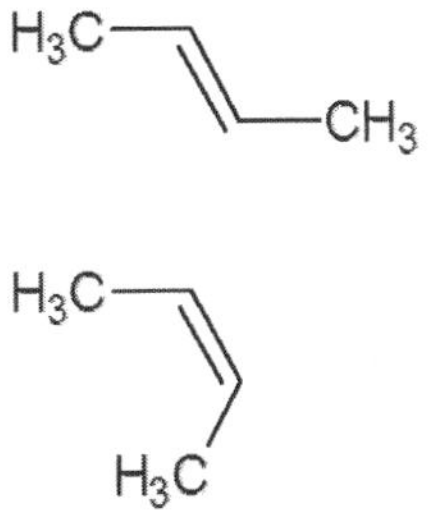

cis and *trans* 2-butene

The other structural isomers of C_4H_8 shown above do not have *cis* and *trans* isomers.

Section 9.7: Cyclic Molecules

9.61

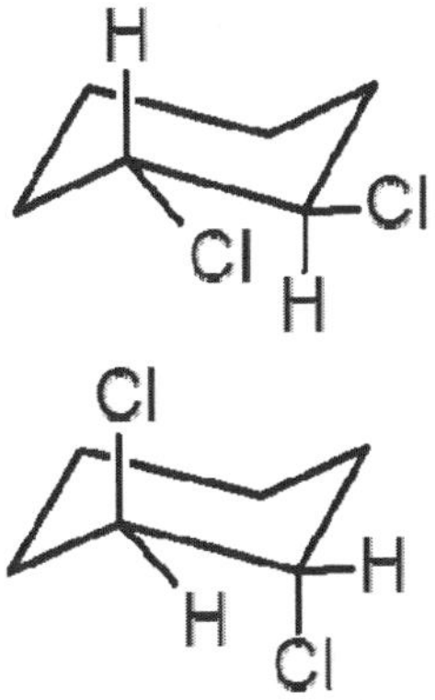

trans-1,2-dichlorocyclohexane in chair conformer. Either both Cl atoms are equatorial (top conformer) or axial (bottom conformer) because on adjacent carbon atoms the two axial positions are on opposite sides of the ring—the two equatorial are necessarily also on opposite sides of the ring. The *trans* isomer has the two Cl atoms on opposite sides of the ring. The diequatorial form is lower in energy.

9.63

ring flip

9.65

trans-1,2-dimethylcyclohexane in its more stable chair conformer. The two methyl groups (larger than the H atom substituents) are in more accommodating equatorial positions. Because this is *trans*, methyl groups on adjacent carbon atoms must be the same position—i.e., either both axial or both equatorial.

9.67

N-Methylpiperidine

The most stable conformer of *N*-methylpiperidine. Clearly the methyl substituent has greater steric requirements than the electron lone pair—the methyl group is in the more accommodating equatorial position.

9.69

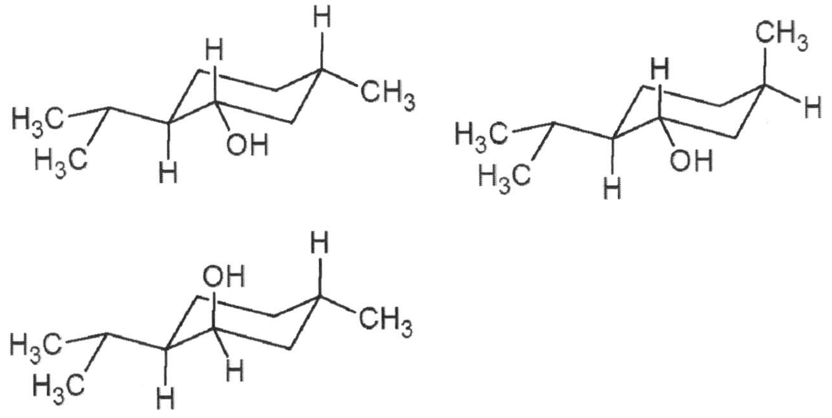

Three of the *cis–trans* isomers of menthol are shown. In the first and second, the hydroxyl group is *trans* to the isopropyl group—it is *cis* in the third isomer. In the first and third, the methyl group is *trans* to the isopropyl group—it is *cis* in the second isomer.

Section 9.8: Stereochemistry

9.71

(a)

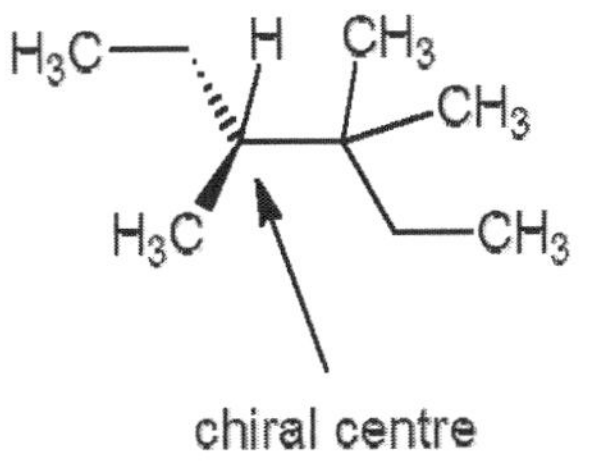

chiral centre

(b)

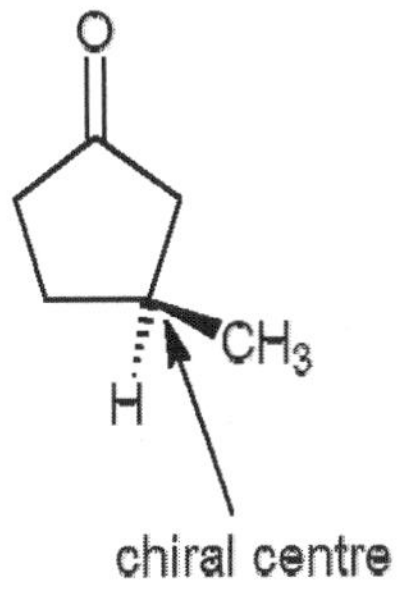

chiral centre

(c)

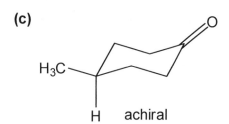

achiral

the ring has a plane of symmetry through C4 and the carbonyl bond

(d) two chiral centres

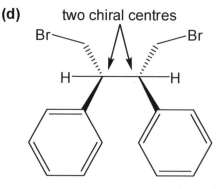

Depending on whether the stereoisomers are configured *S,S* or *R,R*, or *S,R* (*R,S* is the same by symmetry), the molecule is either chiral (former case - *S,S* and *R,R* are enantiomers) or achiral (*S,R* = *R,S* is called a *meso* compound)

(e)

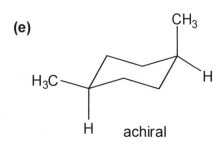

achiral

the ring has a plane of symmetry through C1 and C4

Section 9.9: Optical Activity

9.73

Specific rotation $[\alpha]_D$ equals observed rotation, α, divided by the product of sample concentration c and path length l, measured in g mL^{-1} and dm, respectively.

$[\alpha]_D = \dfrac{\alpha}{c \times l}$, where $\alpha = 2.22°$, $l = 1.00$ cm = 0.100 dm, and $c = 3.00$ g in 5.00 mL = 0.600 g mL^{-1}.

$[\alpha]_D = \dfrac{2.22°}{(0.600 \text{ g mL}^{-1}) \times (0.100 \text{ dm})} = 37.0°$

Section 9.10: Sequence Rules for Specifying Configuration

9.75

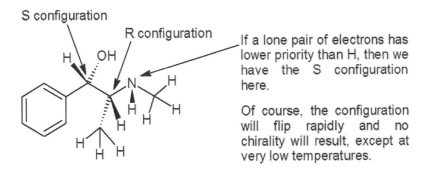

S configuration

R configuration

If a lone pair of electrons has lower priority than H, then we have the S configuration here.

Of course, the configuration will flip rapidly and no chirality will result, except at very low temperatures.

9.77

Sets of substituents are assigned priorities as follows:

(a)

1   2   3   4

(b)

1 –SO$_3$H   2 –SH   3 –OCH$_2$CH$_2$OH   4 –NH$_2$

9.79

(a)   (b)   (c)

The configurations at the chiral centres are

(a) *S* **(b)** *S* **(c)** *S*

Section 9.11: Enantiomers, Diastereomers, and Meso Stereoisomers

9.81

Substances comprised of (2*R*,3*R*)-dihydroxypentane and (2*S*,3*S*)-dihydroxypentane will have the same (not equal to zero) numerical value for their optical rotation except for opposite signs. These stereoisomers are enantiomers. (2*R*,3*S*)-dihydroxypentane and (2*R*,3*R*)-dihydroxypentane are diastereomers of each other, so there will be no predictable relationship between the optical rotation values for substances comprised of these two stereoisomers.

9.83

The stereochemical configurations of the two diastereomers of (2*S*,4*R*)-dibromooctane are (2*R*,4*R*) and (2*S*,4*S*).

Section 9.12: Molecules with More than Two Stereocentres

9.85

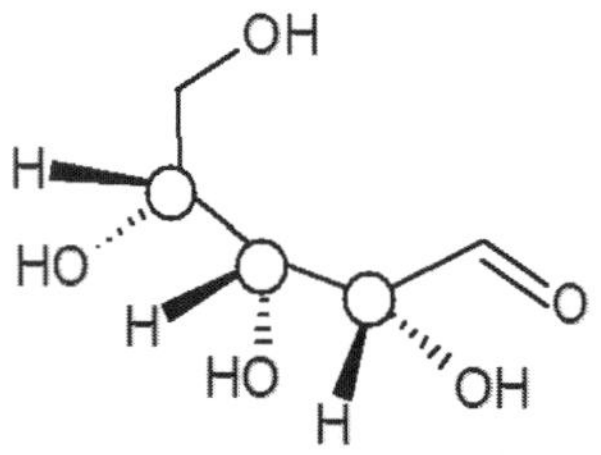

Ribose—three stereocentres indicated with circles. There are $2^3 = 8$ stereoisomers of ribose.

9.87

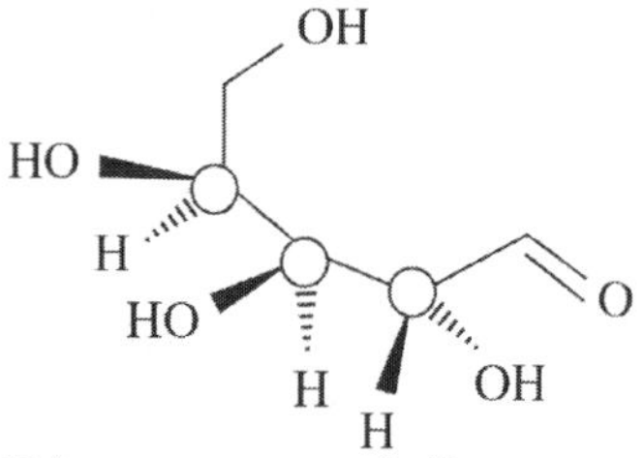

Diastereomer of ribose

Section 9.13: Chiral Environments in Laboratories and Living Systems

9.89

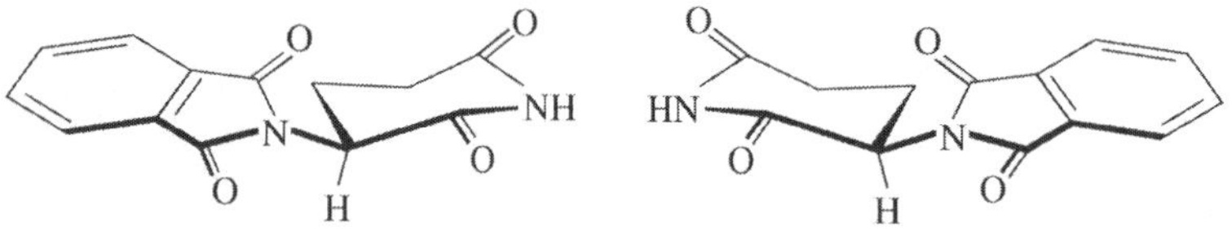

Thalidomide

(a) The lone stereocentre is the C atom bonded to the N atom of the 5-membered ring.
(b) The structure on the right is the R enantiomer—the S enantiomer is the structure on the left.

9.91

Deuterium is a heavier, stable isotope of hydrogen. The shape and polarity of the odorant molecule will not change, as they are approximately the same, although deuterium will make the stronger bond. However, ^{2}H or D, is twice as heavy as ^{1}H (2.014 vs. 1.008) and as such, the vibrational frequencies of the H- and D- substituted molecules would be greatly different (deuterium will vibrate at a lower frequency as it has a much longer amplitude).

SUMMARY AND CONCEPTUAL QUESTIONS

9.93

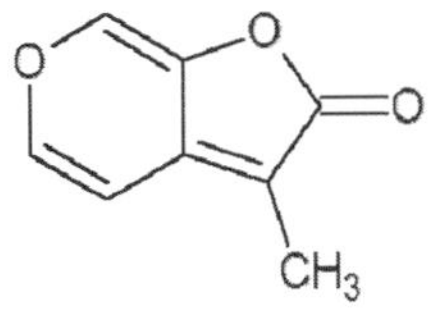

Gavinone is expected to show 8 peaks in its ^{13}C NMR spectrum—all the carbon atoms are non-equivalent.

9.95

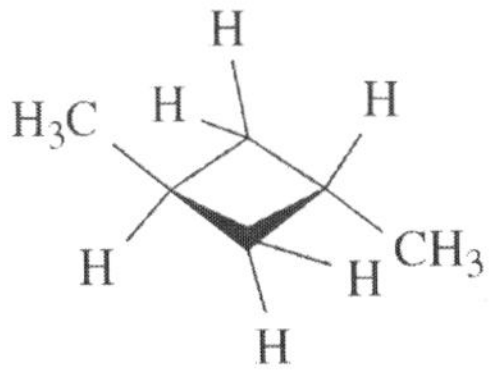

trans-1,3-dimethylcyclobutane

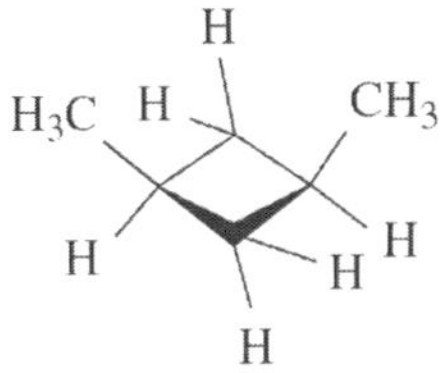

cis-1,3-dimethylcyclobutane—a stereoisomer of *trans*-1,3-dimethylcyclobutane

9.97

Anything with a plane of symmetry is NOT chiral—otherwise it is chiral.

(a) A basketball is NOT chiral

(b) A wine glass is (generally—depending on the pattern—if it is patterned) NOT chiral

(c) An ear is chiral

(d) A snowflake is NOT chiral

(e) A coin is chiral—the images imprinted on the sides of the coin are generally not symmetrical

(f) Scissors are chiral if the thumb handle is bigger (or in any way different) than the finger handle—note that the way the blades slide past each other breaks the plane of symmetry of the handles

9.99

(a) A chiral chloroalkane, $C_5H_{11}Cl$

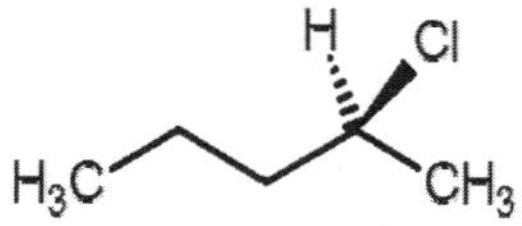

(b) A chiral alcohol, $C_6H_{14}O$

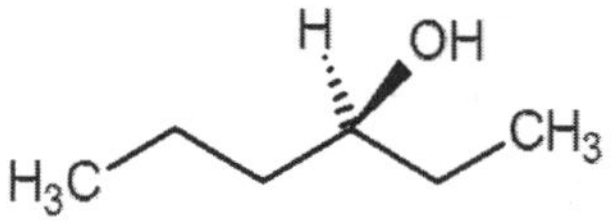

(c) A chiral alkene, C_6H_{12}

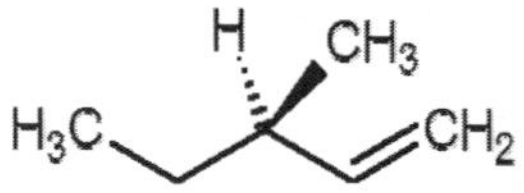

(d) A chiral alkane, C_8H_{18}

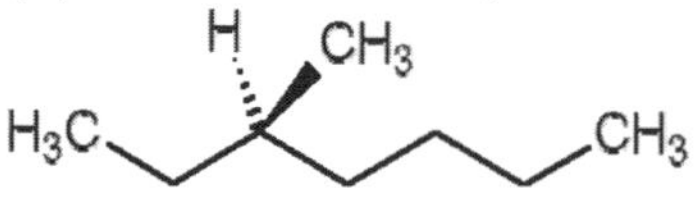

9.101

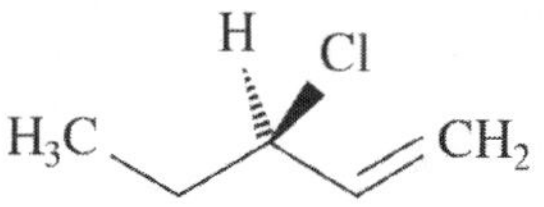

A tetrahedral representation of (*R*)-3-chloropent-1-ene.

9.103

(a)

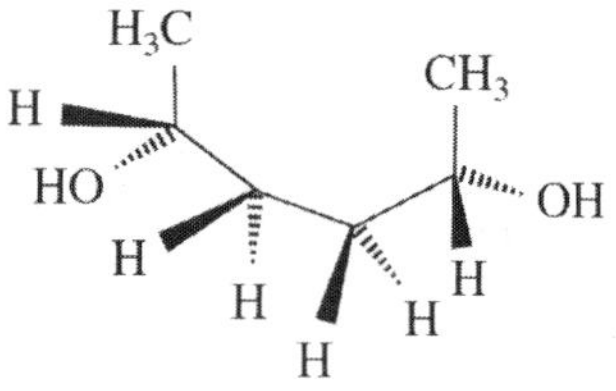

the meso form—(*R,S*) from left to right—of (2*R*, 5*S*)-2,5-hexanediol
The plane of this page is the plane of symmetry.

(b)

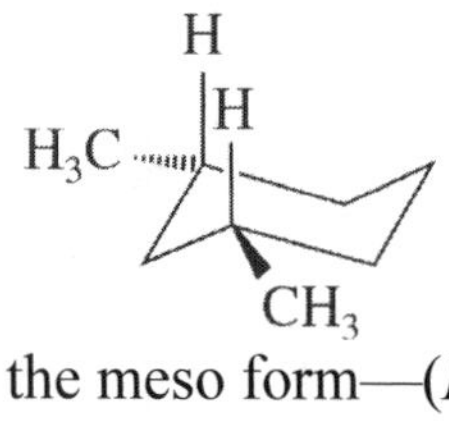

the meso form—(*R,S*) —of (1*R*, 3*S*)-1,3-dimethylcylohexane
The plane of symmetry that goes through C2 and C5 and the H atoms connected to each of them.

(c)

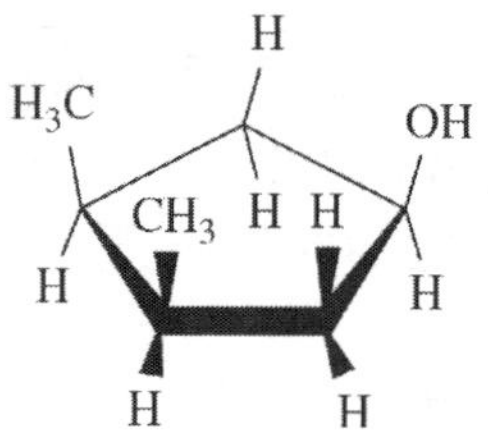

the meso form—(R,S)—of ($3R,4S$)-3,4-dimethylcyclopentan-1-ol

The plane of symmetry that goes through C1 and its OH and H group and in between C3 and C4.

9.105

Fischer projections are useful in this problem—you will see this later. The orientation of the molecules is indicated here by bolded and hashed bonds.

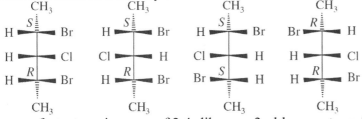

There are four stereoisomers of 2,4-dibromo-3-chloropentane. The two on the right are a pair of enantiomers—substances containing only one of these compounds would be optically active. The two on the left are optically inactive meso compounds—these are diastereomers.

9.107

(a) *cis*-1,3-Dibromocyclohexane and *trans*-1,4-dibromocyclohexane are constitutional isomers—specifically, positional isomers.

(b) 2,3-Dimethylhexane and 2,5,5-trimethylpentane are also constitutional—specifically, positional – isomers.

(c)

are identical—one rotates into the other.

9.109

is the enantiomer of the 2-chlorobutane stereoisomer in Problem 9.108.

CHAPTER 10
Modelling Bonding in Molecules

IN-CHAPTER EXERCISES

Exercise 10.1—Drawing Lewis Structures

Ammonium ion, NH_4^+
Number of valence electrons = 5 (from N) + 4×1 (from H) − 1 (net +1 charge)
$$= 8 \text{ electrons (4 electron pairs)}$$
Use all four electron pairs to form bonds between N and each of the four H's.

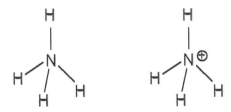

We are finished with the structure on the left. Here, all atoms have a filled valence shell—the N has 8 valence electrons.

Later in the chapter (Section 10.6), you will learn how to attribute the net +1 charge to the N atom, as shown on the right. This is not necessarily a completely accurate description of the distribution of charge. However, it is easy to do—as you will see later—and it gives a first guess at the charge distribution.

CO molecule
Number of valence electrons = 4 (from C) + 6 (from O)
$$= 10 \text{ electrons (5 electron pairs)}$$
Use one electron pair to form a bond between C and O.

   C—O

Place three pairs of electrons on the O atom—as lone pairs—to give O its octet.
There is one pair left. It goes to the C atom. We now have

   :C——Ö:

Here, the C atom only has four electrons in its valence shell. Shift two of the O atom lone pairs into the bonding region of CO to give the C atom its octet—a triple bond between C and O is formed in the process. The result is:

   :C≡O: :C≡O:

The structure on the right has "formal charges" included (see later in the chapter).

Sulfate ion, SO$_4{}^{2-}$

Number of valence electrons = 6 (from S) + 4×6 (from the 4 O's) + 2 (net − 2 charge)
= 32 electrons (16 electron pairs)

Use four electron pairs to form bonds between the S atom and each of the four O atoms.

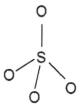

$16 - 4 = 12$ electron pairs remain. Place three pairs of electrons on each O atom—as lone pairs—to give each O atom its octet.
$12 - 4 \times 3 = 0$ electron pairs remain. We now have:

Here, each atom has its octet. Later you will learn how S atoms, being in the third period, are capable of expanding their valence shells beyond the octet. For now, we stop here.

Exercise 10.5—Lewis Structures of Oxoacids and Their Anions

Number of valence electrons = 5 (from P) + 2×1 (from H) + 4×6 (from O) + 1 (net −1 charge)
= 32 electrons (16 electron pairs)

Skeletal structure:

Six electron pairs are used as single bonds here.
$16 - 6 = 10$ electrons pairs remain. Two of the O atoms need three pairs to complete their octet, while the other two only need two pairs. Altogether, $2 \times 3 + 2 \times 2 = 10$ electron pairs are needed to give every atom a filled valence shell—the P atom already has an octet.

Exercise 10.11—Deciding Which Resonance Structures Are Most Important

These are the only electron distributions in which the valence shell of each O atom is full. These distributions of electrons in an ozone molecule are equivalent—by symmetry. They make equal contributions to the actual distribution, which is neither of these, but with π electrons delocalized over the two O–O bonds.

Exercise 10.13—Predicting the Shapes of Molecules

Lewis structure of dichloromethane. It has four electron regions about the central carbon atom—all single bonds. These bonds are tetrahedrally oriented. If the molecule was a regular tetrahedron, the Cl–C–Cl bond angle would be 109.5°. Because the Cl atoms are bigger than the H atoms, we might predict the bond angle to be a little bigger than 109.5°—and experimental measurements tell us that it is.

Exercise 10.15—Predicting the Shapes of Molecules and Ions

(a) Phosphate ion, PO_4^{3-}, is tetrahedral. It has four electron regions—3 single bonds and a double bond.

(b) Phosphoric acid molecule, H_3PO_4, is also tetrahedral around the P atom. It has four electron regions—3 single bonds and a double bond.

(c) Sulfate ion, SO_4^{2-}, is tetrahedral. It has four electron regions around the S atom—2 single bonds and 2 double bonds.

(d) A sulfite ion, SO_3^{2-}, is trigonal pyramidal. It has four electron regions around the S atom—2 single bonds, a double bond, and a lone pair—which are oriented tetrahedrally.

(e) Ethanol is tetrahedral at the carbon atoms and bent at the oxygen atom (4 electron regions, two of which are non-bonding).

(f) Acetone, $CH_3C(O)CH_3$, is trigonal planar at the carbonyl carbon atom (three electron regions, a double bond and two single bonds), and tetrahedral at the methyl carbon atoms.

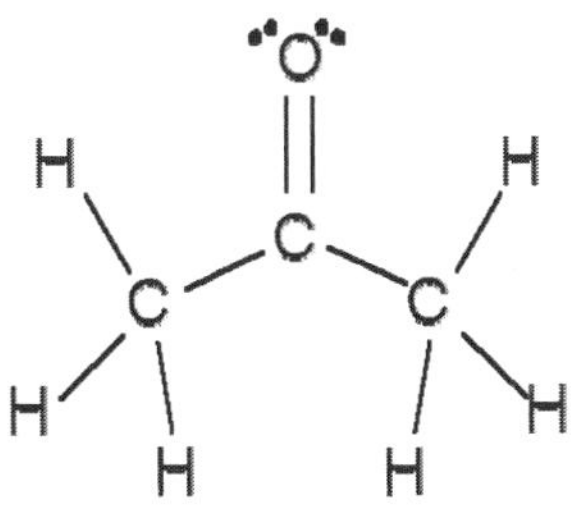

Exercise 10.19—Bonding in a Molecule of Acetone

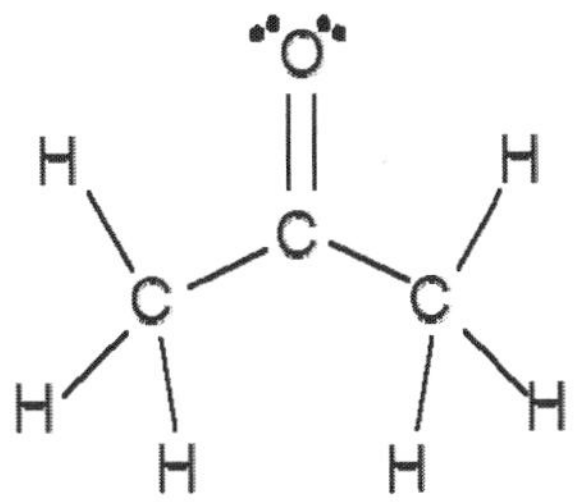

Lewis structure of acetone

The methyl C atoms have the tetrahedral distribution of electron regions. We presume sp^3 hybridization: Three of the sp^3 hybrid orbitals overlap with s orbitals on H atoms to form the C–H bonds, while the fourth makes a bond with the carbonyl C atom.

The carbonyl C atom has a trigonal-planar arrangement of electron regions—consequently, we use sp^2 hybridization here. The C–C bonds are formed from carbonyl-carbon atom sp^2 hybrid orbitals, and a methyl-carbon sp^3 hybrid orbital. The remaining sp^2 hybrid orbital forms the σ bond to oxygen.

The oxygen atom has three electron regions—the trigonal-planar arrangement—as depicted above. This O atom uses one sp^2 orbital to make the σ bond to C. The remaining two sp^2 orbitals accommodate the lone pairs.

The C=O double bond is formed from the unhybridized p orbitals on each atom—both atoms are sp^2 hybridized with one unhybridized p orbital. This p-p bond is the C–O π bond.

Since the O atom is terminal, it could also be described in terms of sp hybridization. In this case, one sp hybrid orbital forms the σ bond to C, while the other accommodates a lone pair. One of the unhybridized p orbitals on O (the p orbital perpendicular to the plane of the molecule) forms the π bond to carbon, while the other accommodates the remaining lone pair of electrons.

Exercise 10.21—Bonding and Hybridization

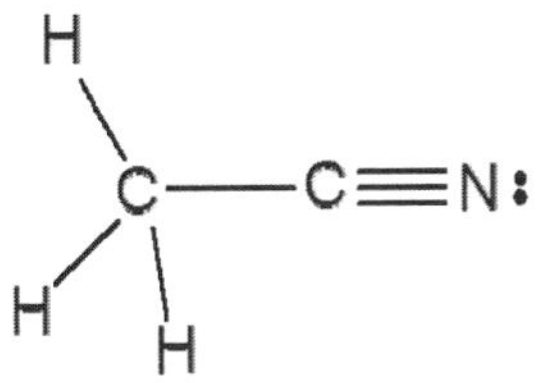

Lewis structure of acetonitrile

The H–C–H, H–C–C, and C–C–N bond angles are predicted to be 109.5°, 109.5°, and 180°, respectively.

The methyl C atom has the tetrahedral distribution of electron regions—presumed sp^3 hybridization. Three of the sp^3 orbitals overlap with s orbitals on H atoms to form the C–H bonds, while the fourth makes a bond with the nitrile C atom.

The nitrile C and N atoms have a linear arrangement of two electron regions. We presume sp hybridization. One sp orbital from each atom is used to make the C–N σ bond. The remaining sp orbital on the nitrile C atom forms the bond with the methyl C atom, while the remaining sp orbital on the N atom accommodates the lone pair.

The two remaining p orbitals on the nitrile C and N atoms overlap in pairs to form the two C–N π bonds.

Exercise 10.25—Deducing Electron Configuration and Bond Order

Electron configurations

$H_2^+ : (\sigma_{1s})^1$

$H_2^-: (\sigma_{1s})^2(\sigma_{1s}^*)^1$

$He_2^+: (\sigma_{1s})^2(\sigma_{1s}^*)^1$

Bond order $= \frac{1}{2} \times$ (# of electrons in bonding MOs $-$ # of electrons in antibonding MOs)

$\qquad = \frac{1}{2} \times (1 - 0) = \frac{1}{2}$ \qquad for H_2^+

$\qquad = \frac{1}{2} \times (2 - 1) = \frac{1}{2}$ \qquad for He_2^+

$\qquad = \frac{1}{2} \times (2 - 1) = \frac{1}{2}$ \qquad for H_2^-

H_2^+ has a bond order of ½. It should exist. But, it is weakly bound. He_2^+ and H_2^- have the same bond order: although they have more electrons in bonding orbitals, but also have an electron in an antibonding orbital.

Exercise 10.27—MO Electron Configuration for Homonuclear Diatomic Ions

Electron configuration of ground state O_2^+ ion (11 **valence** electrons):

$\qquad$ [core electrons]$(\sigma_{2s})^2(\sigma_{2s}^*)^2(\pi_{2p_y})^2(\pi_{2p_z})^2(\sigma_{2p_x})^2(\pi_{2p_y}^*)^1$

Bond order $= \frac{1}{2} \times (2 + 2 + 2 + 2 - 2 - 1) = 2\frac{1}{2}$

O_2^+ ion: it has an unpaired electron.

REVIEW QUESTIONS

Section 10.1: Observe, Measure, and Imagine

10.31

Yes, this statement is consistent with the view of chemistry as a human activity. Chemical structures, properties, and reactions are models that we construct to attempt to explain the behaviour of chemical species. These models can then be applied to solve real world problems in science and technology.

10.33

This statement applies to the material in Chapter 10 in that our chemical models, while requiring some imagination to visualize, must be based on sound experimental evidence. Without this experimental support, a model is purely speculative and is of limited usefulness in chemistry.

Section 10.2: Covalent Bonding in Molecules

10.35

Consider the formation of a covalent bond between two hydrogen atoms. Before the bond has formed, the two hydrogen atoms are attracted to each other due to the attraction between the proton of one atom and the electron of the other atom (and vice versa). The two atoms therefore approach each other, and the force of attraction becomes stronger as the distance between the atoms decreases. As the two atoms get closer, however, the repulsion between the two protons becomes more and more significant. Beyond a certain distance, this repulsion increases very sharply. If the atoms get too close together, this proton–proton repulsion pushes them away from each other, back to a stable distance at which the *net attractive force* is maximized. At this distance, the atoms are covalently bonded.

Section 10.3: Lewis Structures

10.37

Elements in group 13 have 3 valence electrons. To obey the octet rule, such elements must form four bonds. Three of these bonds are formed by sharing each of the valence electrons with another atom. This gives the atom 6 valence electrons. To get 8, the atom must form a fourth "coordinate bond" wherein both electrons come from a different atom. B and Al can form four bonds in this way. These species typically have a net negative charge. They are the products of a Lewis acid—the electron-deficient 6 valence electron B or Al—and a (typically negatively charged) Lewis base (e.g., F^- or Cl^-).

Elements in group 14 have 4 valence electrons. In order to obey the octet rule, these elements generally form four covalent bonds and have a formal charge of zero in their covalently bonded forms. The most notable example from this group is carbon, which almost always forms four stable covalent bonds.

Elements in group 15 have 5 valence electrons. To attain a full octet, these elements generally form three covalent bonds, with a lone pair completing the octet.

Elements in group 16 have 6 valence electrons. These elements can attain a full octet by forming two covalent bonds, with two lone pairs completing the octet.

Elements in group 17 have 7 valence electrons. They can form an octet by forming a single bond.

10.39

(a) NF_3 molecules
Number of valence electrons = 5 (from N) + 3×7 (from F)
= 26 electrons (13 electron pairs)
Use three electron pairs to form bonds between N and each of the three F's.
13 − 3 = 10 electron pairs remain.
Use 3 × 3 = 9 electron pairs to complete the octet about each of the three F's. The remaining pair of electrons is given to N to complete its octet.

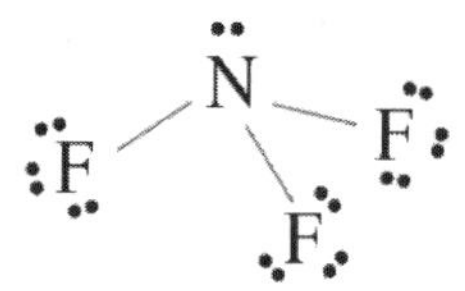

(b) ClO_3^- ions

Number of valence electrons = 7 (from Cl) + 3×6 (from the 3 O's) + 1 (net −1 charge)

= 26 electrons (13 electron pairs)

Use three electron pairs to form bonds between Cl and each of the three O's.

13 − 3 = 10 electron pairs remain.

Use 3 × 3 = 9 electron pairs to complete the octet about each of the three O's. The remaining pair of electrons is given to Cl to complete its octet.

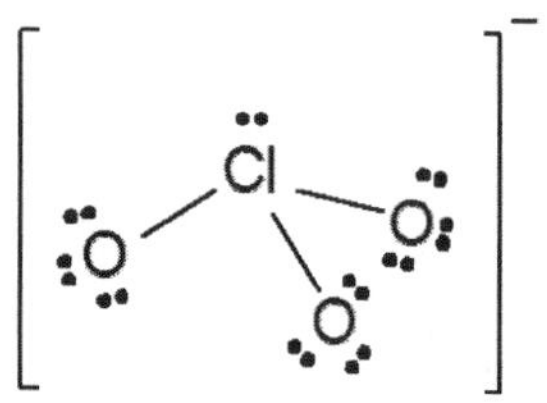

If you have studied formal charge (see Section 10.6), then you know that this structure can be taken further to get:

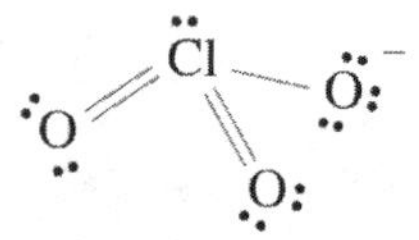

The structure with single bonds has a formal charge of +2 on Cl and −1 on each O atom. Forming a double bond by moving a lone pair on O into a bonding pair between O and Cl raises the formal charge on the O to zero and lowers the formal charge on Cl. Doing this twice reduces the formal charge on Cl to zero and minimizes formal charges overall. This is the preferred structure.

(c) HOBr molecules

Number of valence electrons = 7 (from Br) + 6 (from O) + 1 (from H)

= 14 electrons (7 electron pairs)

Use two electron pairs to form the H–O and O–Br bonds.

7 − 2 = 5 electron pairs remain.

Use 3 electron pairs to complete the octet about Br, and 2 electron pairs to complete the octet about O.

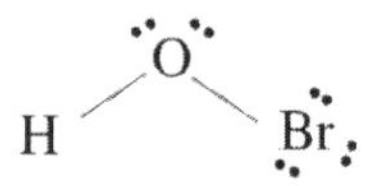

(d) SO_3^{2-} ions

Number of valence electrons = 6 (from S) + 3×6 (from the 3 O's) + 2 (net − 2 charge)

= 26 electrons (13 electron pairs)

Use three electron pairs to form bonds between S and each of the three O's.

$13 - 3 = 10$ electron pairs remain.

Use $3 \times 3 = 9$ electron pairs to complete the octet about each of the three O's. The remaining pair of electrons is given to S to complete its octet.

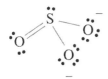

If you have studied formal charge (see Section 10.6), then you know that this structure can be taken further to get:

The structure with single bonds has a formal charge of +1 on S and −1 on each O atom. Forming a double bond by moving a lone pair on an O into a bonding pair between that O and S raises the formal charge on that O to zero and lowers the formal charge on S. Doing this once reduces the formal charge on S to zero and minimizes formal charges overall. This is the preferred structure.

10.41

:Cl:

Cl — B — Cl

:Cl⁻ — B — N⁺ — H

Cl — H

Section 10.4: Resonance and Delocalized Electron Models

10.43

(a) SO_2

(b) NO_2^-

Note that NO_2^- cannot have octets on all atoms—it has an odd number of valence electrons (= 5 + $2 \times 6 = 17$).

We choose N to be missing the one electron because it is less electronegative than O.

(c) SCN⁻

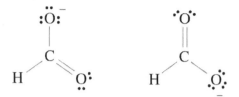

10.45

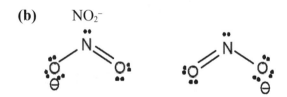

Although in each of these Lewis structures, one C–O bond is a single bond, and the other is a double bond, they are equivalent, with delocalization of the π electrons, so that bond has three electrons.

$$\text{Average C–O bond order} = \frac{(2 + 1)}{2} = \frac{3}{2}$$

10.47

(a) NO_2^+

Formal charges are zero on O atoms and +1 on N atom.
Check:
For the equivalent O atoms, formal charge $= 6 - 4 - 2 = 0$
For N, formal charge $= 5 - 0 - 4 = +1$

(b) NO_2^-

The N has a formal charge of zero—group 15, valence of 3. The formal charges on the two equivalent (although not shown as equivalent in these Lewis structures) O atoms are zero and −1.
Check:
formal charge (double-bonded O atom) $= 6 - 4 - 2 = 0$
formal charge (single-bonded O atom) $= 6 - 6 - 1 = -1$

$$\text{Average formal charge on O atoms} = \frac{(0 - 1)}{2} = -\frac{1}{2}$$

(c) NF₃

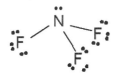

All atoms have zero formal charge. A group 17 element with a valence of 1 and a group 15 element with a valence of 3.

(d) HNO₃

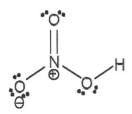

Formal charges are as shown. Note the +1 on the tetravalent N atom (group 15) and −1 on the univalent O (group 16).

10.49

If an H⁺ ion attaches to NO₂⁻ ion (to form the acid HNO₂), it attaches to an O atom—not N. This is consistent with the negative charge localized on the O atoms—as indicated in these resonance structures.

Section 10.5: Spatial Arrangement of Atoms within Molecules

10.51

(a) CO₂

$$:O = C = O:$$

linear
There are two electron regions about the carbon atom. The electron geometry and the molecular geometry (shape of the molecule) are both linear since there are no lone pairs on the C atom.

(b) NO₂⁻

bent
There are three electron regions about the nitrogen atom. The electron geometry is trigonal planar, while the molecular geometry (shape of the ion) is bent—there is one lone pair on the central atom.

(c) O_3

bent
There are three electron regions about the oxygen atom. The electron geometry is trigonal planar, while the molecular geometry (shape of the molecule) is bent—there is one lone pair on the central atom. This molecule is isoelectronic with NO_2^-. They have the same shape and Lewis structure—aside from the formal charge on the central atom.

(d) ClO_2^-

bent
There are four electron regions about the oxygen atom. The electron geometry is tetrahedral, while the molecular geometry (shape of the ion) is bent—there are two lone pairs on the central atom. This ion is probably more bent than O_3 molecules or NO_2^- ions, because there are two lone pairs on the central atom in this case.

10.53

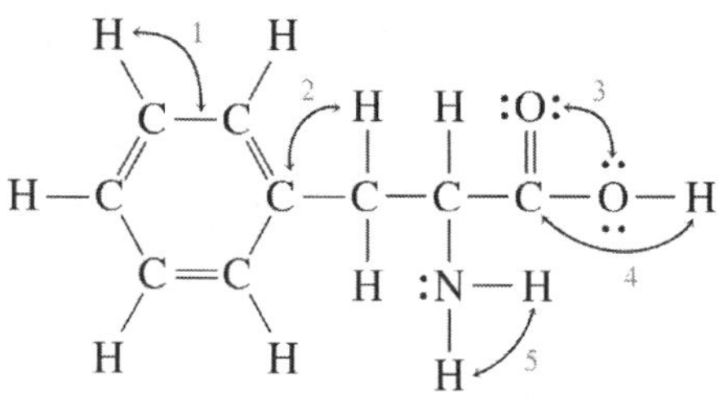

Phenylalanine bond angles:
(1) 120°   (2) 109.5°   (3) 120°
(4) about 109.5° (5) about 109.5°

The CH_2–$CH(NH_2)$–$COOH$ chain is not linear because the two tetrahedral carbon atoms in the chain (–CH_2– and –$CH(NH_2)$–) have bond angles of 109.5°, the C=O carbon has a bond angle of 120°, and the C–O–H has a bond angle of about 109.5°.

Section 10.6: The Valence Bond Model of Covalent Bonding

10.55

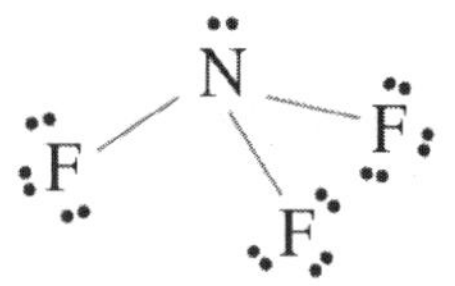

Lewis structure of NF_3

The molecular geometry (shape of the molecule) is trigonal pyramidal—there is a lone pair—while the electron geometry about the N atom is tetrahedral.

A tetrahedral distribution of electron regions implies sp^3 hybridization at nitrogen. Three of the sp^3 hybrid orbitals overlap with sp orbitals on F atoms—we assume the F atoms to be sp hybridized. The remaining nitrogen sp^3 hybrid orbital accommodates the lone pair on the N atom. The remaining one sp and two p orbitals on each F atom accommodate the lone pairs—three on each F atom. We can equally well invoke sp^3 hybridization on the F atoms (with pairs of electrons in three of the hybrid orbitals, and a single electron in the other). Overlap of the half-filled sp^3 hybrid orbital on each F atom with an sp^3 hybrid orbital on the N atom gives rise to a single bond, and there are three residual electron pairs in sp^3 hybrid orbitals.

10.57

(a) The carbon atoms in dimethyl ether, H_3COCH_3, use sp^3 hybridized orbitals on the C atoms to form σ bonds with H atoms, and with the O atom. The oxygen atom in dimethyl ether, H_3COCH_3 uses two sp^3 hybridized orbitals to form σ bonds with the C atom.
The remaining two sp^3 orbitals on the O atom accommodate the two lone pairs.

(b) The carbon atoms in propene, $H_3CCH=CH_2$ use sp^3 hybridized orbitals on the methyl C atoms to form σ bonds with H atoms, and with the middle C atom. The vinyl C atoms (i.e., the doubly bonded C atoms) use sp^2 hybridized orbitals to form σ bonds with each other, with H atoms and with the methyl C atom. Use the remaining unhybridized p orbital on each of the two vinyl C atoms to form a π bond.

(c) The alkyl C atom (i.e., the C atom bonded to N) and the N atom in glycine use sp^3 hybridized orbitals to form σ bonds with H atoms, and with each other. The N atom has a lone pair in the remaining sp^3 orbital. The carboxylic C atom uses sp^2 orbitals to form σ bonds with the other C atom and the two O atoms. Use the remaining p orbital on the carboxylic C and O atoms to form a π bond.

10.59

(a) SO_2
The O–S–O bond angle is about 120º. The sulfur atom uses sp^2 orbitals to form two σ bonds—one with each O atom. The other sp^2 orbital accommodates a lone pair. We can describe the Lewis structure with only one double S–O bond with the remaining unhybridized p orbital. It forms a π bond with just one of the O atoms.
However, the structure with two S-O double bonds minimizes formal charges and, as such, is preferred. To describe the bonding in this case requires the use of d orbitals on the S atom—since in this structure, an S atom accommodates 10 electrons in its valence shell.

(c) SO_3^{2-}
The O–S–O bond angle is 109.5º. The sulfur atom uses sp^3 orbitals to form three σ bonds—one with each O atom. The other sp^3 orbital accommodates a lone pair on S. In accord with the formal charge minimized structure, we can imagine simultaneous sideways overlap of the p orbital on the S atom with a p orbital on each of two of the O atoms.

(b) SO_3
The O–S–O bond angle is 120°.
The sulfur atom uses sp^2 orbitals to form
three σ bonds—one with each O atom.
In accord with the formal charge
minimized structure, we can imagine
simultaneous sideways overlap of the p
orbital on the S atom with a p orbital on
each of the O atoms.

(d) SO_4^{2-}

The O–S–O bond angle is 109.5°.
The sulfur atom uses sp^3 orbitals to
form four σ bonds—one with each O
atom.
In accord with the formal charge
minimized structure, we can imagine
simultaneous sideways overlap of the p
orbital on the S atom with a p orbital on
each of two of the O atoms.

10.61

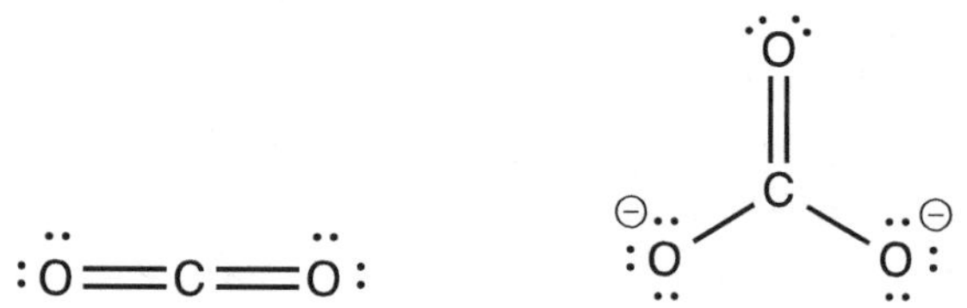

In linear CO_2 the carbon atom uses two sp orbitals to form σ bonds with each O atom. Each of the two unhydridized p orbitals forms a π bond with an O atom. In CO_3^{2-}, there are two single C–O bonds and one double C–O bond, giving rise to three equivalent resonance structures. The carbon atom uses three sp^2 orbitals to form σ bonds with each of the three O atoms. The unhydridized p orbital forms a π bond with an O atom. Since it can be any one of the three, we get three equivalent resonance structures. This differs from the delocalized electron picture in which the unhydridized p orbital forms a degree of π bond with all three O atoms.

10.63

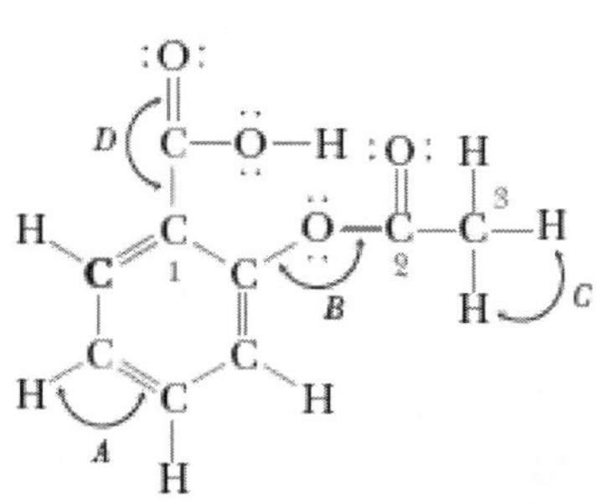

(a) The angles *A, B, C,* and *D* are about 120°, 109.5°, 109.5°, and 120°, respectively.
(b) According to the valence bond model, carbon atoms 1, 2, and 3 are sp^2, sp^2, and sp^3 hybridized, respectively.

Section 10.7: Molecular Orbital Theory of Bonding in Molecules

10.65

(a)

$$:\overset{\cdot\cdot}{\underset{\cdot\cdot}{\overset{-}{O}}}\!\!-\!\!\overset{\cdot\cdot}{\underset{\cdot\cdot}{\overset{-}{O}}}:$$

The bond order in O_2^{2-} is 1.

(b) The molecular orbital theory electron configuration for O_2^{2-} is (using unhybridized orbitals)

[core electrons]$(\sigma_{2s})^2(\sigma_{2s}^*)^2(\pi_{2p_y})^2(\pi_{2p_z})^2(\sigma_{2p_x})^2(\pi_{2p_y}^*)^2(\pi_{2p_z}^*)^2$

bond order $= \frac{1}{2} \times (2 + 2 + 2 + 2 - 2 - 2 - 2) = 1$

(c) For O_2^{2-}, the valence bond description and the molecular orbital description predict the same bond order and the same magnetic behaviour (i.e., diamagnetic—NOT paramagnetic—there are no unpaired electrons in either description).

10.67

Molecular orbital; theory electron configurations (core electrons omitted):

(a) NO

$(\sigma_{2s})^2(\sigma_{2s}^*)^2(\pi_{2p_y})^2(\pi_{2p_z})^2(\sigma_{2p_x})^2(\pi_{2p_y}^*)^1$

an unpaired electron $\rightarrow$ paramagnetic

HOMO $= \pi_{2p_y}^*$ or $\pi_{2p_z}^*$ $\leftarrow$ same energy

(b) OF⁻

$(\sigma_{2s})^2(\sigma_{2s}^*)^2(\pi_{2p_y})^2(\pi_{2p_z})^2(\sigma_{2p_x})^2(\pi_{2p_y}^*)^2(\pi_{2p_z}^*)^2$

no unpaired electrons $\rightarrow$ NOT paramagnetic—i.e., diamagnetic

HOMO $= \pi_{2p_y}^*$ or $\pi_{2p_z}^*$ $\leftarrow$ same energy

(c) O_2^{2-}

$(\sigma_{2s})^2(\sigma_{2s}^*)^2(\pi_{2p_y})^2(\pi_{2p_z})^2(\sigma_{2p_x})^2(\pi_{2p_y}^*)^2(\pi_{2p_z}^*)^2$

no unpaired electrons $\rightarrow$ NOT paramagnetic—i.e., diamagnetic

HOMO $= \pi_{2p_y}^*$ or $\pi_{2p_z}^*$ $\leftarrow$ same energy

(d) Ne₂⁺

$(\sigma_{2s})^2(\sigma_{2s}^*)^2(\pi_{2p_y})^2(\pi_{2p_z})^2(\sigma_{2p_x})^2(\pi_{2p_y}^*)^2(\pi_{2p_z}^*)^2(\sigma_{2p_x}^*)^1$

an unpaired electron $\rightarrow$ paramagnetic

HOMO $= \sigma_{2p_x}^*$

(e) CN

$(\sigma_{2s})^2(\sigma_{2s}^*)^2(\pi_{2p_y})^2(\pi_{2p_z})^2(\sigma_{2p_x})^1$

an unpaired electron $\rightarrow$ paramagnetic

HOMO $= \sigma_{2p_x}$

SUMMARY AND CONCEPTUAL QUESTIONS

10.69

NF_3 has four electron pairs about the central atom, and a trigonal pyramidal shape.

OCl_2 has four electron pairs about the central atom, and a bent shape.
In each case, the bond angles are near 109.5° (the angle between tetrahedrally distributed electron regions).

10.71

The average C–O bond order in the formate ion (HCO_2^- is 1.5)—there are two equivalent C–O bonds represented as a single and a double bond in the two resonance structures.
In a methanol molecule (CH_3OH), the C–O bond is just a single bond—there is just one structure.
In a carbonate ion (CO_3^{2-}) there are three equivalent C–O bonds—represented as a double bond and two single bonds in the resonance structures. The average bond order is 4/3 = 1.3333.
The formate ion has the strongest C–O bond, which is expected to be the shortest.
Methanol molecules have the weakest C–O bond, which is expected to be the longest.

10.73

The NO_2^+ ion is linear. Its O–N–O bond angle is 180°.
The NO_2^- ion is bent. Its O–N–O bond angle is about 120°.
The NO_2^+ ion has the larger O–N–O bond angle.

10.75

    Acrylonitrile
(a) Angles 1, 2, and 3 equal about 120 °, 180°, and 120°, respectively.
(b) The carbon–carbon double bond is shorter.
(c) The carbon–carbon double bond is stronger.
(d) The C–N triple bond is the most polar—largest difference in electronegativity.

10.77

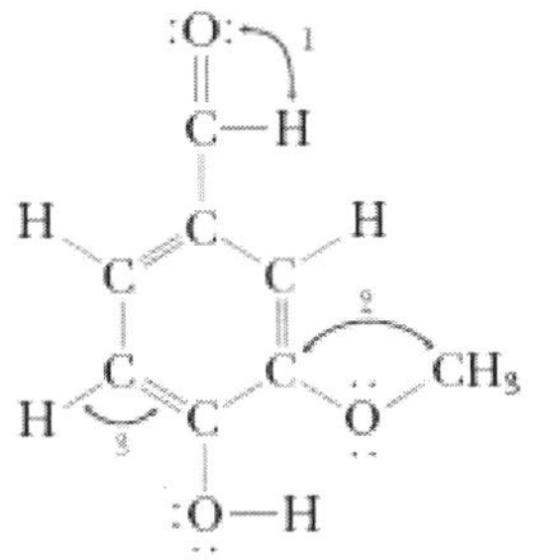

(a) Bond angles 1, 2 and 3 are about 120°, 109.5°, and 120°, respectively.
(b) The shortest carbon-oxygen bond is the carbonyl C=O double bond.
(c) The most polar bond in the molecule is the O–H bond.

10.79

(a) The geometry about the boron atom in BF_3 is trigonal planar. The geometry about the boron atom in $H_3N{\rightarrow}BF_3$ is tetrahedral.
(b) In BF_3, the valence orbitals of boron are sp^2 hybridized.
In $H_3N{\rightarrow}BF_3$, the valence orbitals of boron are sp^3 hybridized.
(c) We expect the hybridization of boron to change when this "coordinate" bond forms.

10.81

This experimental evidence can be explained by looking at the main contributing resonance structures for the NO_2^- ion vs the HONO molecule.
For the NO_2^- ion, we can draw two main resonance structures, in which the relative locations of the double bond and negatively charged oxygen are reversed.

Because the true structure is the average of the contributing resonance structures, the two bonds are identical. The bond length is shorter than would be expected for a single bond, but longer than a double bond.
For the HONO molecule, the covalent bond between oxygen and hydrogen prevents us from redistributing the electrons to draw a second resonance structure that would be analogous to the second resonance structure of the NO_2^- ion.

As such, we have a true singe bond and a true double bond in this molecule, and the observed bond lengths are consistent with what is expected for these types of bonds.

10.83

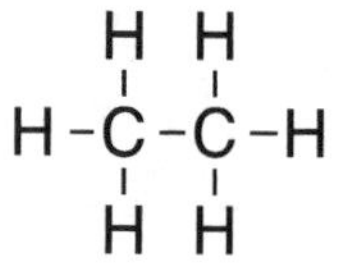

H-C-C-H (ethane, with H atoms above and below each carbon)   H-C≡C-H (ethyne)

ethane    ethyne

The Lewis structures above show that there is one covalent bond between the two carbon atoms of ethane, while there are three covalent bonds between the two carbon atoms of ethyne. In this representation, however, there is no distinction made between bond types, *i.e.*, σ and π bonds. This model is therefore unable to explain the observed bond strengths as all bonds appear to be equal . The valence bond model clearly shows σ and π bonds. Moreover, this model shows that π bonds are weaker than σ bonds: π bonds result from indirect orbital overlap while σ bonds result from direct orbital overlap. Because π bonds are weaker, the overall bond strength is weaker than it would be if all three bonds were σ bonds, and the resulting triple bond is less than three times as strong as the σ bond in ethane.

10.85

Many responses possible. One example: The sp^3 hybridization model for the atomic orbitals of carbon is consistent with the tetrahedral structure of methane.

CHAPTER 11
States of Matter

IN-CHAPTER EXERCISES

Exercise 11.1—Ideal Gas Equation

To use the ideal gas equation with R having units of $(\text{L kPa K}^{-1}\text{ mol}^{-1})$, the pressure must be measured in kPa and the temperature in K. Therefore,

$p = 100$ kPa
$T = 23 + 273.15 = 296.15$ K

Now substitute the values of p, T, n, and R into the ideal gas equation and solve for the volume of gas, V:

$$V = \frac{nRT}{p} = \frac{(1.3 \times 10^3 \text{ mol})(8.314 \text{ L kPa K}^{-1}\text{ mol}^{-1})(296.15 \text{ K})}{(100 \text{ kPa})} = 3.2 \times 10^4 \text{ L}$$

Notice that the units kPa, mol, and K cancel to leave the answer in L.

Exercise 11.5—Ideal Gas Equation

Initial Conditions	Final Conditions
$V_1 = 22$ L	$V_2 = ?$ L
$p_1 = 152$ bar $= 1.52 \times 10^4$ kPa	$p_2 = 1.00$ bar $= 100$ kPa
$T_1 = 31°C$ (304 K)	$T_2 = 22°C$ (295 K)

$$V_2 = \left(\frac{T_2}{p_2}\right) \times \left(\frac{p_1 V_1}{T_1}\right) = V_1 \times \frac{p_1}{p_2} \times \frac{T_2}{T_1}$$

$$= 22 \text{ L} \times \frac{1.52 \times 10^4 \text{ kPa}}{100 \text{ kPa}} \times \frac{295 \text{ K}}{304 \text{ K}}$$

$$= 3250 \text{ L}$$

This is the total volume of gas at the final temperature and pressure—with one additional digit (beyond significant) carried. This gas fills

$$\frac{V_2}{\text{volume of 1 balloon}} = \frac{3250 \text{ L}}{5.0 \text{ L balloon}^{-1}} = 650 \text{ balloons}$$

Exercise 11.7—Molar Mass from Density

The molar mass of the unknown gas is

$$M = \frac{\rho RT}{p} = \frac{(5.02 \text{ g L}^{-1})(8.314 \text{ L kPa K}^{-1}\text{mol}^{-1})(288.15 \text{ K})}{(99.3 \text{ kPa})} = 121 \text{ g mol}^{-1}$$

Exercise 11.11—Partial Pressures

We have 15.0 g of halothane ($C_2HBrClF_3$) vapour and 23.5 g of oxygen gas. This corresponds to

$$n\,(C_2HBrClF_3) = \frac{m}{M} = \frac{15.0\ \cancel{g}}{197.4\ \cancel{g}\ mol^{-1}} = 0.0760\ mol$$

$$n\,(O_2) = \frac{m}{M} = \frac{23.5\ \cancel{g}}{32.00\ \cancel{g}\ mol^{-1}} = 0.734\ mol$$

The total amount of gases (number of moles) is
$$n\,(total) = n\,(C_2HBrClF_3) + n\,(O_2) = 0.810\ mol$$
The gas mixture is in a 5.00 L tank at 25.0 °C (298.15 K).
The ideal gas law gives the total pressure if we substitute $n\,(total)$,
the partial pressure of halothane if we substitute $n\,(C_2HBrClF_3)$, and
the partial pressure of oxygen if we substitute $n\,(O_2)$.

$$p = \frac{nRT}{V} = \frac{(0.810\ \cancel{mol})(8.314\ \cancel{L}\ kPa\ \cancel{K^{-1}}\ \cancel{mol^{-1}})(298.15\ \cancel{K})}{(5.00\ \cancel{L})} = 402\ kPa$$

The partial pressures of halothane and oxygen are determined more easily using mole fractions,

$$x(C_2HBrClF_3) = \frac{n(C_2HBrClF_3)}{n(total)} = \frac{0.0760\ \cancel{mol}}{0.810\ \cancel{mol}} = 0.0938$$

$$x(O_2) = \frac{n(O_2)}{n(total)} = \frac{0.734\ \cancel{mol}}{0.810\ \cancel{mol}} = 0.906 = 1.000 - x(C_2HBrClF_3)$$

From which we get
$$p(C_2HBrClF_3) = x(C_2HBrClF_3) \times p(total) = 0.0938 \times 402\ kPa = 37.7\ kPa$$
and
$$p(O_2) = x(O_2) \times p(total) = 0.906 \times 402\ kPa = 364\ kPa$$

Exercise 11.17—Graham's Law of Effusion

Taking He as the reference, we have

$$\frac{Rate\ of\ effusion\ of\ SF_6}{Rate\ of\ effusion\ of\ He} = \sqrt{\frac{M(He)}{M(SF_6)}} = \sqrt{\frac{4.003\ g\ \cancel{mol^{-1}}}{146.06\ g\ \cancel{mol^{-1}}}} = 0.166$$

and

$$\frac{Rate\ of\ effusion\ of\ N_2}{Rate\ of\ effusion\ of\ He} = \sqrt{\frac{M(He)}{M(N_2)}} = \sqrt{\frac{4.003\ g\ \cancel{mol^{-1}}}{28.02\ g\ \cancel{mol^{-1}}}} = 0.378$$

He has the highest rate of effusion, with N_2 at 37.8% the rate of He, and the much bigger SF_6 molecules at 16.6% that rate.

Exercise 11.21—Relative Strength of Intermolecular Forces

In order of increasing strength, we have
(a) the dispersion forces in liquid O_2 < **(c)** the dipole-induced dipole interactions of O_2 dissolved in H_2O < **(b)** the hydrogen bonding forces in liquid CH_3OH

Exercise 11.25—Vapour Pressure Curves

The vapour pressure of water at 60°C (using Figure 11.19 in the textbook) is about 20 kPa. To achieve this pressure of water in a 5.0 L flask at 60°C (333 K) requires

$$n(H_2O) = \frac{pV}{RT} = \frac{(20 \text{ kPa})(5.0 \text{ L})}{(8.314 \text{ L kPa K}^{-1}\text{mol}^{-1})(333 \text{ K})} = 0.036 \text{ mol}$$

or

$$m(H_2O) = n(H_2O) \times M(H_2O) = (0.036 \text{ mol})(18.02 \text{ g mol}^{-1}) = 0.65 \text{ g}$$

0.50 g of water is not enough to achieve this partial pressure. Instead we achieve only

$$20 \times \frac{0.50 \text{ g}}{0.65 \text{ g}} = 15 \text{ kPa}$$

If we started with 2.0 g of water, 0.65 g of it would evaporate to give a vapour in equilibrium with the remaining 2.0 − 0.65 g = 1.35 g of liquid water. The partial pressure of water would be 20 kPa.

Exercise 11.27—Phase Diagrams

(a) See Figure 11.24 in the textbook.
Because the slope of the liquid-solid equilibrium curve is positive, we conclude that the density of liquid CO_2 is less than that of solid CO_2. If we increase the pressure of a system containing liquid and solid at equilibrium, some of the less dense phase will change to the more dense phase.
(b) At 500 kPa and 0°C, the most stable phase of carbon dioxide is the vapour.
(c) CO_2 cannot be liquefied at 45°C, because this temperature is above the critical temperature. Compressing CO_2 at this temperature produces a dense supercritical fluid.

REVIEW QUESTIONS

Section 11.1: Understanding Gases: Understanding Our World

11.29

In order of increasing temperature, altitudes in the atmosphere are ranked as follows:
$T(80 \text{ km}) < T(25 \text{ km}) < T(110 \text{ km}) < T(5 \text{ km})$

11.31

The temperature stops decreasing, and starts increasing, as we move from the troposphere into the stratosphere. Absorption of UV light by oxygen produces oxygen atoms which, in turn, produce ozone (together with other oxygen molecules). The ozone produced absorbs additional UV light. The energy absorbed increases the local temperature. The density of the atmosphere at the top of the stratosphere and bottom of mesosphere is sufficient to absorb enough UV light to increase the temperature. As we descend into the stratosphere and below, the intensity of incoming UV radiation is diminished due to absorption at higher altitudes. At 1 km above

Earth's surface in the troposphere, there is too little UV radiation to cause significant absorption by oxygen to produce O atoms, and ozone, etc.

11.33

(a) The earth's atmospheric concentration of CO_2 has varied in the following fashion: $c(CO_2)$ was lower 30 000 years ago compared to 125 000 years ago, and then was lower again at 260 000 years ago.

(b) The earth's atmospheric temperature has varied in the following fashion: Temperature was lower 30 000 years ago compared to 125 000 years ago, and then was lower again at 260 000 years ago.

Section 11.2: Relationships among Gas Properties

11.35

$$p = 210 \text{ mm Hg} = \frac{210 \text{ mm Hg}}{760 \text{ mm Hg atm}^{-1}} = 0.276 \text{ atm}$$

$$= 0.276 \text{ atm} \times 101.3 \text{ kPa atm}^{-1} = 30.0 \text{ kPa} = 0.300 \text{ bar}$$

11.37

Initial Conditions	Final Conditions
$V_1 = 25.0 \text{ mL} = 0.0250 \text{ L}$	$V_2 = ?$
$p_1 = 58.2 \text{ kPa}$	$p_2 = 12.6 \text{ kPa}$
$T_1 = 20.5°C$ (293.65 K)	$T_2 = 24.5°C$ (297.65 K)

$$V_2 = \left(\frac{T_2}{p_2}\right) \times \left(\frac{p_1 V_1}{T_1}\right) = V_1 \times \frac{p_1}{p_2} \times \frac{T_2}{T_1}$$

$$= 0.0250 \text{ L} \times \frac{58.2 \text{ kPa}}{12.6 \text{ kPa}} \times \frac{297.65 \text{ K}}{293.65 \text{ K}}$$

$$= 0.117 \text{ L}$$

11.39

Initial Conditions	Final Conditions
$V_1 = 0.40 \text{ L}$	$V_2 = 0.050 \text{ L}$
$p_1 = 1.00 \text{ atm}$	$p_2 = ?$
$T_1 = 15°C$ (288 K)	$T_2 = 77°C$ (350 K)

$$p_2 = \left(\frac{T_2}{V_2}\right) \times \left(\frac{p_1 V_1}{T_1}\right) = p_1 \times \frac{V_1}{V_2} \times \frac{T_2}{T_1}$$

$$= 1.00 \text{ atm} \times \frac{0.40 \text{ L}}{0.050 \text{ L}} \times \frac{350 \text{ K}}{288 \text{ K}}$$

$$= 9.7 \text{ atm}$$

Section 11.3: Different Gases: How Similar? How Different?

11.41

Gas	Molar Volume at 298.15 K (L mol^{-1})	Molar Volume at 448.15 K (L mol^{-1})
H_2	24.804	37.277
CO_2	24.666	37.218
SF_6	24.512	37.157

The molar volumes of the three gases are more similar at 448.15 K than 298.15 K.

Section 11.4: The Ideal Gas Equation

11.43

$$n(C_2H_5OH) = m(C_2H_5OH) / M(C_2H_5OH) = (1.50 \text{ g}) / (46.07 \text{ g mol}^{-1}) = 0.0326 \text{ mol}$$

$$p = \frac{nRT}{V} = \frac{(0.0326 \text{ mol})(8.314 \text{ L kPa K}^{-1}\text{mol}^{-1})(523.15 \text{ K})}{(0.251 \text{ L})} = 565 \text{ kPa}$$

11.45

$$n = \frac{pV}{RT} = \frac{(95.3 \text{ kPa})(0.452 \text{ L})}{(8.314 \text{ L kPa K}^{-1}\text{mol}^{-1})(296 \text{ K})} = 0.0175 \text{ mol}$$

$$M \text{ (unknown)} = m \text{ (unknown)} / n \text{ (unknown)} = (1.007 \text{ g}) / (0.0175 \text{ mol}) = 57.5 \text{ g mol}^{-1}$$

11.47

Initial Conditions	Final Conditions
$V_1 = 1.2 \times 10^4 \text{ m}^3 = 1.2 \times 10^7 \text{ L}$	$V_2 = ?$
$1 \text{ L} = 1 \text{ dm}^3 = 10^{-3} \text{ m}^3$	
$p_1 = 98.3 \text{ kPa}$	$p_2 = 80.0 \text{ kPa}$
$T_1 = 16°C \ (289 \text{ K})$	$T_2 = -33.0°C \ (240 \text{ K})$

$$V_2 = \left(\frac{T_2}{p_2}\right) \times \left(\frac{p_1 V_1}{T_1}\right) = V_1 \times \frac{p_1}{p_2} \times \frac{T_2}{T_1}$$

$$= 1.2 \times 10^7 \text{ L} \times \frac{98.3 \text{ kPa}}{80.0 \text{ kPa}} \times \frac{240 \text{ K}}{289 \text{ K}}$$

$$= 1.22 \times 10^7 \text{ L}$$

Though the pressure dropped by almost 20%, the decrease in temperature compensates and the change in volume is not very big. Nevertheless, there would have to be allowance for expansion of the balloon at higher altitudes.

Section 11.5: The Density of Gases

11.49

$$\frac{n}{V} = \frac{p}{RT} = \frac{25.2 \text{ kPa}}{(8.314 \text{ L kPa K}^{-1}\text{mol}^{-1})(290 \text{ K})} = 1.045 \times 10^{-2} \text{ mol L}^{-1}$$

$$M(\text{organofluorine compound}) = \text{density} \times \left(\frac{n}{V}\right)^{-1} = (0.355 \text{ g L}^{-1}) / (1.045 \times 10^{-2} \text{ mol L}^{-1})$$

$$= 33.97 \text{ g mol}^{-1} = 34 \text{ g mol}^{-1} \text{ (2 significant figures)}$$

11.51

Amount of N_2 required =

$$= n(N_2) = \frac{pV}{RT} = \frac{(132 \text{ kPa})(75.0 \text{ L})}{(8.314 \text{ L kPa K}^{-1}\text{mol}^{-1})(298.15 \text{ K})} = 3.99 \text{ mol}$$

$$2 \text{ NaN}_3(s) \longrightarrow 2 \text{ Na}(s) + 3 \text{ N}_2(g)$$

Amount of NaN_3 required, $n(\text{NaN}_3) = \dfrac{2}{3} \times (\text{amount of } N_2) = 2.66 \text{ mol}$

Mass of NaN_3 required, $m = (2.66 \text{ mol}) \times (65.02 \text{ g mol}^{-1}) = 173 \text{ g}$

11.53

Amount of N_2H_4 to be consumed, $n(\text{N}_2\text{H}_4) = (1.00 \times 10^3 \text{ g}) / (32.05 \text{ g mol}^{-1}) = 31.2 \text{ mol}$

$$N_2H_4(g) + O_2(g) \longrightarrow N_2(g) + 2 \text{ H}_2\text{O}(\ell)$$

Amount of O_2 required, $n(\text{O}_2) = \text{amount of } N_2H_4 = 31.2 \text{ mol}$

$$p = \frac{n(O_2)RT}{V} = \frac{(31.2 \text{ mol})(8.314 \text{ L kPa K}^{-1}\text{mol}^{-1})(296 \text{ K})}{(450 \text{ L})} = 171 \text{ kPa}$$

Section 11.6: Gas Mixtures and Partial Pressures

11.55

We have 1.0 g of H_2 and 8.0 g of Ar. This corresponds to

$$n(H_2) = \frac{m}{M} = \frac{1.0 \text{ g}}{2.016 \text{ g mol}^{-1}} = 0.50 \text{ mol}$$

$$n(Ar) = \frac{m}{M} = \frac{8.0 \text{ g}}{39.95 \text{ g mol}^{-1}} = 0.20 \text{ mol}$$

The total number of moles is

$n(\text{total}) = n(H_2) + n(Ar) = 0.70 \text{ mol}$

The gas mixture is in a 3.0 L tank at 27 °C (300 K).

The ideal gas law gives the total pressure if we substitute $n(\text{total})$,

$$p = \frac{nRT}{V} = \frac{(0.70 \text{ mol})(8.314 \text{ L kPa K}^{-1}\text{ mol}^{-1})(300 \text{ K})}{(3.0 \text{ L})} = 580 \text{ kPa}$$

The partial pressures of H_2 and Ar are determined using mole fractions,

$$x(H_2) = \frac{n(H_2)}{n(\text{total})} = \frac{0.50 \text{ mol}}{0.70 \text{ mol}} = 0.71$$

$$x(Ar) = \frac{n(Ar)}{n(\text{total})} = \frac{0.20 \text{ mol}}{0.70 \text{ mol}} = 0.29$$

From which we get

$$p(H_2) = x(H_2) \times p(\text{total}) = 0.71 \times 580 \text{ kPa} = 412 \text{ kPa} = 4.07 \text{ atm}$$

$$p(Ar) = x(Ar) \times p(\text{total}) = 0.29 \times 580 \text{ kPa} = 168 \text{ kPa} = 1.66 \text{ atm}$$

Section 11.7: The Kinetic-Molecular Theory of Gases

11.61

$$\text{Average speed of a } CO_2 \text{ molecule} = \sqrt{u_{CO_2}^2} = \sqrt{\frac{3RT}{M_{CO_2}}} = \sqrt{\frac{M_{O_2}}{M_{CO_2}}}\sqrt{\frac{3RT}{M_{O_2}}} = \sqrt{\frac{M_{O_2}}{M_{CO_2}}}\sqrt{u_{O_2}^2}$$

$$= \sqrt{\frac{31.9988}{44.0098}} \times 4.28104 \text{ cm s}^{-1} = 3.65041 \text{ cm s}^{-1}$$

11.63

Refer to 11.61.

$$\text{Average speed of CO molecules} = \sqrt{\frac{32.00}{28.01}} \times 4.28 \text{ cm s}^{-1} = 4.58 \text{ cm s}^{-1}$$

Ratio of speed of CO molecules to speed of Ar atoms (at the same temperature)

$$= \sqrt{\frac{M_{Ar}}{M_{CO}}} = \sqrt{\frac{39.95}{28.01}} = 1.194$$

Section 11.8: Diffusion and Effusion

11.67

$$\frac{\text{Rate of effusion of unknown}}{\text{Rate of effusion of He}} = \sqrt{\frac{M(\text{He})}{M(\text{unknown})}} = \frac{1}{3}$$

So,

$$M(\text{unknown}) = 9 \times M(\text{He}) = 36 \text{ g mol}^{-1}$$

11.69

According to the ideal gas law,

$$p = \frac{nRT}{V} = \frac{(8.00 \text{ mol})(8.314 \text{ L kPa K}^{-1}\text{ mol}^{-1})(300.15 \text{ K})}{(4.00 \text{ L})} = 4990 \text{ kPa} = 49.90 \text{ bar}$$

The van der Waals equation gives a better description of chlorine at such a high pressure.

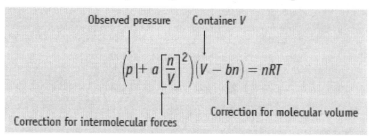

From Table 11.4, the van der Waals' constants for chlorine are
$$a = = 658 \text{ kPa L}^2\text{ mol}^{-2} \quad \text{and} \quad b = 0.0562 \text{ L mol}^{-1}$$
Using these constants and solving for pressure, we get

$$p = \frac{nRT}{V - nb} - a\left(\frac{n}{V}\right)^2$$

$$= \frac{(8.00 \text{ mol})(8.314 \text{ L kPa K}^{-1}\text{ mol}^{-1})(300.15 \text{ K})}{(4.00 \text{ L} - (8.00 \text{ mol})(0.0562 \text{ L mol}^{-1}))} - (658 \text{ L}^2\text{kPa mol}^{-2})\left(\frac{8.00 \text{ mol}}{4.00 \text{ L}}\right)^2$$

$$= 5623 - 2632 \text{ kPa} = 2991 \text{ kPa}$$

Section 11.10: Liquid and Solid States—Stronger Intermolecular Forces

11.71

When solid I_2 dissolves in methanol, CH_3OH, dispersion forces holding I_2 molecules in their lattice positions must be overcome. Hydrogen bonding forces between methanol molecules are disrupted when methanol solvates iodine. The solvation forces between methanol and iodine are dipole-induced dipole interactions.

Section 11.12 Liquids: Properties and Phase Changes

11.75

(a) The equilibrium vapour pressure of water at 60°C is about 19 kPa.
(b) Water has an equilibrium vapour pressure of 80 kPa at about 92°C.
(c) The equilibrium vapour pressure of ethanol is higher than that of water at 70°C (and at all temperatures up to their boiling points).

11.77

The vapour pressure of diethyl ether at 30°C is (from Figure 11.19) about 75 kPa.
To achieve this pressure of diethyl ether in a 0.10 L flask at 30 °C (303 K) requires

$$n((CH_3CH_2)_2O) = \frac{pV}{RT} = \frac{(75\ \text{kPa})(0.10\ \text{L})}{(8.314\ \text{L kPa K}^{-1}\text{mol}^{-1})(303\ \text{K})} = 0.0030\ \text{mol}$$

or

$$m((CH_3CH_2)_2O) = n((CH_3CH_2)_2O) \times M((CH_3CH_2)_2O) = (0.0030\ \text{mol})(74.12\ \text{g mol}^{-1}) = 0.22\ \text{g}$$

1.0 g of diethyl ether is more than enough to achieve this partial pressure. 0.22 g of diethyl ether evaporates, while 0.78 g remains in the liquid phase—the two phases are in equilibrium.
If the flask is placed in an ice bath, the temperature lowers causing the vapour pressure to lower. The gas becomes supersaturated and liquid diethyl ether condenses out of the vapour.

11.81

Oxygen *T-p* phase diagram

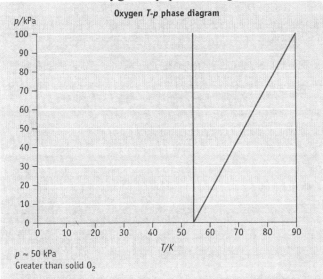

This phase diagram is constructed by (1) connecting ($T = 0$ K, $p = 0$ kPa) to the triple point, ($T = 54.34$ K, $p = 0.267$ kPa), to get the solid-vapour equilibrium curve, (2) connecting the triple point to the normal melting point, ($T = 54.8$ K, $p = 101.3$ kPa), to get the solid–liquid equilibrium curve, and (3) connecting the triple point to the normal boiling point, ($T = 90.18$ K, $p = 101.3$ kPa), to get part of the liquid–gas equilibrium curve. This curve could be taken further if data for the critical point of oxygen were given. The degree of curvature of both the solid–vapour and liquid–vapour curves cannot be qualitatively judged. We will see (Chapter 17) that the curves are such that the data fit a straight-line plot of ln p vs. $1/T$.
At $T = -196$°C (i.e., $T = 77$ K), $p \approx 50$ kPa
The solid–liquid equilibrium line has a positive slope (54.34 K at 0.267 kPa, to 54.8 K at 101.3 kPa). If we increase the pressure on a solid–liquid equilibrium system, we would be taken to the more dense phase—the solid.

Section 11.13: Solids: Properties and Phase Changes

11.83

To heat 5.00 g of solid silver from 25°C to 962°C requires

$q_1 = (0.235 \text{ J g}^{-1}\text{K}^{-1}) \times (5.00 \text{ g}) \times (962 \text{ K} - 25 \text{ K}) = 1100 \text{ J} = 1.10 \text{ kJ}$

Amount of silver $= (5.00 \text{ g}) / (107.9 \text{ g mol}^{-1}) = 0.0463 \text{ mol}$

To melt 5.00 g of solid silver at 962°C requires

$q_2 = (0.0463 \text{ mol}) \times (11.3 \text{ kJ mol}^{-1}) = 0.523 \text{ kJ}$

Total heat required $= q_1 + q_2 = 1.10 + 0.523 \text{ kJ} = 1.623 \text{ kJ}$

Section 11.14: Phase Diagrams

11.87

CH_3Cl can be liquefied at or above room temperature, up to 416 K (143°C) the critical temperature. Above the critical temperature, chloromethane cannot be liquefied—it can only be compressed into a supercritical fluid. Below the triple point temperature, 175.4 K (–97.8°C), chloromethane cannot be liquefied—it solidifies upon compression. However, room temperature is between these limits on liquefaction.

Section 11.15: Polymorphic Forms of Solids

11.89

Ice V only exists over the range of temperatures and pressures shown in the phase diagram (Figure 11.27 of the textbook). The lowest pressure at which Ice V is stable is about 3 Mbar, i.e., about 3 million atmospheres.

SUMMARY AND CONCEPTUAL QUESTIONS

11.91

Initial Conditions	Final Conditions
$V_1 = 0.0255 \text{ L}$	$V_2 = 0.0215 \text{ L}$
p_1 is not given—do not worry	$p_2 = p_1 \quad \leftarrow$ this is all you need to know
$T_1 = 90°C \ (363 \text{ K})$	$T_2 = ?$

$$T_2 = p_2 V_2 \times \left(\frac{T_1}{p_1 V_1} \right) = T_1 \times \frac{p_2}{p_1} \times \frac{V_2}{V_1}$$

$$= 363 \text{ K} \times \frac{\cancel{p_1}}{\cancel{p_1}} \times \frac{0.0215 \ \cancel{L}}{0.0255 \ \cancel{L}}$$

$$= 306 \text{ K} \quad \text{or} \quad 33 \ °C$$

11.93

(a) A 1.0-L flask containing 10.0 g each of O_2 and CO_2 at 25°C has greater partial pressure of O_2 than CO_2. The partial pressures depend on the numbers of moles—10.0 g of the lower molar mass O_2 has more moles than 10.0 g of CO_2.

(b) The lighter O_2 molecules have greater average speed—average speed is proportional to $M^{-1/2}$.

(c) At a given temperature, molecules of all substances have the same average kinetic energy.

11.95

(a) A material is in a steel tank at 100 bar pressure. When the tank is opened to the atmosphere, the material suddenly expands, increasing its volume by 10%. This is NOT a gas. An ideal gas would expand 100 fold if its pressure were changed from 100 bar to 1 bar (atmospheric pressure). Only a liquid or solid is so incompressible as to expand only 10% when relieved of such a large pressure.

(b) A 1.0 mL sample of material weighs 8.2 g. Assuming the temperature is about 25°C, and pressure is about 1 bar, we can say that this is NOT a gas. A density of 8.2 g mL^{-1} = 8200 g L^{-1} is way too high for a gas. An ideal gas at 25°C and 1 bar pressure has the following molar density:

$$\frac{n}{V} = \frac{p}{RT} = \frac{100 \text{ kPa}}{(8.314 \text{ L kPa K}^{-1}\text{mol}^{-1})(298 \text{ K})} = 0.0404 \text{ mol L}^{-1}$$

The ratio of mass density to molar density is the molar mass of the substance.

M (substance) = $(8200$ g $L^{-1})$ / $(0.0404$ mol $L^{-1})$ = 203000 g mol^{-1}

There are no substances with such a huge molar mass that are not solids under ordinary conditions.

(c) There is insufficient information. Chlorine is a transparent and pale green gas. But, there are also transparent and pale green liquids and solids.

(d) A material that contains as many molecules in 1.0 m^3 as the same volume of air, at the same temperature and pressure, is definitely a gas (assuming the temperature and pressure are not unusual). Air is approximately an ideal gas under ordinary conditions. Anything else with the same density under the same conditions is behaving like an ideal gas.

11.97

$$3 N_2O(g) + 4 Na(s) + NH_3(\ell) \longrightarrow NaN_3(s) + 3 NaOH(s) + 2 N_2(g)$$

(a)

Amount of sodium, $n(Na)$ = $(65.0$ g$)$ / $(22.99$ g $mol^{-1})$ = 2.83 mol

Amount of N_2O, $n(N_2O)$ = $\dfrac{pV}{RT}$ = $\dfrac{(215 \text{ kPa})(35.0 \text{ L})}{(8.314 \text{ L kPa K}^{-1}\text{mol}^{-1})(296 \text{ K})}$ = 3.06 mol

To consume 3.06 mol of N_2O requires

$$\frac{4}{3} \times n(N_2O) = \frac{4}{3} \times 3.06 \text{ mol} = 4.08 \text{ mol of Na}$$

Since we only have 2.83 mol of Na, Na is the limiting reactant.

The amount of NaN_3 that can be produced is

$$\frac{1}{4} \times n(\text{Na}) = \frac{1}{4} \times 2.83 \text{ mol} = 0.708 \text{ mol of NaN}_3$$

Mass of NaN$_3$ that can be produced is

$$m(\text{NaN}_3) = (0.708 \text{ mol}) \times (65.02 \text{ g mol}^{-1}) = 46.0 \text{ g}$$

(b)

The structure on the left is the most reasonable, as it has the lowest formal charges—the electron distributions are closest to those of the isolated atoms.

(c) The azide ion is linear.

11.99

For the phase diagram of CO_2, see the solution to Exercise 11.27.
0.61 kPa corresponds to almost zero pressure in the phase diagram. To solidify carbon dioxide requires a temperature no greater than −90°C.

11.101

$$B_2H_6(g) + 3\,O_2(g) \longrightarrow B_2O_3(s) + 3\,H_2O(g)$$

(a) According to increasing average molecular speeds, the gases in this reaction are ordered as follows:

$O_2 \;<\; B_2H_6 \;<\; H_2O$—i.e., in descending order according to molar mass: the heavier the molecules, the less is their average speed.

(b) We do not need to compute the amount of $B_2H_6(g)$. Under the conditions of this question, the partial pressures of $B_2H_6(g)$ and $O_2(g)$ are in the correct stoichiometric ratio—partial pressures are proportional to amounts.
Therefore,

$$\text{Partial pressure of } O_2 = \frac{3}{1} \text{ partial pressure of } B_2H_6 = 102.3 \text{ kPa}$$

This presumes that no reaction between $B_2H_6(g)$ and $O_2(g)$ has taken place.

11.103

$$\text{Amount of air, } n = \frac{pV}{RT} = \frac{(324 \text{ kPa})(17 \text{ L})}{(8.314 \text{ L kPa K}^{-1}\text{mol}^{-1})(298 \text{ K})} = 2.22 \text{ mol}$$

Mass of air, $m = 2.22 \text{ mol} \times 28.96 \text{ g mol}^{-1} = 64.3 \text{ g} = 64 \text{ g}$ (to correct number of significant figures)

11.105

The density of air 20 km above the earth's surface is 92 g m^{-3} = 9.2×10^{-2} g L^{-1}. The pressure and temperature are 5.6 kPa and −63°C, respectively.

(a)

$$\frac{n}{V} = \frac{p}{RT} = \frac{5.6 \text{ kPa}}{(8.314 \text{ L kPa K}^{-1}\text{mol}^{-1})(210 \text{ K})} = 3.21 \times 10^{-3} \text{ mol L}^{-1}$$

$$M(\text{air at 20 km altitude}) = \text{density} \times \left(\frac{n}{V}\right)^{-1} = (9.2 \times 10^{-2} \text{ g L}^{-1}) / (3.21 \times 10^{-3} \text{ mol L}^{-1})$$

$$= 28.7 \text{ g mol}^{-1}$$

Note that we have used molar mass of air = 28.96 g mol^{-1} to solve previous problems. How can the average molar mass of air be different at higher altitude?

The answer is that lighter molecules are distributed to higher altitudes in comparison with heavier molecules. Lighter and heavier molecules (most notably nitrogen and oxygen in the case of air) both have decreasing density with altitude. However, the density of the heavier molecule falls off faster. The result is an average molar mass that decreases slightly with altitude.

(b) The average molar mass is the average of the O_2 and N_2 molar masses—weighted by their mole fractions.

$$\text{Average molar mass} = x(N_2) \times M(N_2) + x(O_2) \times M(O_2)$$

i.e.,

$$28.7 \text{ g mol}^{-1} = x(N_2) \times (28.02 \text{ g mol}^{-1}) + x(O_2) \times (32.00 \text{ g mol}^{-1})$$
$$= (1 - x(O_2)) \times (28.02 \text{ g mol}^{-1}) + x(O_2) \times (32.00 \text{ g mol}^{-1})$$
$$28.7 = x(O_2) \times (32.00 - 28.02) + 28.02$$
$$x(O_2) = (28.7 - 28.02) / (32.00 - 28.02) = 0.17 \quad \leftarrow \text{a lower value than on the surface of Earth}$$
$$x(N_2) = 1 - 0.17 = 0.83$$

11.107

$$p(N_2) = p(\text{total}) - p(O_2) - p(CO_2) - p(H_2O)$$
$$= 33.7 - 4.7 - 1.0 - 6.3 \text{ kPa} = 21.7 \text{ kPa}$$

11.109

(a) At 20°C, the vapour pressure of water is 23.38 mbar = 2.338 kPa.
At 45% relative humidity, the partial pressure of water is 0.45 × 2.338 kPa = 1.05 kPa

$$\frac{n}{V} = \frac{p}{RT} = \frac{1.05 \text{ kPa}}{(8.314 \text{ L kPa K}^{-1}\text{mol}^{-1})(293 \text{ K})} = 4.31 \times 10^{-4} \text{ mol L}^{-1}$$

Density of water = (18.02 g mol^{-1}) × (4.31 × 10^{-4} mol L^{-1}) = 0.00776 g L^{-1} = 7.76 mg L^{-1}

(b) At 0°C, the vapour pressure of water is 6.11 mbar = 0.611 kPa.
At 95% relative humidity, the partial pressure of water is 0.95 × 0.611 kPa = 0.580 kPa

$$\frac{n}{V} = \frac{p}{RT} = \frac{0.580 \text{ kPa}}{(8.314 \text{ L kPa K}^{-1}\text{mol}^{-1})(273 \text{ K})} = 2.56 \times 10^{-4} \text{ mol L}^{-1}$$

Density of water = (18.02 g mol^{-1}) × (2.56 × 10^{-4} mol L^{-1}) = 0.00461 g L^{-1} = 4.61 mg L^{-1}

Therefore, the density of water vapour is higher at 20°C with 45% humidity than at 0°C with 95% humidity.

11.111

Amount density of mercury

$$= \frac{n}{V} = \frac{p}{RT} = \frac{0.225 \times 10^{-3} \text{ kPa}}{(8.314 \text{ L kPa K}^{-1}\text{mol}^{-1})(297 \text{ K})} = 9.11 \times 10^{-8} \text{ mol L}^{-1}$$

Atoms in $1 \text{ m}^3 = (1.00 \times 10^3 \text{ L}) \times (6.022 \times 10^{23} \text{ atoms mol}^{-1}) \times (9.11 \times 10^{-8} \text{ mol L}^{-1})$

$\qquad = 5.49 \times 10^{19} \text{ atoms}$

11.113

The molecules of cooking oil do not form hydrogen bonds with water, and do not have strong dipoles to form significant dipole–dipole interactions.

11.115

(a) The normal boiling point of CCl_2F_2 is the temperature at which the vapour pressure = 1 atm (which is very close to 1 bar, or 100 kPa). The normal boiling point of $CCl_2F_2 = -27°C$.

(b) At 25°C, there is net evaporation of liquid CCl_2F_2 until the dichlorodifluoromethane partial pressure inside the steel cylinder equals about 6.5 bar, the vapour pressure of CCl_2F_2 at 25°C. Here we assume that the cylinder is not so big that 25 kg of CCl_2F_2 is not enough to fill the cylinder to this pressure—i.e., no bigger than about 780 L.

(c) The CCl_2F_2 vapour rushes out of the cylinder quickly at first because of the large pressure imbalance. The flow slows as the inside pressure approaches 1 atm.
The outside of the cylinder becomes icy because when the gas expands, intermolecular forces are overcome—the energy required is drawn from the thermal energy of the gas and cylinder. Expansion of a gas can cause significant cooling. This expansion is driven by the increase in entropy, and the resulting cooling is inexorable.

(d)

(1) Turning the cylinder upside down, and opening the valve, would produce a dangerous situation. The cylinder would behave like a rocket. It would empty quickly though.

(2) Cooling the cylinder to −78°C in dry ice, then opening the valve would allow the cylinder to be emptied safely. It would not happen quickly though—the vapour pressure of CCl_2F_2 at −78°C (not shown in the figure) is quite small.

(3) Knocking the top off the cylinder, valve and all, with a sledge hammer would provide rapid and relatively safe discharge of the cylinder. The flow velocity would be smaller because of the large cross-sectional area of the open top of the cylinder.

11.117

Mass of ethanol, $m = (0.125 \text{ L}) \times (784.9 \text{ g L}^{-1}) = 98.1 \text{ g}$
Amount of ethanol, $n = (98.1 \text{ g}) / (46.07 \text{ g mol}^{-1}) = 2.13 \text{ mol}$
Heat required to evaporate 2.13 mol of ethanol at 25°C is
$\qquad q = (42.32 \text{ kJ mol}^{-1}) \times (2.13 \text{ mol}) = 90.1 \text{ kJ}$

11.119

(a) The structure of aspartame.

(b) Aspartame is capable of hydrogen bonding. Sites of hydrogen bonding are circled on the structure.

CHAPTER 12
Solutions and Their Behaviour

IN-CHAPTER EXERCISES

Exercise 12.3—Using Henry's Law

Henry's law constant for CO_2 in water is 3.36×10^{-4} mol L^{-1} kPa^{-1} (see Table 12.1).
The solubility of CO_2 in water is

$$s = k_H \times p = (3.36 \times 10^{-4} \text{ mol } L^{-1} \text{ kPa}^{-1})(33.4 \text{ kPa}) = 1.12 \times 10^{-2} \text{ mol } L^{-1}$$

This is the equilibrium concentration of CO_2 in water at 25°C when the pressure of $CO_2(g)$ above the aqueous phase is 33.4 kPa.

Exercise 12.5—Temperature Dependence of Solubility

The solubility of Li_2SO_4 in water decreases slightly as we increase the temperature from 10°C.
The amount of solid Li_2SO_4 in the second beaker will increase a little (presuming that the solution was a saturated solution at 10°C).
The solubility of LiCl in water increases slightly as we increase the temperature from 10°C. The amount of solid LiCl in the first beaker will decrease.

Exercise 12.7—Solution Concentration

Amount of sucrose $= (10.0 \text{ g}) / (342.3 \text{ g mol}^{-1}) = 0.0292$ mol
Amount of water $= (250 \text{ g}) / (18.02 \text{ g mol}^{-1}) = 13.9$ mol
The mole fraction of sucrose $=$ (amount of sucrose) / (total amount)
$\quad x = (0.0292 \text{ mol}) / (0.0292 \text{ mol} + 13.9 \text{ mo }) = 0.00210$
The molality of sucrose solution, $m =$ (amount of sucrose) / (mass of water in
kg)$= \dfrac{0.0292 \text{ mol}}{0.250 \text{ kg}} = 0.117$ mol kg^{-1}
The mass percent of sucrose $= 100\% \times$ (mass of sucrose) / (total mass)
$\quad\quad\quad\quad\quad = 100\% \times (10.0 \text{ g}) / (260 \text{ g}) = 3.85\%$

Exercise 12.9—Vapour Pressure of Solutions, Non-volatile Solute

$$n(\text{H}_2\text{O}) = \frac{500.0\ \cancel{g}}{18.02\ \cancel{g}\ \text{mol}^{-1}} = 27.75\ \text{mol}$$

$$n(\text{HOCH}_2\text{CH}_2\text{OH}) = \frac{35.0\ \cancel{g}}{62.07\ \cancel{g}\ \text{mol}^{-1}} = 0.564\ \text{mol}$$

$$\therefore x_{\text{H}_2\text{O}} = \frac{27.75\ \cancel{\text{mol}}}{27.75\ \cancel{\text{mol}} + 0.564\ \cancel{\text{mol}}} = 0.980$$

From Raoult's law,

$$p_{\text{H}_2\text{O}} = x_{\text{H}_2\text{O}} \times p^{\circ}_{\text{H}_2\text{O}} = (0.980)\,(4.76\,\text{kPa}) = 4.66\ \text{kPa}$$

Exercise 12.11—Freezing Point Depression

Determine the freezing point of the ethylene glycol–water solution.
Amount of HOCH2CH2OH, $n = (525\ \text{g})\,/\,(62.07\ \text{g mol}^{-1}) = 8.46\ \text{mol}$
The molality of HOCH2CH2OH $=$ (amount of HOCH2CH2OH) / (mass of water in kg)

$$= \frac{8.46\ \text{mol}}{3.00\ \text{kg}} = 2.82\ \text{mol kg}^{-1}$$

Freezing point depression, $\Delta T_f = K_f \times m = (1.86\ \text{K}\ \cancel{\text{kg}}\ \cancel{\text{mol}}^{-1})\,(2.82\ \cancel{\text{mol}}\ \cancel{\text{kg}}^{-1})$

$$= 5.25\ \text{K} = 5.25\ \text{degrees Celsius}$$

Note that the freezing point depression constant, K_f is characteristic of the solvent—it does not depend on the solute, ethylene glycol.
Therefore, the freezing point of the solution is −5.25°C. The concentration of ethylene glycol is not enough to prevent freezing at −25°C.

Exercise 12.13—Molar Mass from Freezing Point Depression

Molality of aluminon in water $=$

$$m = \frac{\Delta T_f}{K_f} = \frac{0.197\ \cancel{\text{K}}}{1.86\ \cancel{\text{K}}\ \text{kg mol}^{-1}} = 0.106\ \text{mol kg}^{-1}$$

The amount of aluminon in 50.0 g water, n, is $(0.0500\ \cancel{\text{kg}}) \times (0.106\ \text{mol}\ \cancel{\text{kg}}^{-1}) = 0.00530\ \text{mol}$

Molar mass of aluminon, $M =$ (mass of aluminon) / (amount of aluminon)

$$= (2.50\ \text{g})\,/\,(0.00530\ \text{mol}) = 472\ \text{g mol}^{-1}$$

Exercise 12.15—Freezing Points of Solutions of Electrolytes

Amount of NaCl, $n = (25.0\ \text{g})\,/\,(58.44\ \text{g mol}^{-1}) = 0.428\ \text{mol}$
Molality of NaCl $=$ (amount of NaCl) / (mass of water, in kg)

$$= \frac{0.428\ \text{mol}}{0.525\ \text{kg}} = 0.815\ \text{mol kg}^{-1}$$

Assuming that the solute is a non-electrolyte

$$\Delta T_f \text{ (non-electrolyte)} = K_f \times m = (1.86 \text{ K kg mol}^{-1})(0.815 \text{ mol kg}^{-1}) = 1.52 \text{ K}$$

$$\Delta T_f \text{ (actual)} = i \times \Delta T_f \text{ (non-electrolyte)} = 1.85 \times 1.52 \text{ K} = 2.81 \text{ K}$$

The freezing point of this solution is $-2.81°C$.

Exercise 12.17—Molar Mass from Osmotic Pressure

$$\text{Molar concentration, } c = \frac{\Pi}{RT} = \frac{0.248 \text{ kPa}}{(8.314 \text{ L kPa K}^{-1} \text{ mol}^{-1})(298 \text{ K})} = 1.00 \times 10^{-4} \text{ mol L}^{-1}$$

In 100 mL of solution, amount of polyethylene, $n = (1.00 \times 10^{-4} \text{ mol L}^{-1}) \times (0.100 \text{ L}) = 1.00 \times 10^{-5} \text{ mol}$

Average molar mass of polyethylene $= (1.40 \text{ g}) / (1.00 \times 10^{-5} \text{ mol}) = 1.40 \times 10^{5} \text{ g mol}^{-1}$

Exercise 12.19—Osmotic Pressure of Electrolyte Solutions

The osmotic pressure of human blood (37°C), treated as a 0.154 mol L^{-1} NaCl solution, is

$$\Pi = i \times cRT = 1.9 \times (0.154 \text{ mol L}^{-1}) (8.314 \text{ L kPa K}^{-1} \text{ mol}^{-1}) (310 \text{ K})$$

$$= 754 \text{ kPa}$$

This is a very high pressure. Human cells burst from the osmotic pressure of water entering the cell—if the cell is exposed to pure water.

REVIEW QUESTIONS

Section 12.1: The Killer Lakes of Cameroon

12.21

(a) No, CO_2 is not normally poisonous to humans.
(b) The concentration in the air that night at ground level was so high that people suffocated (starved of oxygen).
(c) The amount of CO_2 expelled when we open a bottle of carbonated drink quickly dissipates into the atmosphere, and its concentration is very low comparatively.
(d) CO_2 is a greenhouse gas.

12.23

Dissolution of CO_2 in water

$$CO_2(g) \underset{}{\overset{H_2O(\ell)}{\rightleftharpoons}} CO_2(aq)$$

Formation of aquated carbonic acid

$$CO_2(aq) + H_2O(\ell) \rightleftharpoons H_2CO_3(aq)$$

Ionization of $H_2CO_3(aq)$ as a weak diprotic acid

$$H_2CO_3(aq) + H_2O(\ell) \rightleftharpoons HCO_3^-(aq) + H_3O^+(aq)$$

$$HCO_3^-(aq) + H_2O(\ell) \rightleftharpoons CO_3^{2-}(aq) + H_3O^+(aq)$$

When CO_2 is dissolved in water, the water becomes more acidic, which results in a multitude of negative consequences. Some negative consequences are as follows: interference with corals' deposition, the inability of organisms to use $CaCO_3$ to build shells, food supplies/food chain interruptions and breeding ground endangerment, plant life, tourism, and dwindled supply of marine-derived natural medicinal products.

Section 12.2: Solutions and Solubility

12.25

$CO_2(g)$ molecules that are dissolved may become aquated or react with water to reversibly form H_2CO_3. CO_2 (O=C=O) has two potential sites with 4 lone pairs of electrons to participate in hydrogen bonding with 4 molecules of water forming a shell around the molecule ($CO_2(aq)$).

12.27

The solubility of a substance in a solvent at a given temperature is the concentration of the substance in a saturated solution. The concentration of sugar in water for the given conditions is 40 g L^{-1}. If all 20 g of sugar dissolved, then it is soluble in water. There is not enough information to determine if this is a saturated solution or not.

Section 12.3: Enthalpy Change of Solution: Ionic Solutes

12.29

(a) The dissolution of solid potassium nitrate in water is endothermic, so the flask will feel cold to the touch.

(b) The dissolution of $KNO_3(s)$ can be thought of as two parts:

Step 1: endothermic separation of ions

$$KNO_3(s) \rightarrow K^+(g) + NO_3^-(g) \quad \Delta_{lattice}H$$

Step 2: exothermic aquation of ions

$$K^+(g) + NO_3^-(g) \xrightarrow{H_2O(\ell)} K^+(aq) + NO_3^-(aq) \quad \Delta_{aq}H$$

$$\Delta_{sol}H = \Delta_{lattice}H + \Delta_{aq}H$$

Since the flask felt cold, the endothermic term (lattice enthalpy) must be greater than the aquation enthalpy.

(c) For many ionic compounds, $\Delta_{sol}H$ is a small difference between two large enthalpy changes and is greatly affected by ion size and charge. Smaller ions will interact more strongly with water resulting in a larger aquation enthalpy. Even if energy is absorbed during the dissolving process, the net result is favourable. The term "going uphill" refers to endothermic reactions where energy is required to overcome a barrier—in this case to dissolution.

Section 12.4: Factors Affecting Solubility: Pressure and Temperature

12.31

Table 12.1 gives the Henry's law constant for CO_2 in water as 3.36×10^{-4} mol L^{-1} kPa^{-1}. The partial pressure of CO_2 inside the can (above the aqueous solution) is

$(0.0506$ mol $L^{-1}) / (3.36 \times 10^{-4}$ mol L^{-1} kPa$^{-1}) = 151$ kPa

Section 12.5: More Units of Solute Concentration

12.33

Sea water has a sodium ion concentration of 1.08×10^4 ppm.
The mass of 1 L of sea water is 1000 mL $\times$ 1.05 g mL^{-1} = 1050 g
The mass of NaCl in 1 L of sea water is   1050 g $\times$ 1.08 $\times$ 10^4 $\times$ 10^{-6} = 11.34 g

12.35

The order of the steps taken are indicated as (1), (2), and (3).

Compound	Molality	Mass Percent	Mole Fraction
KNO_3		10.0%	
Suppose you have 1.00 kg of solution	(2) amount of KNO_3 $= (100 \text{ g}) / (101.11 \text{ g mol}^{-1})$ $= 0.989$ mol of KNO_3 mass of water $1000 - 100$ g $= 900$ g $= 0.900$ kg **molality of KNO_3** = $\dfrac{0.989 \text{ mol}}{0.900 \text{ kg}} = 1.10 \text{ mol kg}^{-1}$	(1) 0.100×1.00 kg $= 0.100$ kg of KNO_3 $= 100$ g	(3) amount of water $= (900 \text{ g}) / (18.02 \text{ g mol}^{-1}$ $= 49.9$ mol **mole fraction of KNO_3** = $\dfrac{0.989 \text{ mol}}{0.989 + 49.9 \text{ mol}} = 0.019$
CH_3CO_2H	0.0183 mol kg^{-1}		
Suppose you have 1.00 kg of water $= 1000$ g of water	(1) amount of CH_3CO_2H $= (0.0183 \text{ mol kg}^{-1}) \times 1.00$ kg $= 0.0183$ mol	(2) mass of CH_3CO_2H $= (0.0183 \text{ mol}) \times (60.05$ g mol^{-1}) $= 1.10$ g **mass percent of CH_3CO_2H** = $100\% \times \dfrac{1.10 \text{ g}}{1000 \text{ g} + 1.10 \text{ g}}$ $= \mathbf{0.110\%}$	(3) amount of water $= (1000 \text{ g}) / (18.02 \text{ g mol}^-$ $= 55.5$ mol **mole fraction of CH_3CO_2H** $=$ $\dfrac{0.0183 \text{ mol}}{0.0183 + 55.5 \text{ mol}}$ $= \mathbf{0.000330}$
$HOCH_2CH_2$ OH		18.0%	
Suppose you have 1.00 kg of solution	(2) amount of $HOCH_2CH_2OH$ $= (180 \text{ g}) / (62.07 \text{ g mol}^{-1})$ $= 2.90$ mol of $HOCH_2CH_2OH$ mass of water $1000 - 180$ g $= 820$ g $= 0.820$ kg **molality of $HOCH_2CH_2OH$** = $\dfrac{2.90 \text{ mol}}{0.820 \text{ kg}} = 3.54 \text{ mol kg}^{-1}$	(1) 0.180×1.00 kg $= 0.180$ kg of $HOCH_2CH_2OH$ $= 180$ g	(3) amount of water $= (820 \text{ g}) / (18.02 \text{ g mol}^{-1}$ $= 45.5$ mol **mole fraction of $HOCH_2CH_2OH$** = $\dfrac{2.90 \text{ mol}}{2.90 + 45.5 \text{ mol}} = 0.0599$

12.41

Sea water has a lithium ion concentration of 0.18 ppm.
The mass of Li$^+$ ions in 1 kg of sea water is
$$1000 \text{ g} \times 0.18 \times 10^{-6} = 1.8 \times 10^{-4} \text{ g}$$
Amount of Li$^+$ ions in 1 kg of sea water $= (1.8 \times 10^{-4} \text{ g}) / (6.941 \text{ g mol}^{-1}) = 2.6 \times 10^{-5}$ mol
Molality of Li$^+$ ions $= (2.6 \times 10^{-5} \text{ mol}) / (1.000 \text{ kg}) = 2.6 \times 10^{-5} \text{ mol kg}^{-1}$

Section 12.6: Colligative Properties

12.43

According to Raoult's law, the water vapour pressure = (mole fraction of water) × (vapour pressure of pure water)

Therefore,

Mole fraction of water = (water vapour pressure) / (vapour pressure of pure water)
   = (60.93 kPa) / (70.10 kPa) = 0.8692

Amount of water = (2000 g) / (18.02 g mol^{-1}) = 111 mol

Mole fraction of ethylene glycol (EG) = 1.0000 − 0.8692 = 0.1308

Amount of EG = (mole fraction of EG) × (total amount)

i.e., $n_{EG} = x_{EG} (n_{EG} + n_{water})$

So, $n_{EG} = \dfrac{x_{EG} n_{water}}{(1 - x_{EG})} = \dfrac{0.1308 \times 111.\,mol}{0.8692} = 16.7\,mol$

Mass of EG added = (16.7 mol) × (62.07 g mol^{-1}) = 1037 g = 1040 g (to three significant figures)

12.45

Amount of I$_2$ = (105 g) / (126.9 g mol^{-1}) = 0.827 mol

Amount of CCl$_4$ = (325 g) / (153.81 g mol^{-1}) = 2.11 mol

The vapour pressure of the solution comes almost entirely from the volatile carbon tetrachloride.

Vapour pressure of CCl$_4$ = (mole fraction of CCl$_4$) × (vapour pressure of pure CCl$_4$)

$= \dfrac{2.11\ \cancel{mol}}{2.11\ \cancel{mol} + 0.827\ \cancel{mol}} \times 70.80\ kPa = 0.718 \times 70.80\ kPa = 50.8\ kPa$

12.47

Amount of sucrose, n = (15.0 g) / (342.3 g mol^{-1}) = 0.0438 mol

Molality of sucrose, m = (0.0438 mol) / (0.225 kg) = 0.195 mol kg^{-1}

Freezing point depression =

$\Delta T_f = K_f \times m = 1.86\ K\ \cancel{kg}\ \cancel{mol}^{-1} \times 0.195\ \cancel{mol}\ \cancel{kg}^{-1} = 0.363\ K$

The freezing point of the solution is −0.362°C

12.49

The molality of the naphthalene (in biphenyl) solution is

$m = \dfrac{\Delta T_f}{K_f} = \dfrac{(70.03 - 69.40)\,K}{8.00\ K\,kg\,mol^{-1}} = 0.079\ mol\ kg^{-1}$

Amount of naphthalene added to the 10.0 g of biphenyl = 0.079 mol kg^{-1} × (0.010 kg) = 0.00079 mol

Molar mass of naphthalene = (0.100 g) / (0.00079 mol) = 127 g mol^{-1}

SUMMARY AND CONCEPTUAL QUESTION

12.55

With the same amounts, $CaCl_2$ lowers the freezing point of the water almost 1.5 times as much as the dissolved NaCl because there are 3 mol of aquated ions for every mol of $CaCl_2$ dissolved, and only 2 mol of aquated ions for every mol of NaCl dissolved.

CHAPTER 13
Dynamic Chemical Equilibrium

IN-CHAPTER EXERCISES

Exercise 13.1—Expressing the Form of the Reaction Quotient

(a) $PCl_5(g) \rightleftharpoons PCl_3(g) + Cl_2(g)$

$$Q = \frac{[PCl_3][Cl_2]}{[PCl_5]}$$

(b) $CO_2(g) + C(s) \rightleftharpoons 2\,CO(g)$

$$Q = \frac{[CO]^2}{[CO_2]}$$

The solid reactant, C(s), does not appear in the reaction quotient.

(c) $Cu(NH_3)_4^{2+}(aq) \rightleftharpoons Cu^{2+}(aq) + 4\,NH_3(aq)$

$$Q = \frac{[Cu^{2+}][NH_3]^4}{[Cu(NH_3)_4^{2+}]}$$

(d) $CH_3COOH(aq) + H_2O(\ell) \rightleftharpoons CH_3COO^-(aq) + H_3O^+(aq)$

$$Q = \frac{[CH_3COO^-][H_3O^+]}{[CH_3COOH]}$$

Remember that the liquid reactant, $H_2O(\ell)$, does not appear in the reaction quotient.

Exercise 13.7—Estimation of Equilibrium Constants

$C_6H_{10}I_2\,(in\,CCl_4) \rightleftharpoons C_6H_{10}\,(in\,CCl_4) + I_2\,(in\,CCl_4)$
The order of the steps taken are indicated as (1), (2), etc.

	$C_6H_{10}I_2$	$\rightleftharpoons$	C_6H_{10}	+	I_2
Initial / (mol L^{-1})	0.050 mol / 1.00 L $=\ 0.050\ mol\ L^{-1}$		0		0
Change / (mol L^{-1})	(2) the change in $[C_6H_{10}I_2]$ $=\ -0.035$		(3) the change in $[C_6H_{10}]$ $=\ +0.035$		(1) the change in $[I_2]$ $=\ +0.035$
Equilibrium / (mol L^{-1})	(4) $0.050 - 0.035$ $=\ 0.015$		(5) $0 + 0.035$ $=\ 0.035$		$0 + 0.035$ $=\ 0.035$

Having determined the equilibrium concentrations, we can evaluate the reaction quotient, which equals K at equilibrium.

$$Q = \frac{[C_6H_{10}][I_2]}{[C_6H_{10}I_2]} = \frac{0.035 \times 0.035}{0.015} = 0.082 = K$$

Exercise 13.9—Calculating Equilibrium Concentrations

$$H_2(g) + I_2(g) \rightleftharpoons 2HI(g)$$

At some temperature, $K = 33$.

The ICE table is constructed in terms of an unknown extent of reaction, x, that we must solve for using the given equilibrium constant.

	H_2	+	I_2	$\rightleftharpoons$	2 HI
Initial / (mol L^{-1})	6.00×10^{-3}		6.00×10^{-3}		0
Change / (mol L^{-1})	$-x$		$-x$		$+2x$
Equilibrium / (mol L^{-1})	$6.00 \times 10^{-3} - x$		$6.00 \times 10^{-3} - x$		$2x$

The equilibrium constant is determined by the equilibrium concentrations, expressed here in terms of the unknown, x.

$$K = \frac{[HI]^2_{eqm}}{[H_2]_{eqm}[I_2]_{eqm}} = \frac{(2x)^2}{(6.00 \times 10^{-3} - x)(6.00 \times 10^{-3} - x)} = 33$$

Taking the square root of both sides of this equation gives

$$\frac{2x}{(6.00 \times 10^{-3} - x)} = \sqrt{33} = 5.74$$

$$2x = 5.74\,(6.00 \times 10^{-3} - x) = 34.5 \times 10^{-3} - 5.74x$$

$$7.74x = 34.5 \times 10^{-3}$$

$$x = 4.46 \times 10^{-3}$$

Now that we have the extent of reaction, x, we can calculate all of the reactant and product concentrations at equilibrium.

$[H_2] = [I_2] = 6.00 \times 10^{-3} - x = 6.00 \times 10^{-3} - 4.46 \times 10^{-3} = 1.54 \times 10^{-3}$ mol L^{-1}

$[HI] = 2x = 2 \times 4.46 \times 10^{-3} = 8.92 \times 10^{-3}$ mol L^{-1}

Exercise 13.11—Reaction Equations and Equilibrium Constants

$$A + B \xrightleftharpoons{K1} 2C$$
$$2A + 2B \xrightleftharpoons{K2} 4C$$

The correct answer is **(b)** $K_2 = K_1^2$

Doubling all stoichiometric coefficients doubles all exponents in the reaction quotient expression— squaring the reaction quotient.

Exercise 13.15—Effect of Concentration Changes on Equilibrium

We will need the equilibrium constant. Use the given concentrations for the initial equilibrium mixture.

$$K = \frac{[\text{isobutane}]_{\text{eqm}}}{[\text{butane}]_{\text{eqm}}} = \frac{0.50}{0.20} = 2.5$$

Now, we set up an ICE table with the initial data corresponding to after increasing [isobutane] by 2.00.

	butane	$\rightleftharpoons$	**isobutane**
Initial / (mol L^{-1})	0.20		0.50 + 2.00 = 2.50 Note that by increasing [isobutane] by 2.00, the mixture is taken away from equilibrium.
Change / (mol L^{-1})	$+x$		$-x$
Equilibrium / (mol L^{-1})	$0.20 + x$		$2.50 - x$

The equilibrium constant is determined by the equilibrium concentrations, expressed here in terms of the unknown, x.

$$K = \frac{[\text{isobutane}]_{\text{eqm}}}{[\text{butane}]_{\text{eqm}}} = \frac{2.50 - x}{0.20 + x} = 2.5$$

Solving for x:
$$2.50 - x = 2.5\,(0.20 + x) = 0.50 + 2.5\,x$$

$$3.5\,x = 2.00$$

$$x = 0.57$$

The new equilibrium concentrations are
$$[\text{butane}] = 0.20 + x = 0.20 + 0.57 = 0.77 \text{ mol L}^{-1}$$
$$[\text{isobutane}] = 2.50 - x = 2.50 - 0.57 = 1.93 \text{ mol L}^{-1}$$

Exercise 13.17—Dependence of Equilibrium Constant on Temperature

(a) The equilibrium concentration of NOCl decreases if the temperature of the system is increased.
$$2\,\text{NOCl(g)} \rightleftharpoons 2\,\text{NO(g)} + \text{Cl}_2(\text{g}) \qquad \Delta_{\text{rxn}}H^\circ = +77.1 \text{ kJ mol}^{-1}$$

(i) Initially, the mixture is at equilibrium. So, $Q = K$. When temperature increases, the equilibrium constant generally changes. For an endothermic reaction such as this, the equilibrium constant increases. If there were no net reaction, we would have $Q < K'$ where K' is the new equilibrium constant. *Net* reaction must proceed in the direction that forms more NO(g) and Cl_2(g) to return to equilibrium—by increasing Q up to K'.

(ii) The reaction proceeds in the forward—endothermic—direction after the temperature is increased. The endothermic direction is the direction that consumes heat and lowers temperature, countering the increase in temperature in accord with Le Chatelier's principle.

(b) The equilibrium concentration of SO_3 decreases if the temperature is increased.

$$2\,SO_2(g) + O_2(g) \rightleftharpoons 2\,SO_3(g) \qquad \Delta_{rxn}H^\circ = -198\ kJ\ mol^{-1}$$

(i) Initially, the mixture is at equilibrium, so, $Q = K$. For an exothermic reaction, the equilibrium constant decreases. Now, $Q > K'$ where K' is the new equilibrium constant. To return to equilibrium, Q must decrease—net reaction proceeds in the direction to form more SO_2(g) and O_2(g) to return to equilibrium—decreasing Q down to K'.

(ii) The reaction proceeds in the reverse—endothermic—direction after the temperature is increased. The endothermic direction is the direction that absorbs heat and lowers temperature, countering the increase in temperature in accord with Le Chatelier's principle.

REVIEW QUESTIONS

Section 13.1: Air into Bread

13.19

Nitrogen from Haber-Bosch process is used to make about half of the world's synthetic nitrogen fertilizer. We consume this nitrogen directly by eating chemically fertilized plants, or indirectly by eating animals that have consumed chemically fertilized plants. Most of the food in the human diet is derived from these sources.

13.21

The work of Fritz Haber, through the Haber-Bosch process, resulted in dramatically increased food production, allowing the global population to reach six billion. Haber's later work on explosives and chemical weapons led to enormous loss of life in WWI and WWII. This contrast shows the profound connection between human motivations and the applications of chemistry.

13.23

$$\text{number of nitrogen atoms} = (2\ kg)\left(\frac{1000\ g}{1\ kg}\right)\left(\frac{1\ mol}{14.01\ g}\right)\left(\frac{6.02\times10^{23}\ atoms}{1\ mol}\right) = 9\times10^{25}\ atoms$$

Section 13.2: Reaction Mixtures in Dynamic Chemical Equilibrium

13.25

Net rate of people entering = (rate of people entering) – (rate of people leaving)

Imagine that, over a one hour time period, ten people enter the theatre while three people leave the theatre. The net rate of people entering is therefore:

Net rate of people entering = 10 people/hour– 3 people/hour = 7 people/hour

We therefore say that over the given time period, the net rate is seven people entering per hour.

Section 13.3: The Reaction Quotient and the Equilibrium Constant

13.27

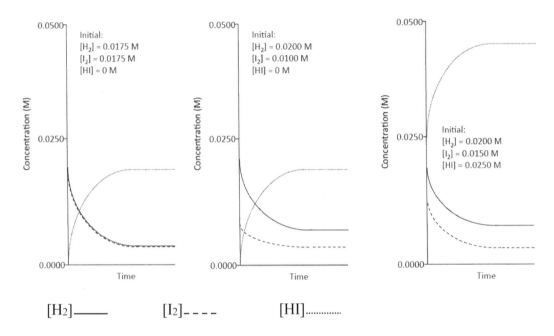

[H₂]———   [I₂]- - - -   [HI]..............

13.29

$$N_2(g) + O_2(g) \rightleftharpoons 2\,NO(g)$$

$$Q = \frac{[NO]^2}{[N_2][O_2]} = \frac{(4.2 \times 10^{-3})^2}{(0.50)(0.25)} = 1.4 \times 10^{-4} < 4.0 \times 10^{-4} = K$$

Because $Q < K$, net reaction proceeds in the direction that forms more NO(g), to increase Q up to K.

Section 13.4: Quantitative Aspects of Equilibrium Constants

13.31

$$PCl_5(g) \rightleftharpoons PCl_3(g) + Cl_2(g)$$

$$Q = \frac{[PCl_3][Cl_2]}{[PCl_5]} = \frac{(1.3 \times 10^{-2})(3.9 \times 10^{-3})}{(4.2 \times 10^{-5})} = 1.2 = K$$

since the mixture is at equilibrium.

13.35

	Br₂(g)	⇌	**2 Br(g)**
Initial / (mol L⁻¹)	0.086 / 1.26 = 0.068		0
Change / (mol L⁻¹)	(2) = −x = −0.00125		(1) + 0.0025 = 2x
Equilibrium / (mol L⁻¹)	(3) 0.068 − 0.00125 = 0.067		3.7% of 0.068 = 0.037 × 0.068 = 0.0025

Having determined the equilibrium concentrations, we can evaluate the reaction quotient which equals K at equilibrium.

$$K = \frac{[\text{Br}]^2_{\text{eqm}}}{[\text{Br}_2]_{\text{eqm}}} = \frac{(0.0025)^2}{(0.067)} = 9.3 \times 10^{-5}$$

13.39

$$\text{H}_2(\text{g}) + \text{CO}_2(\text{g}) \rightleftharpoons \text{H}_2\text{O}(\text{g}) + \text{CO}(\text{g})$$

(a) $K = \dfrac{[\text{H}_2\text{O}]_{\text{eqm}}[\text{CO}]_{\text{eqm}}}{[\text{H}_2]_{\text{eqm}}[\text{CO}_2]_{\text{eqm}}} = \dfrac{(0.11/1.0)(0.11/1.0)}{(0.087/1.0)(0.087/1.0)} = 1.6$

(b)

	H₂(g)	+	**CO₂(g)**	⇌	**H₂O(g)**	+	**CO(g)**
Initial / (mol L⁻¹)	0.050 / 2.0 = 0.025		0.050 / 2.0 = 0.025		0		0
Change / (mol L⁻¹)	−x		−x		+x		+x
Equilibrium / (mol L⁻¹)	0.025 − x		0.025 − x		x		x

The equilibrium constant is determined by the equilibrium concentrations, expressed here in terms of the unknown, x.

$$K = \frac{[\text{H}_2\text{O}]_{\text{eqm}}[\text{CO}]_{\text{eqm}}}{[\text{H}_2]_{\text{eqm}}[\text{CO}_2]_{\text{eqm}}} = \frac{x\,x}{(0.025 - x)(0.025 - x)} = 1.6$$

Taking the square root of both sides of this equation gives

$$\frac{x}{(0.025 - x)} = \sqrt{1.6} = 1.26$$

$$x = 1.26\,(0.025 - x) = 0.0315 - 1.26x$$

$$2.26x = 0.0315$$

$$x = 0.014$$

The new equilibrium concentrations of $H_2O(g)$ and $CO(g)$ are both 0.014 mol L^{-1}, and the amounts (in mol) are 0.028 for both $H_2O(g)$ and $CO(g)$.

13.41

	I_2(aq)	$\rightleftharpoons$	I_2(CCl₄)
Initial / (mol L^{-1})	amount = 0.0340 g / 253.81 g mol^{-1} = 0.000134 mol concentration = 0.000134 mol / 0.1000 L = 0.00134		0
Change / (mol L^{-1})	$-x$		$+x$
Equilibrium / (mol L^{-1})	$0.00134 - x$		x

The equilibrium constant is given by

$$K = \frac{[I_2(CCl_4)]_{eqm}}{[I_2(aq)]_{eqm}} = \frac{x}{0.00134 - x} = 85.0$$

Solve this equation for x.

$$x = 85.0\,(0.00134 - x) = 0.114 - 85.0\,x$$

$$86.0\,x = 0.114$$

$$x = 0.00133$$

The equilibrium concentrations are

$[I_2(aq)]_{eqm} = 0.00134 - x = 0.00001$ mol L^{-1}

$[I_2(CCl_4)]_{eqm} = x = 0.00133$ mol L^{-1}

To the precision of the given data, only 0.00001 mol L^{-1} concentration of I_2 remains in the aqueous layer.

The corresponding fraction of the total concentration of I_2 is 0.00001 / 0.00134 = 0.007. This same fraction applies to the initial mass of I_2, to give the mass of I_2 remaining in the aqueous layer. Note that it is unnecessary to convert the aqueous concentration back into an amount and then a mass.

Mass of I_2 remaining in the aqueous layer = 0.007 × 0.0340 g = 0.0002 g.

13.43

	2 NH₃(g)	$\rightleftharpoons$	N₂(g)	+	3 H₂(g)
Initial / (mol L^{-1})	3.60 / 2.00 = 1.80		0		0
Change / (mol L^{-1})	$-2\,x$		$+x$		$+3\,x$
Equilibrium / (mol L^{-1})	$1.80 - 2\,x$		x		$3\,x$

The equilibrium constant is given by

$$K = \frac{[N_2]_{eqm}[H_2]^3_{eqm}}{[NH_3]^2_{eqm}} = \frac{x\,(3x)^3}{(1.80-2x)^2} = 6.3$$

Solve this equation for x. First take the square root of both sides

$$\frac{\sqrt{27}x^2}{1.80-2x} = \frac{5.196\,x^2}{1.80-2x} = \sqrt{6.3} = 2.5$$

$$5.196\,x^2 = 2.5\,(1.80-2x)$$

$$5.196\,x^2 + 5.0x - 4.5 = 0$$

$$x = \frac{-5.0 \pm \sqrt{5.0^2 + 4 \times 5.196 \times 4.5}}{2 \times 5.196}$$

$$x = 0.57 \quad \text{or} \quad -1.5$$

The latter solution gives negative $N_2(g)$ and $H_2(g)$ concentrations. Only the former solution gives admissible concentrations:

$[NH_3]_{eqm} = 1.80 - 2\,x = 1.80 - 1.14 = 0.66$ mol L^{-1}

$[N_2]_{eqm} = x = 0.57$ mol L^{-1}

$[H_2]_{eqm} = 3\,x = 1.71$ mol L^{-1}

The total concentration of gases is the sum of the three gas concentrations.

Total gas concentration $= 0.66 + 0.57 + 1.71 = 2.94$ mol L^{-1}

Pressure is determined from concentration using the ideal gas law:

$$p = \frac{n}{V}RT = \left(2.94\ \text{mol}\ L^{-1}\right)\left(8.314\ L\ \text{kPa}\,K^{-1}\,\text{mol}^{-1}\right)\left(723\ K\right)$$

$$= 17670\ \text{kPa} = 17.7\ \text{MPa}\ \text{(to three significant digits)}$$

Section 13.5: Reaction Equations and Equilibrium Constants

13.45

The second equation is double the inverse of the first. Therefore, K for the second reaction is the inverse of the square of K for the first reaction.

$$K = 1/(6.66 \times 10^{-12})^2 = 1/4.44 \times 10^{-23} = 2.25 \times 10^{22}$$

13.47

$N_2(g) + O_2 \rightleftharpoons 2\,NO(g) \qquad K = 1.7 \times 10^{-3}$ at 2300 K

(a) For $\frac{1}{2}N_2(g) + \frac{1}{2}O_2(g) \rightleftharpoons NO(g) \qquad K = (1.7 \times 10^{-3})^{1/2} = 0.041$

(b) For $2\,NO(g) \rightleftharpoons N_2(g) + O_2(g) \qquad K = (1.7 \times 10^{-3})^{-1} = 588 = 590$ (to two significant digits)

Section 13.6: Disturbing Reaction Mixtures at Equilibrium

13.49

We will need the equilibrium constant. Use the given concentrations for the initial equilibrium mixture.

$$K = \frac{[\text{isobutane}]_{\text{eqm}}}{[\text{butane}]_{\text{eqm}}} = \frac{2.5 \text{ mol L}^{-1}}{1.0 \text{ mol L}^{-1}} = 2.5$$

(a) Now, we set up an ICE table with the initial data corresponding to just after increasing [isobutane] by 0.50.

	butane	$\rightleftharpoons$	isobutane
Initial / (mol L^{-1})	1.0		2.5 + 0.50
Change / (mol L^{-1})	$-x$		$+x$
Equilibrium / (mol L^{-1})	$1.0 - x$		$3.0 + x$

We have used $+x$ even though we expect the reaction to shift in the other direction. In this case, solving for x will give a negative value. This shows that you can get the right answer even if you do not anticipate correctly the direction of net reaction.

The equilibrium constant is determined by the equilibrium concentrations, expressed here in terms of the unknown, x.

$$K = \frac{[\text{isobutane}]_{\text{eqm}}}{[\text{butane}]_{\text{eqm}}} = \frac{3.0 + x}{1.0 - x} = 2.5$$

Solving for x:

$$3.0 + x = 2.5\,(1.0 - x) = 2.5 - 2.5\,x$$

$$3.5\,x = -0.5$$

$$x = -0.14$$

The new equilibrium concentrations are

$$[\text{butane}] = 1.0 - x = 1.0 + 0.14 = 1.1 \text{ mol L}^{-1}$$
$$[\text{isobutane}] = 3.0 + x = 3.0 - 0.14 = 2.9 \text{ mol L}^{-1}$$

(b) Now, we set up an ICE table with the initial data corresponding to just after increasing [butane] by 0.50.

	butane	$\rightleftharpoons$	isobutane
Initial / (mol L^{-1})	1.0 + 0.50 = 1.5		2.5
Change / (mol L^{-1})	$-x$		$+x$
Equilibrium / (mol L^{-1})	$1.5 - x$		$2.5 + x$

$$K = \frac{[\text{isobutane}]_{\text{eqm}}}{[\text{butane}]_{\text{eqm}}} = \frac{2.5 + x}{1.5 - x} = 2.5$$

Solving for x:

$$2.5 + x = 2.5\,(1.5 - x) = 3.75 - 2.5\,x$$

$$3.5\,x = 1.25$$

$$x = 0.36$$

The new equilibrium concentrations are

$$[\text{butane}] = 1.5 - x = 1.5 - 0.36 = 1.1 \text{ mol L}^{-1}$$
$$[\text{isobutane}] = 2.5 + x = 2.5 + 0.36 = 2.9 \text{ mol L}^{-1}$$

We end up with the same concentrations as in part (a) at the level of significant figures used.

13.51

$$N_2O_3(g) \rightleftharpoons NO(g) + NO_2(g) \qquad \Delta_r H^\circ = 40.5 \text{ kJ mol}^{-1}$$

(a) If more N_2O_3 (a reactant gas) is added, then the reaction quotient Q decreases—the concentration of $N_2O_3(g)$ is in the denominator of the reaction quotient. Therefore, there is net reaction in the direction that forms more $NO(g)$ and $NO_2(g)$, increasing Q until again $Q = K$. The net reaction does not consume all of the added $N_2O_3(g)$. So, all three gases increase in concentration, compared with the concentrations before more $N_2O_3(g)$ was added.

(b) If more NO_2 (a product gas) is added, then the reaction quotient Q increases—the concentration of $NO_2(g)$ appears in the numerator of the reaction quotient. Therefore, there is net reaction in the direction that produces more $N_2O_3(g)$, decreasing Q until again $Q = K$. Again, all three gases increase in concentration, compared with the concentrations before more $NO_2(g)$ was added. This is the case whenever reactant or product gas is added.

(c) If the volume of the reaction flask is increased, the concentrations of all gases decrease by the same factor. Since the sum of the coefficients in the equation of the two product gases, is bigger than the coefficient of the one reactant gas, the reaction quotient Q decreases (the numerator has more decreasing factors—by one). Therefore, there is net reaction in the direction that forms more $NO(g)$ and $NO_2(g)$, increasing Q until again $Q = K$.
Reactant gas concentration decreases in this case—because of the increase in volume and again because of the net reaction. Product gas concentrations decrease due to the volume increase, but then increase upon net forward reaction. They do not, however, return to their original concentrations. Altogether, there is a net decrease in all gas concentrations.

(d) If the temperature is lowered, the equilibrium constant K decreases because the reaction is endothermic. Upon decreasing K, by lowering the temperature, we have $Q > K$. Therefore, there is net reaction in the direction that produces more $N_2O_3(g)$, decreasing Q until $Q =$ (the new) K. Here, the only change in gas concentrations is due to the net reaction. Product gases decrease in concentration, while the reactant gas increases in concentration.

13.53

$$BaCO_3(s) \rightleftharpoons BaO(s) + CO_2(g) \qquad \text{is endothermic}$$

(a) More BaCO₃(s) is added. (i) no net reaction, the system remains at equilibrium.
   Adding a solid does not change Q.
(b) More CO₂(g) is added. (ii) net reaction to form more BaCO₃(s)
   Adding $CO_2(g)$ brings about an increase of Q, so net reaction happens to decrease Q till
   $Q = K$ again.
(c) More BaO(s) is added. (i) no net reaction, the system remains at equilibrium
   Adding a solid does not change Q.
(d) The temperature is raised. (iii) net reaction to form more BaO(s) and CO₂(g).
   Increasing temperature increases K for the endothermic direction. There will be net reaction
   in the direction that increases Q until $Q = $ new K.
(e) The volume of the reaction vessel is increased. (iii) net reaction to form more BaO(s)
   and CO₂(g).
   Increasing volume decreases the concentration of $CO_2(g)$—the only gas in the reaction—and
   so decreases Q.
   There will be net reaction in the direction that increases $Q = [CO_2(g)]$ back to its
   equilibrium value.

13.55 $PCl_5(g) \rightleftharpoons PCl_3(g) + Cl_2(g)$

First, get the equilibrium constant.
Amount of $PCl_5(g)$ = 3.120 g / 208.22 g mol^{-1} = 0.01498 mol
Amount of $PCl_3(g)$ = 3.845 g / 137.32 g mol^{-1} = 0.02800 mol
Amount of $Cl_2(g)$ = 1.787 g / 70.90 g mol^{-1} = 0.02520 mol
   Extra amount of $Cl_2(g)$ = 1.418 g / 70.90 g mol^{-1} = 0.02000 mol
These amounts are within a 1.00 L flask—the concentrations have the same numerical values
except that they have 3 significant figures. We carry 4 figures because these calculations can
amplify round-off errors.

$$K = \frac{[PCl_3]_{eqm}[Cl_2]_{eqm}}{[PCl_5]_{eqm}} = \frac{0.02800 \times 0.02520}{0.01498} = 0.04710$$

	PCl₅(g)	$\rightleftharpoons$	**PCl₃(g)**	**+**	**Cl₂(g)**
Initial / (mol L^{-1})	0.01498		0.02800		0.02520 + 0.02000 = 0.04520
Change / (mol L^{-1})	$+x$		$-x$		$-x$
Eqm / (mol L^{-1})	$0.01498 + x$		$0.02800 - x$		$0.04520 - x$

The equilibrium constant is given by

$$K = \frac{[PCl_3]_{eqm}[Cl_2]_{eqm}}{[PCl_5]_{eqm}} = \frac{(0.02800 - x)(0.04520 - x)}{(0.01498 + x)} = 0.04710$$

Solve this equation for x.

$$x^2 - (0.02800 + 0.04520)\,x + 0.02800 \times 0.04520 = 0.04710\,(0.01498 + x)$$

$$x^2 - 0.07320\,x + 0.001266 = 0.0007056 + 0.04710\,x$$

$$x^2 - 0.12030\,x + 0.000560 = 0$$

$$x = \frac{0.12030 \pm \sqrt{0.12030^2 - 4 \times 0.000560}}{2}$$

$$x = 0.1154 \quad \text{or} \quad 0.004851$$

The former solution gives negative $PCl_3(g)$ and $Cl_2(g)$ concentrations. Only the latter solution gives admissible concentrations:

$[PCl_5]_{eqm} = 0.01498 + x = 0.01498 + 0.00485 = 0.0198 \text{ mol L}^{-1}$
$[PCl_3]_{eqm} = 0.02800 - x = 0.02800 - 0.00485 = 0.0232 \text{ mol L}^{-1}$
$[Cl_2]_{eqm} = 0.04520 - x = 0.04520 - 0.00485 = 0.0404 \text{ mol L}^{-1}$

SUMMARY AND CONCEPTUAL QUESTIONS

13.57

	$2\ CH_3COOH(g)$	$\rightleftharpoons$	$(CH_3COOH)_2(g)$
Initial / (mol L^{-1})	5.4×10^{-4}		0
Change / (mol L^{-1})	$-2x$		$+x$
Eqm / (mol L^{-1})	$5.4 \times 10^{-4} - 2x$		x

$$K = \frac{[\text{dimer}]_{eqm}}{[\text{monomer}]_{eqm}^2} = \frac{x}{(5.4 \times 10^{-4} - 2x)^2} = 3.2 \times 10^4$$

Solving for x:

$$\frac{1}{3.2 \times 10^4}\,x = (5.4 \times 10^{-4})^2 - 2 \times 2 \times 5.4 \times 10^{-4}\,x + 4\,x^2$$

$$3.1 \times 10^{-5}\,x = 2.9 \times 10^{-7} - 2.16 \times 10^{-3}\,x + 4\,x^2$$

$$4\,x^2 - 2.19 \times 10^{-3}\,x + 2.9 \times 10^{-7} = 0$$

$$x = \frac{2.19 \times 10^{-3} \pm \sqrt{(2.19 \times 10^{-3})^2 - 4 \times 4 \times 2.9 \times 10^{-7}}}{2 \times 4}$$

$$x = 3.2 \times 10^{-4} \quad \text{or} \quad 2.2 \times 10^{-4}$$

The former solution produces a negative monomer concentration. We use the latter solution. The new equilibrium concentrations are

$$[\text{monomer}] = 5.4 \times 10^{-4} - 2x = 5.4 \times 10^{-4} - 2 \times 2.2 \times 10^{-4} = 1.0 \times 10^{-4} \text{ mol L}^{-1}$$
$$[\text{dimer}] = x = 2.2 \times 10^{-4} \text{ mol L}^{-1}$$

(a) Percentage of monomer converted to dimer =
100% × (initial amount of monomer − final amount of monomer) / (initial amount of monomer)
= 100% × $(5.4 \times 10^{-4} - 1.0 \times 10^{-4}) / (5.4 \times 10^{-4})$ = 81%

(b) The reaction producing the dimer involves only the making of bonds—it must be exothermic. If the temperature is increased, the equilibrium constant will decrease and there will be net reaction in the direction that decreases Q—by formation of more monomer.

13.59

(a) $SO_2Cl_2(g) \rightleftharpoons SO_2(g) + Cl_2(g)$ $K = 0.045$ at $375°C$
Amount of SO_2Cl_2 = 6.70 g / 134.96 g mol^{-1} = 0.0496 mol
placed in a 1.00 L flask → 0.0496 mol L^{-1}

	$SO_2Cl_2(g)$	$\rightleftharpoons$	$SO_2(g)$	+	$Cl_2(g)$
Initial / (mol L^{-1})	0.0496		0		0
Change / (mol L^{-1})	$-x$		$+x$		$+x$
Eqm / (mol L^{-1})	$0.0496 - x$		x		x

The equilibrium constant is given by

$$K = \frac{[SO_2]_{eqm}[Cl_2]_{eqm}}{[SO_2Cl_2]_{eqm}} = \frac{x \, x}{(0.0496 - x)} = 0.045$$

Solve this equation for x.
$$x^2 = 0.045 \, (0.0496 - x) = 0.0022 - 0.045x$$

$$x^2 + 0.045x - 0.0022 = 0$$

$$x = \frac{-0.045 \pm \sqrt{0.045^2 + 4 \times 0.0022}}{2}$$

$$x = 0.030 \quad \text{or} \quad -0.075$$

The latter solution gives negative $SO_2(g)$ and $Cl_2(g)$ concentrations. We use the former solution.
$[SO_2Cl_2]_{eqm} = 0.0496 - x = 0.0496 - 0.030 = 0.020$ mol L^{-1}
$[SO_2]_{eqm} = x = 0.030$ mol L^{-1}
$[Cl_2]_{eqm} = x = 0.030$ mol L^{-1}
Fraction of $SO_2Cl_2(g)$ dissociated
= (initial amount of $SO_2Cl_2(g)$ − final amount of $SO_2Cl_2(g)$) / (initial amount of $SO_2Cl_2(g)$)

$$= (0.0496 - 0.020) / 0.0496 = 0.60$$

(b)
Initial concentration of $Cl_2(g) =$

$$\frac{n}{V} = \frac{p}{RT} = \frac{101.3 \text{ kPa}}{(8.314 \text{ L kPa mol}^{-1} \text{ K}^{-1})(648 \text{ K})} = 0.0188 \text{ mol L}^{-1}$$

	$SO_2Cl_2(g)$	$\rightleftharpoons$	$SO_2(g)$	+	$Cl_2(g)$
Initial / (mol L^{-1})	0.0496		0		0.0188
Change / (mol L^{-1})	$-x$		$+x$		$+x$
Eqm / (mol L^{-1})	$0.0496 - x$		x		$0.0188 + x$

The equilibrium constant is given by

$$K = \frac{[SO_2]_{eqm}[Cl_2]_{eqm}}{[SO_2Cl_2]_{eqm}} = \frac{x(0.0188 + x)}{(0.0496 - x)} = 0.045$$

Solve this equation for x.

$$x^2 + 0.0188\,x = 0.045\,(0.0496 - x) = 0.0022 - 0.045\,x$$

$$x^2 + 0.0638\,x - 0.0022 = 0$$

$$x = \frac{-0.0638 \pm \sqrt{0.0638^2 + 4 \times 0.0022}}{2}$$

$$x = 0.025 \quad \text{or} \quad -0.089$$

The latter solution gives negative $SO_2(g)$ and $Cl_2(g)$ concentrations. We use the former solution.
$[SO_2Cl_2]_{eqm} = 0.0496 - x = 0.0496 - 0.025 = 0.025 \text{ mol L}^{-1}$
$[SO_2]_{eqm} = x = 0.025 \text{ mol L}^{-1}$
$[Cl_2]_{eqm} = 0.0188 + x = 0.0188 + 0.025 = 0.044 \text{ mol L}^{-1}$
Fraction of $SO_2Cl_2(g)$ dissociated
    = (initial amount of $SO_2Cl_2(g)$ − final amount of $SO_2Cl_2(g)$) / (initial amount of $SO_2Cl_2(g)$)
    $= (0.0496 - 0.025) / 0.0496 = 0.50$

(c) The fractions of $SO_2Cl_2(g)$ dissociated in part (b) is less than that in part (b), in agreement with Le Chatelier's principle. There was net reaction in the direction to counter the increase in chlorine gas concentration.

Chapter 14
Acid-Base Equilibria in Aqueous Solution

IN-CHAPTER EXERCISES

Exercise 14.1—Brønsted-Lowry Model of Acids and Bases

(a) $HCOOH(aq) + H_2O(\ell) \rightleftharpoons HCOO^-(aq) + H_3O^+(aq)$
  acid   base   conj. base   conj. acid

(b) $NH_3(aq) + H_2S(aq) \rightleftharpoons NH_4^+(aq) + HS^-(aq)$
  base   acid   conj. acid   conj. base

(c) $HSO_4^-(aq) + OH^-(aq) \rightleftharpoons SO_4^{2-}(aq) + H_2O(\ell)$
  acid   base   conj. base   conj. acid

Exercise 14.9—Acid and Base Ionization Constant

(a) pK_a for aquated benzoic acid, $C_6H_5COOH(aq)$, is
$$pK_a = -\log_{10}K_a = -\log_{10}(6.3 \times 10^{-5}) = -(-4.2) = 4.2$$

(b) $pK_a = 2.87$ for aquated chloroacetic acid, $ClCH_2COOH(aq)$.
Since $2.87 < 4.2$ and (like pH) a smaller pK_a corresponds to a stronger acid, aquated chloroacetic acid is a stronger acid than aquated benzoic acid.

Exercise 14.11—Ionization Constants of Weak Acids and Their Conjugate Bases

The conjugate base of lactic acid—i.e., aquated lactate ions—has
$$K_b = \frac{K_w}{K_a} = \frac{10^{-14}}{1.4 \times 10^{-4}} = 7.1 \times 10^{-11}$$
It fits between dihydrogenphosphate, $H_2PO_4^-(aq)$, ions and fluoride, $F^-(aq)$, ions in Table 14.5.

Exercise 14.13—Acid-Base Character of Solutions of Salts

(a) An aqueous KBr solution has a pH of 7. KOH is a strong base—$K^+(aq)$ ions have no appreciable reaction with water. HBr is a strong acid—its conjugate base, $Br^-(aq)$ ion, is a *very* weak base with insignificant reaction with water.

(b) An aqueous NH_4NO_3 solution has a pH < 7—it is acidic. $NO_3^-(aq)$ ions undergo negligible reaction as a base with water—they are the conjugate base of a strong acid (HNO_3). $NH_4^+(aq)$ ions, on the other hand, are a weak acid.

(c) An aqueous $AlCl_3$ solution has a pH < 7—it is acidic. $Cl^-(aq)$ ions undergo negligible basic reaction with water—they are the conjugate base of a strong acid (HCl). Although $Al^{3+}(aq)$ ions have no protons, they are a weak acid because the Al^{3+} ions strongly attract the six water molecules bound to each, (because of the large +3 charge and the small size of the cation), withdrawing electrons from the water molecules, so that these more easily have a proton removed than do the water molecules not bound to the Al^{3+} cations.

(d) An aqueous Na_2HPO_4 solution has pH > 7—it is basic. $Na^+(aq)$ ions have no appreciable reaction with water, whereas $HPO_4^{2-}(aq)$ ions are a weak base. $HPO_4^{2-}(aq)$ ions can also act as an acid. However, the K_a value of 3.6×10^{-13} is much smaller than its K_b value of 1.6×10^{-7}.

Exercise 14.15—Lewis Acids and Bases

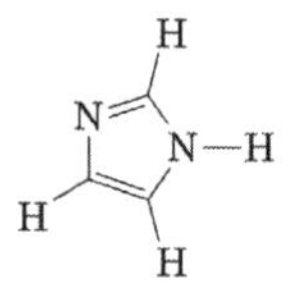

Imidazole

(a) The H atom attached to an N atom has the greatest concentration of positive charge. As such, this is the most acidic hydrogen atom.

(b) The unprotonated ring nitrogen has the greatest concentration of negative charge. It is the most basic nitrogen atom in imidazole. This N atom is the site most likely to attract $H^+(aq)$ ions.

Exercise 14.17—Lewis Acids and Bases

(a) CH_3CH_2OH

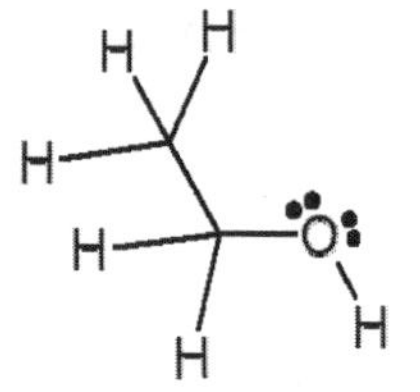

The Lewis structure of ethanol shows lone pairs of electrons on the O atom that can be donated to bond formation with a Lewis acid. It is a Lewis base. There are no places in this structure that can accept a lone pair of electrons, and the H atoms bonded to alkyl carbons are not very acidic.

(b) $(CH_3)_2NH$

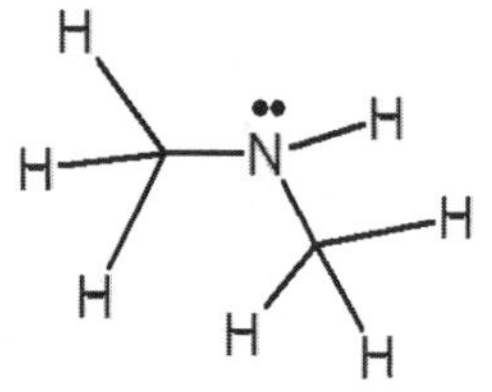

The Lewis structure of dimethyl amine shows a lone pair of electrons on the N atom that can be donated to bond formation. It is a Lewis base.

(c) Br^-

The bromide ion clearly has lone pairs that it can donate to bond formation. It is a Lewis base. However, because it is the conjugate base of a strong acid, we know that it is a very weak base in the Brønsted-Lowry sense.

(d) $(CH_3)_3B$

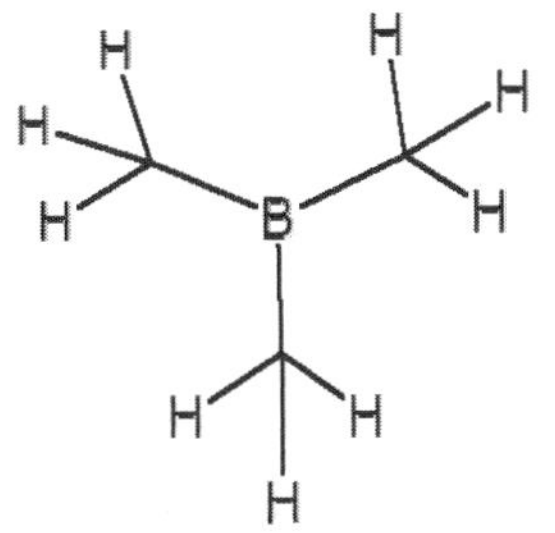

Trimethyl borane is a Lewis acid. The boron atom has only 6 electrons in its valence shell. It can accept a lone pair of electrons from a Lewis base. Trimethyl borane has no lone pairs available to be donated—it cannot act as a Lewis base.

(e) H_3CCl

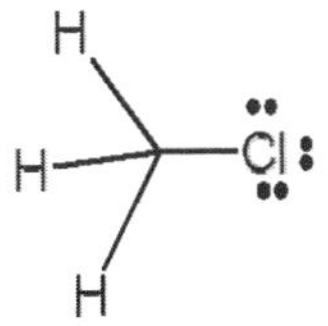

Methyl chloride has lone pairs on the Cl atom, so has the potential to react as a Lewis base.

(f) $(CH_3)_3P$

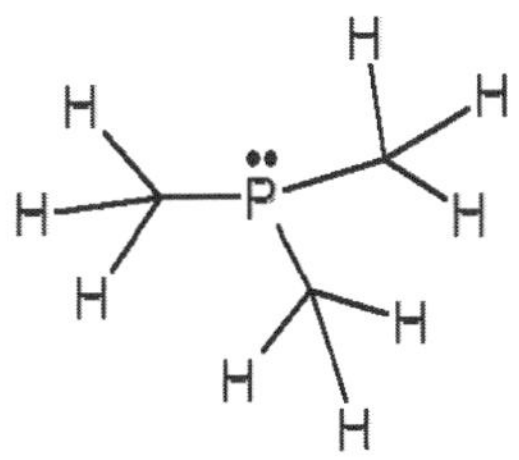

Trimethylphosphine has a lone pair of electrons on the P atom that can be donated to bond formation. It is a Lewis base.

Exercise 14.23—Magnitude of Ionization Constant and Percentage Ionization

For benzoic acid, $K_a = 6.3 \times 10^{-5}$. The quadratic equation derived from assuming that x mol L^{-1} of the weak acid ionizes before equilibrium is achieved is

$$x^2 + (6.3 \times 10^{-5})\, x - (6.3 \times 10^{-6}) = 0$$

Solving the quadratic equation, gives $x = 0.0025$
The equilibrium H_3O^+ concentration is

$$[H_3O^+] = x = 0.0025 \text{ mol L}^{-1}$$

and $\quad pH = -\log_{10}[H_3O^+] = -\log_{10}(0.0025) = 2.61$

and $\quad$ % of benzoic acid ionized $= 100\% \times \dfrac{x}{0.10} = 100\% \times \dfrac{0.0025}{0.10} = 2.5\%$

Note that the pH comes out as 2.60 if computed as shown above. The value 2.61 results if you carry an extra digit from the previous calculation—i.e., use $x = 0.00248 \text{ mol L}^{-1}$.

For hypochlorous acid, $K_a = 3.5 \times 10^{-8}$. The quadratic equation derived is

$$x^2 + (3.5 \times 10^{-8})\,x - (3.5 \times 10^{-9}) = 0$$

Solving the quadratic equation, gives $x = 5.9 \times 10^{-5}$

The equilibrium H_3O^+ concentration is

$$[H_3O^+] = x = 5.9 \times 10^{-5} \text{ mol L}^{-1}$$

and $\quad pH = -\log_{10}[H_3O^+] = -\log_{10}(5.9 \times 10^{-5}) = 4.23$

and $\quad$ % of hypochlorous acid ionized $= 100\% \times \dfrac{x}{0.10} = 100\% \times \dfrac{5.9 \times 10^{-5}}{0.10} = 5.9 \times 10^{-2}\%$

For ammonium ions, $K_a = 5.6 \times 10^{-10}$. The quadratic equation derived is

$$x^2 + (5.6 \times 10^{-10})\,x - (5.6 \times 10^{-11}) = 0$$

Solving the quadratic equation, gives $x = 7.5 \times 10^{-6}$

The equilibrium $H_3O^+(aq)$ ion concentration is

$$[H_3O^+] = x = 7.5 \times 10^{-6} \text{ mol L}^{-1}$$

and $\quad pH = -\log_{10}[H_3O^+] = -\log_{10}(7.5 \times 10^{-6}) = 5.13$

and $\quad$ % of ammonium reacted $= 100\% \times \dfrac{x}{0.10} = 100\% \times \dfrac{7.5 \times 10^{-6}}{0.10} = 7.5 \times 10^{-3}\%$

Exercise 14.25—Dependence of Percentage Ionization on Solution Concentration

Here, only the initial acid concentration is varied. The acid is propanoic acid with $K_a = 1.3 \times 10^{-5}$ at 25°C.

For initial acid concentration, 1.00 mol L^{-1}, the quadratic equation derived from assuming x mol L^{-1} ionization takes the form

$$x^2 + K_a\,x - c_{initial}K_a = 0$$

$$x^2 + 1.3 \times 10^{-5}\,x - 1.3 \times 10^{-5} = 0$$

Solving the quadratic equation, gives $x = 3.6 \times 10^{-3}$

The equilibrium $H_3O^+(aq)$ ion concentration is

$$[H_3O^+] = x = 3.6 \times 10^{-3} \text{ mol L}^{-1}$$

and $\quad pH = -\log_{10}[H_3O^+] = -\log_{10}(3.6 \times 10^{-3}) = 2.44$

and $\quad$ % of propanoic acid ionized $= 100\% \times \dfrac{x}{1.00} = 100\% \times \dfrac{3.6 \times 10^{-3}}{1.00} = 0.36\%$

For initial acid concentration, $1.00 \times 10^{-2} \text{ mol L}^{-1}$, the quadratic equation is of the form

$$x^2 + 1.3 \times 10^{-5}\,x - 1.3 \times 10^{-7} = 0$$

Solving the quadratic equation, gives $x = 3.5 \times 10^{-4}$

The equilibrium $H_3O^+(aq)$ ion concentration is

$$[H_3O^+] = x = 3.5 \times 10^{-4} \text{ mol L}^{-1}$$

and $\quad pH = -\log_{10}[H_3O^+] = -\log_{10}(3.5 \times 10^{-4}) = 3.45$

and $\quad$ % of propanoic acid ionized $= 100\% \times \dfrac{x}{0.0100} = 100\% \times \dfrac{3.5 \times 10^{-4}}{0.0100} = 3.5\%$

For initial acid concentration, 1.00×10^{-4} mol L^{-1}, the quadratic equation derived is

$$x^2 + 1.3 \times 10^{-5} x - 1.3 \times 10^{-9} = 0$$

Solving the quadratic equation, gives $x = 3.0 \times 10^{-5}$
The equilibrium H_3O^+(aq) ion concentration is

$$[H_3O^+] = x = 3.0 \times 10^{-5} \text{ mol L}^{-1}$$

and $\quad pH = -\log_{10}[H_3O^+] = -\log_{10}(3.0 \times 10^{-5}) = 4.52$

and $\quad$ % of propanoic acid ionized $= 100\% \times \dfrac{x}{0.000100} = 100\% \times \dfrac{3.0 \times 10^{-5}}{0.000100} = 30\%$

Exercise 14.29—pH of a Solution of a Polyprotic Acid

Oxalic acid is a diprotic acid with $K_{a1} = 5.6 \times 10^{-2}$ and $K_{a2} = 5.4 \times 10^{-5}$.
Because the second ionization constant is 1000 times smaller than the first, we can neglect the second ionization for the purpose of computing pH of solution.

ICE table for stage one ionization of oxalic acid.

Concentrations (mol L^{-1})	(COOH)$_2$(aq)	+	H$_2$O(ℓ)	⇌	HOOCCOO$^-$(aq)	+	H$_3$O$^+$(aq)
Initial	0.10				0		0
Change	$-x$				$+x$		$+x$
Equilibrium	$0.10 - x$				x		x

At equilibrium the condition $Q = K$ must be satisfied:

$$Q = \frac{[\text{oxalate}^-][H_3O^+]}{[\text{oxalic acid}]} = \frac{(x)(x)}{0.10 - x} = K_{a1} = 5.6 \times 10^{-2}$$

From this we obtain the quadratic equation

$$x^2 + (5.6 \times 10^{-2}) x - (5.6 \times 10^{-3}) = 0$$

Solving the quadratic equation, gives $x = 0.047$
We now have the equilibrium concentrations

$$[\text{oxalate}^-] = [H_3O^+] = x = 0.047 \text{ mol L}^{-1}$$
$$[\text{oxalic acid}] = 0.10 - x = 0.10 - 0.047 = 0.053 \text{ mol L}^{-1}$$

and $\quad pH = -\log_{10}[H_3O^+] = -\log_{10}(0.047) = 1.33$

Exercise 14.31—Acidity of Solutions

(a) To get the pH in cases A and C, requires equilibrium calculations. B requires no equilibrium calculation because HCl is a strong acid.

A. 0.005 mol L^{-1} formic acid ($K_a = 1.8 \times 10^{-4}$) has pH $= 3.06$. This is derived from the quadratic equation

$$x^2 + (1.8 \times 10^{-4}) x - (9.0 \times 10^{-7}) = 0$$
$$\rightarrow x = 0.00086$$
$$[H_3O^+] = x = 0.00086 \text{ mol L}^{-1}$$

$$pH = -\log_{10}[H_3O^+] = -\log_{10}(0.00086) = 3.06$$

B. 0.001 mol L^{-1} hydrochloric acid (strong acid) has $[H^+] = 0.001$ mol L^{-1} and pH $= 3$

C. 0.003 mol L^{-1} carbonic acid solution ($K_a = 4.2 \times 10^{-7}$) has pH $= 4.46$

Consideration of only the first step of ionization leads to the quadratic equation:

$$x^2 + (4.2 \times 10^{-7})\,x - (1.26 \times 10^{-9}) = 0$$
$$\rightarrow x = 3.7 \times 10^{-5}$$
$$[H_3O^+] = x = 3.5 \times 10^{-5} \text{ mol } L^{-1}$$
$$pH = -\log_{10}[H_3O^+] = -\log_{10}(3.5 \times 10^{-5}) = 4.46$$

According to increasing acidity in terms of pH (i.e., decreasing pH) we have
C < A < B

The pH calculation for case C could have been skipped because there are fewer moles of carbonic acid and it is a weaker acid—also, the second dissociation occurs to a negligible extent unless base is added.

(b) To rank the three solutions according to amount of NaOH solution that can be consumed, it is sufficient to compare the amounts of ionizable protons available for reaction with OH^-(aq) ions—regardless of the amounts of H^+(aq) ions that are in solution before any base is added. Therefore, according to amount of OH^-(aq) ions that will react with 1 L of each solution, we have

B (0.001 mol) < A (0.005 mol) < C (0.006 mol)

Exercise 14.35—pH-Dependent Speciation of Polyprotic Acid Species

(a) At pH 6.0

(i) $\dfrac{[H_3PO_4]}{[H_2PO_4^-]} = \dfrac{[H_3O^+]}{K_{a1}} = \dfrac{[H_3O^+]}{7.5 \times 10^{-3}} = \dfrac{10^{-6}}{7.5 \times 10^{-3}} = \dfrac{1.33 \times 10^{-4}}{1}$

(ii) $\dfrac{[H_2PO_4^-]}{[HPO_4^{2-}]} = \dfrac{[H_3O^+]}{K_{a2}} = \dfrac{[H_3O^+]}{6.2 \times 10^{-8}} = \dfrac{10^{-6}}{6.2 \times 10^{-8}} = \dfrac{16}{1}$

(iii) $\dfrac{[HPO_4^{2-}]}{[PO_4^{3-}]} = \dfrac{[H_3O^+]}{K_{a3}} = \dfrac{[H_3O^+]}{3.6 \times 10^{-13}} = \dfrac{10^{-6}}{3.6 \times 10^{-13}} = \dfrac{2.78 \times 10^6}{1}$

At pH $= 6.0$, the dominant species is $H_2PO_4^-$(aq) ions.

(b) At pH 9.0

(i) $\dfrac{[H_3PO_4]}{[H_2PO_4^-]} = \dfrac{[H_3O^+]}{K_{a1}} = \dfrac{[H_3O^+]}{7.5 \times 10^{-3}} = \dfrac{10^{-9}}{7.5 \times 10^{-3}} = \dfrac{1.33 \times 10^{-7}}{1}$

(ii) $\dfrac{[H_2PO_4^-]}{[HPO_4^{2-}]} = \dfrac{[H_3O^+]}{K_{a2}} = \dfrac{[H_3O^+]}{6.2 \times 10^{-8}} = \dfrac{10^{-9}}{6.2 \times 10^{-8}} = \dfrac{1.6 \times 10^{-2}}{1}$

(iii) $\dfrac{[HPO_4^{2-}]}{[PO_4^{3-}]} = \dfrac{[H_3O^+]}{K_{a3}} = \dfrac{[H_3O^+]}{3.6 \times 10^{-13}} = \dfrac{10^{-9}}{3.6 \times 10^{-13}} = \dfrac{2.78 \times 10^3}{1}$

At pH $= 9.0$, the dominant species is HPO_4^{2-}(aq) ions.

Exercise 14.39—pH of Buffer Solutions

Amount of benzoic acid, $n(HA) = 2.00 \text{ g} / 122.12 \text{ g mol}^{-1} = 0.0164 \text{ mol}$
Amount of sodium benzoate, $n(A^-) = 2.00 \text{ g} / 144.10 \text{ g mol}^{-1} = 0.0139 \text{ mol}$
dissolved to make 1.00 L of solution
$[C_6H_5COOH] = 0.0164 \text{ mol L}^{-1}$
$[C_6H_5COO^-] = 0.0139 \text{ mol L}^{-1}$
Recognizing that we have a buffer solution, we can apply the buffer equation 14.1:

$$[H_3O^+] = \frac{[C_6H_5COOH]}{[C_6H_5COO^-]} \times K_a = \frac{0.0164 \text{ mol L}^{-1}}{0.0139 \text{ mol L}^{-1}} \times 6.3\times10^{-5} = 7.4\times10^{-5}$$

pH = 4.13
Check that you necessarily should obtain the same answer by using the alternative buffer equation 14.4.

Exercise 14.41—Designing Buffer Solutions

To make a buffer solution at a pH near 9, we use a weak acid (and its conjugate base) with pK_a near 9.
(a) HCl and NaCl would not make a buffer solution since HCl is a strong acid.
(b) NH3 and NH4Cl would make a buffer solution near pH = 9
    $K_a(HA) = K_a(NH_4^+) = 5.6 \times 10^{-10}$ and p$K_a = 9.25$
(c) CH3COOH and NaCH3COO would not make a buffer solution near pH = 9
    $K_a(HA) = K_a(CH_3COOH) = 1.8 \times 10^{-5}$ and p$K_a = 4.74$
    Acetic acid is essentially fully ionized this far above pH = pK_a, so has no capacity to buffer against the addition of bases.

Exercise 14.47—Acid-Base Titration

$CH_3COOH(aq) + OH^-(aq) \rightarrow CH_3COO^-(aq) + H_2O(\ell)$
Amount of $OH^-(aq)$ ions consumed in titration $= 0.02833 \text{ L} \times 0.953 \text{ mol L}^{-1} = 0.0270 \text{ mol}$
    $=$ total amount of $CH_3COOH(aq)$ in the vinegar solution
Therefore, the concentration of acetic acid in the vinegar sample, $c(CH_3COOH)$
$= (0.0270 \text{ mol}) / (0.0250 \text{ L}) = 1.08 \text{ mol L}^{-1}$
and the mass of acetic acid in vinegar sample $= 0.0270 \text{ mol} \times 60.05 \text{ g mol}^{-1} = 1.62 \text{ g}$

Exercise 14.53—Titration of a Weak Base with a Strong Acid

Adding 75.0 mL of 0.100 mol L^{-1} HCl solution to 100.0 mL of 0.100 mol L^{-1} ammonia solution
    Amount of $H_3O^+(aq)$ ions added $= 0.0750 \text{ L} \times 0.100 \text{ mol L}^{-1} = 0.00750 \text{ mol}$
    Initial amount of $NH_3(aq) = 0.1000 \text{ L} \times 0.100 \text{ mol L}^{-1} = 0.0100 \text{ mol}$
Since the amount of added acid does not exceed the amount of weak base, we end up with a buffer solution. $NH_3(aq)$ molecules will react with essentially all of the added $H_3O^+(aq)$ ions. Upon adding 0.00750 mol $H_3O^+(aq)$ ions, the amounts of $NH_3(aq)$ molecules and $NH_4^+(aq)$ ions change to
    Amount of $NH_3(aq)$ molecules $= 0.0100 - 0.00750 \text{ mol} = 0.00250 \text{ mol}$
    Amount of $NH_4^+(aq)$ ions $= 0.00750 \text{ mol}$

$$[H_3O^+] = \frac{[NH_4^+]}{[NH_3]} \times K_a(NH_4^+) = \frac{0.00750}{0.00250} \times (5.6 \times 10^{-10}) = 1.68 \times 10^{-9}$$

$$pH = 8.77$$

REVIEW EXERCISES

Section 14.1: How Do You Like Your Acids: Ionized or Un-ionized?

14.55

(a) The pH in the stomach is ~2. Since pH < pKa, furosemide will remain as an uncharged, protonated species. In the stomach, the drug passes through the lipid membrane into the bloodstream quickly.
(b) The pH in the intestine is 5. Since pH > pKa, furosemide will exist as a negatively charged species. In the intestine, transfer across the lipid membrane will be inefficient and most will be excreted.
(c) The pH of human blood is 7.4. Furosemide will exist as a negatively charged species and will behave the same as in the intestine.
     The pharmacologically active species is the uncharged, protonated species. Furosemide has an overall bioavailability of ~60%.

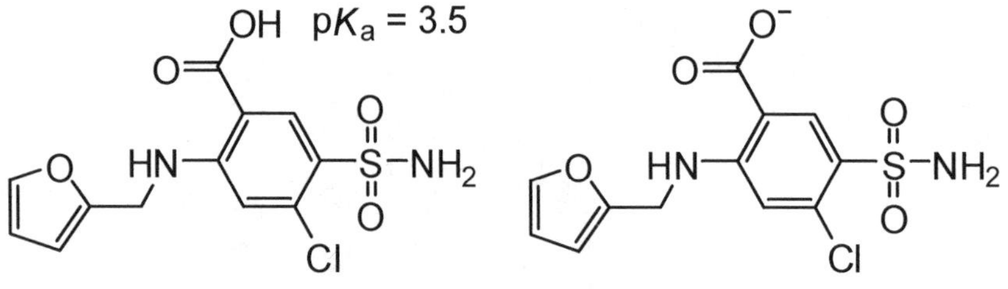

furosemide, pH = 2   furosemide, pH = 5, 7.4

14.57

When the drug is administered intravenously (blood, pH 7.4 < pKa), the concentration of [HA] is slightly more than [A$^-$] (they are approximately equal). Therefore, only approximately half of the drug can pass freely through the lipid membrane. When the dose is given orally, it must pass through the stomach (pH 2) and intestine (pH 5). In these spaces, the drug exists predominantly as an uncharged, protonated species that passes freely through the lipid membrane. The patient will eventually get to sleep, however the uptake of the drug and the length of time before the drug is effective is unpredictable.

Section 14.2: The Brønsted-Lowry Model of Acids and Bases

14.59

$HPO_4{}^{2-}$(aq) ions acting as an acid:

$$HPO_4{}^{2-}(aq) + H_2O(\ell) \rightleftharpoons PO_4{}^{3-}(aq) + H_3O^+(aq)$$

$HPO_4{}^{2-}$(aq) ions acting as a base:

$$HPO_4{}^{2-}(aq) + H_2O(\ell) \rightleftharpoons H_2PO_4{}^-(aq) + OH^-(aq)$$

Section 14.3: Water and the pH Scale

14.61

Assuming that the acidity of the water is due to a strong acid, 1000× dilution of this very low concentration of H_3O^+(aq) ions would mean that the concentration of H_3O^+(aq) ions from the acid solute would be insignificant compared with that in the pure water with which it is diluted.

To a good approximation, $[H_3O^+] = 1 \times 10^{-7}$ mol L^{-1}

More accurately, $[H_3O^+] = 5 \times 10^{-9}$ mol L^{-1} (from acid solute) $+ 1 \times 10^{-7}$ mol L^{-1} (in pure water)
$= 1.05 \times 10^{-7}$ mol L^{-1}

14.63

$$pH = -\log_{10}[H_3O^+] = -\log_{10}(0.0075) = 2.12$$
$$[OH^-] = 10^{-14} / [H_3O^+] = 10^{-14} / 0.0075 = 1.3 \times 10^{-12} \text{ mol } L^{-1}$$

Section 14.4: Relative Strengths of Weak Acids and Bases

14.67

$$pK_a = -\log_{10}(K_a) = -\log_{10}(6.5 \times 10^{-5}) = 4.19$$

14.69

(a) H_2SO_4 is a stronger acid than H_2SO_3
(b) Benzoic acid, C_6H_5COOH, is stronger than acetic acid—it has a larger pK_a value.
(c) Since acetic acid is stronger than boric acid, its conjugate base is weaker than dihydrogenborate—the conjugate base of boric acid.
(d) Ammonia is a stronger base than acetate. It appears lower on the right side of the table.
(e) The conjugate acid of acetate (i.e., acetic acid) is stronger than the conjugate acid of ammonia (ammonium).

14.71

(c) $HClO(aq)$ is the weakest acid among $HSO_4^-(aq)$, $CH_3COOH(aq)$ and $HClO(aq)$. So it has the strongest conjugate base.

14.73

A solution of $K_2CO_3(s)$ is basic because the salt ionizes on dissolving to form $K^+(aq)$ ions and $CO_3^{2-}(aq)$ ions, and some of the $CO_3^{2-}(aq)$ ions react with water to form $HCO_3^-(aq)$ and $OH^-(aq)$

$$CO_3^{2-}(aq) + H_2O(\ell) \rightleftharpoons HCO_3^-(aq) + OH^-(aq)$$

Section 14.5: The Lewis Model of Acids and Bases

14.75

The BH_3 fragment is a Lewis acid. Boron has 6 electrons in its valence shell. It accepts a lone pair of electrons from the N atom in $(CH_3)_3N$ to form the complex.

Section 14.6: Equilibria in Aqueous Solutions of Weak Acids or Bases

14.77

$pH = 11.70$
$[H_3O^+] = 10^{-11.70} = 2.00 \times 10^{-12}$ mol L^{-1}
$[OH^-] = 10^{-14}/[H_3O^+] = 10^{-14}/(2.00 \times 10^{-12}) = 5.00 \times 10^{-3}$ mol L^{-1}
This OH^- concentration equals the extent of the reaction, x mol L^{-1}

$$CH_3NH_2(aq) + H_2O(\ell) \rightleftharpoons CH_3NH_3^+(aq) + OH^-(aq)$$

Since the initial $CH_3NH_2(aq)$ concentration is 0.065 mol L^{-1}, in terms of x, the equilibrium concentrations are
$[CH_3NH_3^+] = [OH^-] = x = 5.00 \times 10^{-3}$ mol L^{-1}
$[CH_3NH_2] = 0.065 - x = 0.065 - 5.00 \times 10^{-3} = 0.060$ mol L^{-1}
Therefore,

$$K_b = \frac{\left[CH_3NH_3^+\right]\left[OH^-\right]}{\left[CH_3NH_2\right]} = \frac{(5.00 \times 10^{-3})^2}{0.060} = 4.17 \times 10^{-4}$$

14.79

The quadratic equation for the extent of reaction, x, in the case of a 0.10 mol L^{-1} aqueous solution of hydrofluoric acid ($K_a = 7.0 \times 10^{-4}$) is

$$x^2 + 7.0 \times 10^{-4}\,x - 7.0 \times 10^{-5} = 0$$

Solving the quadratic equation, gives $x = 8.0 \times 10^{-3}$
The equilibrium $H_3O^+(aq)$ ion concentration is
$[H_3O^+] = x = 8.0 \times 10^{-3}$ mol L^{-1}
and $pH = -\log_{10}[H_3O^+] = -\log_{10}(8.0 \times 10^{-3}) = 2.10$

and $\quad$ % of HF ionized $= 100\% \times \dfrac{x}{0.10} = 100\% \times \dfrac{8.0 \times 10^{-3}}{0.10} = 8.0\%$

14.81

Recognize that this is a buffer solution and use Equation 14.3.

$$[H_3O^+] = \frac{c(NH_4^+)}{c(NH_3)} \times K_a = \frac{0.20 \text{ mol L}^{-1}}{0.20 \text{ mol L}^{-1}} \times (5.6 \times 10^{-10}) = 5.6 \times 10^{-10} \text{ mol L}^{-1}$$

$pH = -\log_{10}(5.6 \times 10^{-10}) = 9.25$

14.83

This solution is analogous to 14.77.

$$CH_3NH_2(aq) + H_2O(\ell) \rightleftharpoons CH_3NH_3^+(aq) + OH^-(aq)$$

$K_b = 4.2 \times 10^{-4}$ at 25 °C

Consider a 0.25 mol L^{-1} solution of methylamine. Presume that x mol L^{-1} ionizes at equilibrium, and express the equilibrium concentrations of species in terms of x in the reaction quotient, Q. The derived quadratic equation takes the form $\quad x^2 + (4.2 \times 10^{-4})x - 1.05 \times 10^{-4} = 0$

with solution $\quad x = 0.010$ mol L^{-1}

Now we can write the equilibrium concentrations:

$$[CH_3NH_3^+] = [OH^-] = x = 0.010 \text{ mol L}^{-1}$$
$$[CH_3NH_2] = 0.25 - x = 0.25 - 0.010 = 0.24 \text{ mol L}^{-1}$$

$$[H_3O^+] = \frac{K_w}{[OH^-]} = \frac{1.0 \times 10^{-14}}{0.010} = 1.0 \times 10^{-12} \text{ mol L}^{-1}$$

Finally, $\quad pH = -\log_{10}(1.0 \times 10^{-12}) = 12$

$\qquad pOH = 14 - pH = 2$

% of CH$_3$NH$_2$ ionized $= 100\% \times \dfrac{x}{0.25} = 100\% \times \dfrac{0.010}{0.25} = 4.0\%$

14.85

K_a for H$_2$SO$_3$(aq) is 1.2×10^{-2}

For initial acid concentration, 0.45 mol L^{-1}, the quadratic equation takes the form

$\qquad x^2 + (1.2 \times 10^{-2})x - (5.4 \times 10^{-3}) = 0$

Solving the quadratic equation, gives $x = 6.8 \times 10^{-2}$

The equilibrium H$_3$O$^+$(aq) ion concentration is

$\qquad [H_3O^+] = x = 6.8 \times 10^{-2}$ mol L^{-1}

(a) $\qquad pH = -\log_{10}[H_3O^+] = -\log_{10}(6.8 \times 10^{-2}) = 1.17$

$\quad$ and

(b) $\qquad [HSO_3^-] = x = 6.8 \times 10^{-2}$ mol L^{-1}

14.87

$H_2NCH_2CH_2NH_2$,(aq) can interact with water in two steps, forming OH^-(aq) ions in each step ($K_{b1} = 8.5 \times 10^{-5}$, $K_{b1} = 2.7 \times 10^{-8}$)
We need consider only the first ionization step of ethylenediamine with water.
For initial base concentration $c = 0.15$ mol L^{-1}, the quadratic equation takes the form
$$x^2 + 8.5 \times 10^{-5} x - 1.28 \times 10^{-5} = 0$$
Solving the quadratic equation, gives $x = 3.5 \times 10^{-3}$
The equilibrium OH^-(aq) ion concentration is
$$[OH^-] = x = 3.5 \times 10^{-3} \text{ mol } L^{-1}$$
and $\text{pOH} = -\log_{10}[OH^-] = -\log_{10}(3.5 \times 10^{-3}) = 2.46$
$$[H_2NCH_2CH_2NH_3^+] = x = 3.5 \times 10^{-3} \text{ mol } L^{-1}$$
The concentration of $H_3NCH_2CH_2NH_3^{2+}$(aq) ions can be computed by considering that the second stage of weak acid ionization occurs in an environment dominated by the concentrations of OH^-(aq) and $H_2NCH_2CH_2NH_3^+$(aq) ions produced in the first stage—and which will be negligibly altered as a result of the very weak second stage.

$$K_{b2} = \frac{[H_3NCH_2CH_2NH_3^{2+}][OH^-]}{[H_2NCH_2CH_2NH_3^+]}$$

$$= \frac{[H_3NCH_2CH_2NH_3^{2+}](3.5\times10^{-3})}{3.5\times10^{-3}} = [H_3NCH_2CH_2NH_3^{2+}] = 2.7\times10^{-8} \text{ mol } L^{-1}$$

This very low concentration justifies our earlier assumption that the extent of the second ionization step is very small, and negligible in terms of calculating the concentrations of species from the first step.

Section 14.7 Speciation: Relative Concentrations of Species

14.89

$\text{p}K_a(C_6H_5COOH) = -\log_{10}(6.3 \times 10^{-5}) = 4.20$
$\text{p}K_a(HCN) = -\log_{10}(6.2 \times 10^{-10}) = 9.21$
A weak acid is mostly in its protonated form until pH is increased to $\text{pH} = \text{p}K_a$, at which [weak acid] = [conjugate base]
At pH < 4.20, both C_6H_5COOH(aq) and HCN(aq) are mostly in their protonated form. At pH > 9.21, both mostly exist as the deprotonated species. In the range 4.20 < pH < 9.21, the dominant species are deprotonated $C_6H_5COO^-$(aq) ions and protonated HCN(aq) molecules.

14.91

(a) At
   (i) pH = 4, A^{2-}(aq) ions are at the highest concentration
   (ii) pH = 6, A^{2-}(aq) ions are at the highest concentration—by a wider margin
   (iii) pH = 8, A^{2-}(aq) ions are at the highest concentration—by an even wider margin
(b) $\text{p}K_{a1}$ and $\text{p}K_{a2}$ can be seen in the plot.

$[H_2A(aq)] = [HA^-(aq)]$ when $pH = pK_{a1}$. So, pK_{a1} = about 1.4
$[HA^-(aq)] = [A^{2-}(aq)]$ when $pH = pK_{a2}$. So, pK_{a2} = about 3.6

(c) The best match in tables of common diprotic acids is oxalic acid for which $pK_{a1} = 1.23$ and $pK_{a2} = 4.19$.
$K_{a1} = 5.9 \times 10^{-2}$ and $K_{a2} = 6.5 \times 10^{-5}$ for oxalic acid.
These pK_a values are not so close to those as in the speciation plot to identify this as oxalic acid, however.

(d) At $pH = 1$, we mostly have the fully protonated form, $H_2A(aq)$, whose potential Lewis base sites (lone pairs of electrons) are "occupied" by protons. There is, however, a significant $HA^-(aq)$ ion concentration which is capable of acting as a Lewis base and complexing metal ions. At $pH = 7$, there is only the fully deprotonated form, $A^{2-}(aq)$ ions, which is an even stronger Lewis base, readily able to complex metal ions.

Section 14.8: Acid-Base Properties of Amino Acids and Proteins

14.93

(a) $K_{a1} = 10^{-2.3}$ and $K_{a2} = 10^{-9.7}$
   (i) $[H_2Ala^+]/[HAla] = [H_3O^+]/K_{a1} = 10^{-1.3} / 10^{-2.3} = 10/1$ at pH 1.3
   (ii) $[H_2Ala^+]/[HAla] = [H_3O^+]/K_{a1} = 10^{-4.3} / 10^{-2.3} = (1 \times 10^{-2})/1$ at pH 4.3
   (iii) $[HAla]/[Ala^-] = [H_3O^+]/K_{a2} = 10^{-4.3} / 10^{-9.7} = 10^{5.4} = (2.5 \times 10^5)/1$ at pH 4.3
   (iv) $[HAla]/[Ala^-] = [H_3O^+]/K_{a2} = 10^{-10.7} / 10^{-9.7} = 0.10/1$ at pH 10.7

(b) Alanine zwitterion, HAla:

(c) The isoelectric pH is the pH at which more of the alanine is present as the zwitterion, HAla, than at any other pH. This is analogous to the situation for glycine, represented in Figure 14.16, and occurs at
$pH = \frac{1}{2}(pK_{a1} + pK_{a2}) = \frac{1}{2}(2.3 + 9.7) = 6.0$

Section 14.9: Controlling pH: Buffer Solutions

14.95

We can use the buffer equation 14.4, expressed in terms of amounts:
Amount of $HCO_3^-(aq)$ ions, $n(HCO_3^-)$ = amount of $NaHCO_3$ added = 15.0 g / 84.00 g mol^{-1} = 0.179 mol
Amount of $CO_3^{2-}(aq)$ ions, $n(CO_3^{2-})$ = amount of Na_2CO_3 added = 18.0 g / 105.99 g mol^{-1} = 0.170 mol

$$[H_3O^+] = \frac{n(HCO_3^-)}{n(CO_3^{2-})} \times K_{a2}(H_2CO_3) = \frac{0.179 \text{ mol}}{0.170 \text{ mol}} \times (4.8 \times 10^{-11}) = 5.05 \times 10^{-11} \text{ mol L}^{-1}$$

$pH = 10.30$

14.97

Amount of $H_2PO_4^-$(aq) ions = amount of KH_2PO_4(s) = 1.360 g / 136.09 g mol^{-1} = 0.009993 mol

Amount of HPO_4^{2-}(aq) ions = amount of Na_2HPO_4(s) = 5.677 g / 141.96 g mol^{-1} = 0.03999 mol

(a) Use the buffer equation 14.4, expressed in terms of amounts

$$[H_3O^+] = \frac{n(H_2PO_4^-)}{n(HPO_4^{2-})} \times K_{a2}(H_3PO_4) = \frac{0.009993 \; \cancel{mol}}{0.03999 \; \cancel{mol}} \times (6.2\times10^{-8}) = 1.55\times10^{-8} \; mol \; L^{-1}$$

pH = 7.81

(b) To decrease the pH by 0.5, to 7.31, we would need

$$[H_3O^+] = 10^{-7.31} = 4.90\times10^{-8} = \frac{n(H_2PO_4^-)}{n(HPO_4^{2-})} \times 6.2\times10^{-8}$$

$$\frac{n(H_2PO_4^-)}{n(HPO_4^{2-})} = \frac{4.90\times10^{-8}}{6.2\times10^{-8}} = \frac{0.79}{1}$$

The amount of HPO_4^{2-}(aq) ions is not changed, so we need

$$n(H_2PO_4^-) = 0.79 \times n(HPO_4^{2-}) = 0.79 \times 0.03999 \, mol = 0.032 \, mol$$

Total amount of KH_2PO_4(s) dissolved = 0.032 mol × 136.09 g mol^{-1} = 4.35 g

Additional amount of KH_2PO_4(s) needed = 4.35 g – 1.36 g = 2.99 g

14.99

We need pH = 4.60 – i.e., $[H_3O^+]$ = 2.51 × 10^{-5} mol L^{-1}

(a) For benzoic acid, K_a = 10$^{-4.20}$ = 6.31 × 10^{-5}

We would need to dissolve in the same solution amounts of benzoic acid and sodium benzoate such that

$$[H_3O^+] = 2.51\times10^{-5} = \frac{n(C_6H_5COOH)}{n(C_6H_5COO^-)} \times K_a = \frac{n(C_6H_5COOH)}{n(C_6H_5COO^-)} \times (6.31\times10^{-5})$$

$$\frac{n(C_6H_5COOH)}{n(C_6H_5COO^-)} = \frac{2.51\times10^{-5}}{6.31\times10^{-5}} = \frac{0.40}{1}$$

(b) For acetic acid, pK_a = 4.74 and K_a = 10$^{-4.74}$ = 1.82 × 10^{-5}

We would need to dissolve in the same solution amounts of acetic acid and sodium acetate such that

$$[H_3O^+] = 2.51\times10^{-5} = \frac{n(CH_3COOH)}{n(CH_3COO^-)} \times K_a = \frac{n(CH_3COOH)}{n(CH_3COO^-)} \times (1.82\times10^{-5})$$

$$\frac{n(CH_3COOH)}{n(CH_3COO^-)} = \frac{2.51\times10^{-5}}{1.82\times10^{-5}} = \frac{1.38}{1}$$

(c) Propanoic acid has pK_a = 4.89, and K_a = 10$^{-4.89}$ = 1.29 × 10^{-5}

We would need to dissolve in the same solution amounts of propanoic acid and sodium propanoate such that

$$[H_3O^+] = 2.51 \times 10^{-5} = \frac{n(CH_3CH_2COOH)}{n(CH_3CH_2COO^-)} \times K_a = \frac{n(CH_3CH_2COOH)}{n(CH_3CH_2COO^-)} \times (1.29 \times 10^{-5})$$

$$\frac{n(CH_3CH_2COOH)}{n(CH_3CH_2COO^-)} = \frac{2.51 \times 10^{-5}}{1.29 \times 10^{-5}} = \frac{1.95}{1}$$

14.101

(a) Before HCl is dissolved in the buffer solution

$n(NH_4^+) = 0.125$ mol and $n(NH_3) = 0.500$ L $\times$ 0.500 mol L^{-1} = 0.250 mol

$$[H_3O^+] = \frac{n(NH_4^+)}{n(NH_3)} \times K_a(NH_4^+) = \frac{0.125 \text{ mol}}{0.250 \text{ mol}} \times (5.6 \times 10^{-10}) = 2.80 \times 10^{-10} \text{ mol L}^{-1}$$

pH = 9.55

(b) If 0.0100 mol of HCl is added to the solution, the added H$^+$(aq) ions are essentially all "mopped up" by the base, NH$_3$(aq) of the conjugate acid-base pair. The new amounts of NH$_4^+$(aq) ions and NH$_3$(aq) molecules are

$n(NH_4^+) = 0.125$ mol + 0.0100 mol = 0.135 mol

$n(NH_3) = 0.250$ mol − 0.0100 mol = 0.240 mol

$$[H_3O^+] = \frac{n(NH_4^+)}{n(NH_3)} \times K_a(NH_4^+) = \frac{0.135 \text{ mol}}{0.240 \text{ mol}} \times (5.6 \times 10^{-10}) = 3.15 \times 10^{-10} \text{ mol L}^{-1}$$

pH = 9.50

Section 14.10: Acid-Base Titrations

14.103

To exactly neutralize the OH$^-$(aq) ions from 1.45 g of NaOH(s) requires
1.45 g / 40.00 g mol^{-1} = 0.0363 mol of H$^+$(aq) ions—which is the amount in
0.0363 mol / 0.812 mol L^{-1} = 0.0446 L = 44.6 mL of the HCl solution.

14.105

Amount of OH$^-$(aq) ions required = 0.0291 L $\times$ 0.513 mol L^{-1} = 0.0149 mol
This is the amount of ionizable protons on the acid that can be neutralized. The molar mass of the acid per mole of ionizable protons is
    0.956 g / 0.0149 mol = 64.16 g mol^{-1}
The molar mass of citric acid = 192.124 g mol^{-1} = 2.994 $\times$ 64.16 g mol^{-1}
Since citric acid is triprotic, it has three acidic H's. Its molar mass per acidic H matches that of the unknown.

14.107

(a) Amount of ammonia = amount of HCl solution required to reach the equivalence point
= 0.03678 L × 0.0105 mol L^{-1} = 0.000386 mol
Concentration of ammonia solution, $c(NH_3)$ = 0.000386 mol / 0.0250 L = 0.0154 mol L^{-1}

(b) At the equivalence point, we have an ammonium chloride solution with
c = 0.000386 mol / total volume of solution after titration
= 0.000386 mol / (0.03678 + 0.0250 L) = 0.000386 mol / (0.0618 L) = 0.00625 mol L^{-1}
K_a for ammonium is 5.6×10^{-10}

Concentrations (mol L^{-1})	NH$_4^+$(aq)	+	H$_2$O(ℓ)	⇌	NH$_3$(aq)	+	H$_3$O$^+$(aq)
Initial	0.00625				0		0
Change	$-x$				$+x$		$+x$
Equilibrium	0. 00625 − x				x		x

Setting Q (at equilibrium) = K_a gives the quadratic equation,
$$x^2 + (5.6 \times 10^{-10})\, x - (3.5 \times 10^{-12}) = 0$$
Solving the quadratic equation, gives $x = 1.9 \times 10^{-6}$
So $[H_3O^+]$ = x = 1.9×10^{-6} mol L^{-1}
$[NH_4^+]$ = $0.00625 - 1.9 \times 10^{-6}$ = 0.00625 mol L^{-1}
$[OH^-]$ = 10^{-14} / 1.9×10^{-6} = 5.3×10^{-9} mol L^{-1}

(c) pH = $-\log_{10}(1.9 \times 10^{-6})$ = 5.72

14.109

(a) We have a 0.10 mol L^{-1} ammonia solution
K_b for ammonia is 1.8×10^{-5}

Concentrations (mol L^{-1})	NH$_3$(aq)	+	H$_2$O(ℓ)	⇌	NH$_4^+$(aq)	+	OH$^-$(aq)
Initial	0.10				0		0
Change	$-x$				$+x$		$+x$
Equilibrium	0. 10 − x				x		x

Setting Q (at equilibrium) = K_b gives the quadratic equation,
$$x^2 + (1.8 \times 10^{-5})\, x - (1.8 \times 10^{-6}) = 0$$
Solving the quadratic equation, gives $x = 1.3 \times 10^{-3}$
So $[OH^-]$ = x = 1.3×10^{-3} mol L^{-1}
$[NH_4^+]$ = 1.3×10^{-3} mol L^{-1}
$[NH_3]$ = $0.10 - 1.3 \times 10^{-3}$ mol L^{-1} = 0.10 mol L^{-1}
$[H_3O^+]$ = 10^{-14} / 1.3×10^{-3} = 7.7×10^{-12} mol L^{-1}
pH = $-\log_{10}(7.7 \times 10^{-12})$ = 11.11

(b) At the equivalence point, we have an ammonium chloride solution with
$c(NH_4^+) = (0.0250\ L \times 0.10\ mol)/$ total volume of solution after titration $= 0.050$ mol L^{-1}

Note that the volume of HCl solution added to reach the equivalence point $= 25.0$ mL because $c(HCl)$ of the HCl solution is the same as $c(NH_3)$ of the ammonia solution.

K_a for ammonium is 5.6×10^{-10}

Concentrations (mol L^{-1})	NH_4^+(a)	+	H_2O(ℓ)	⇌	NH_3(aq)	+	H_3O^+(aq)
Initial	0.050				0		0
Change	$-x$				$+x$		$+x$
Equilibrium	$0.050 - x$				x		x

Setting Q (at equilibrium) $= K_a$ gives the quadratic equation,
$$x^2 + (5.6 \times 10^{-10})\,x - (2.8 \times 10^{-11}) = 0$$

Solving the quadratic equation, gives $x = 5.3 \times 10^{-6}$

So
$$[H_3O^+] = x = 5.3 \times 10^{-6}\ mol\ L^{-1}$$
$$[NH_4^+] = 0.050 - 5.3 \times 10^{-6}\ mol\ L^{-1} = 0.050\ mol\ L^{-1}$$
$$[NH_3] = 5.3 \times 10^{-6}\ mol\ L^{-1}$$
$$pH = -\log_{10}(5.3 \times 10^{-6}) = 5.28$$

(c) At the halfway point of the titration, the NH_3 and NH_4^+ concentrations are equal and
$$pH = pK_a(NH_4^+) = 9.25$$

Note that by using pK_a for ammonium here means we are thinking of the halfway point of the reverse titration. To treat the halfway point of the titration of ammonia means setting $pOH = pK_b$ of ammonia, then getting $pH = 14 - pOH$. This gives the same result.

(d) We need an indicator with pK_a near the equivalence point pH—i.e., $pK_a \approx 5.28$.
Indicators which change colour in the neighbourhood of pH = 5 include congo red, bromocresol green, and methyl red

(e) Adding 5.00 mL of 0.10 mol L^{-1} HCl solution to 25.0 mL of 0.10 mol L^{-1} ammonia solution
amount of H_3O^+ added $= 0.005\ L \times 0.10\ mol\ L^{-1} = 0.0005$ mol
initial amount of $NH_3 = 0.025\ L \times 0.10\ mol\ L^{-1} = 0.0025$ mol
final amount of $NH_4^+ = 0.0005$ mol
final amount of $NH_3 = 0.0025 - 0.0005\ mol = 0.0020$ mol
The pH at this point is

$$pH = pK_a(NH_4^+) + \log_{10}\frac{n(NH_3)}{n(NH_4^+)} = 9.25 + \log_{10}\frac{0.0020}{0.0005}$$

$$= 9.25 + 0.60 = 9.85$$

Adding 15.00 mL of 0.10 mol L^{-1} HCl solution to 25.0 mL of 0.10 mol L^{-1} ammonia solution
amount of H_3O^+ added $= 0.015\ L \times 0.10\ mol\ L^{-1} = 0.0015$ mol
initial amount of $NH_3 = 0.0025$ mol
final amount of $NH_4^+ = 0.0015$ mol
final amount of $NH_3 = 0.0025 - 0.0015\ mol = 0.0010$ mol

The pH at this point is

$$pH = pK_a(NH_4^+) + \log_{10}\frac{n(NH_3)}{n(NH_4^+)} = 9.25 + \log_{10}\frac{0.0010}{0.0015}$$

$$= 9.25 - 0.18 = 9.07$$

Adding 20.00 mL of $0.10\ mol\ L^{-1}$ HCl solution to 25.0 mL of $0.10\ mol\ L^{-1}$ ammonia solution

amount of H_3O^+ added $= 0.020\ L \times 0.10\ mol\ L^{-1} = 0.0020\ mol$

initial amount of $NH_3 = 0.0025\ mol$

final amount of $NH_4^+ = 0.0020\ mol$

final amount of $NH_3 = 0.0025 - 0.0020\ mol = 0.0005\ mol$

The pH at this point is

$$pH = pK_a(NH_4^+) + \log_{10}\frac{n(NH_3)}{n(NH_4^+)} = 9.25 + \log_{10}\frac{0.0005}{0.0020}$$

$$= 9.25 - 0.60 = 8.65$$

Adding 22.00 mL of $0.10\ mol\ L^{-1}$ HCl solution to 25.0 mL of $0.10\ mol\ L^{-1}$ ammonia solution

amount of H_3O^+ added $= 0.022\ L \times 0.10\ mol\ L^{-1} = 0.0022\ mol$

initial amount of $NH_3 = 0.0025\ mol$

final amount of $NH_4^+ = 0.0022\ mol$

final amount of $NH_3 = 0.0025 - 0.0022\ mol = 0.0003\ mol$

The pH at this point is

$$pH = pK_a(NH_4^+) + \log_{10}\frac{n(NH_3)}{n(NH_4^+)} = 9.25 + \log_{10}\frac{0.0003}{0.0022}$$

$$= 9.25 - 0.87 = 8.38$$

Adding 30.00 mL of $0.10\ mol\ L^{-1}$ HCl solution to 25.0 mL of $0.10\ mol\ L^{-1}$ ammonia solution

amount of H_3O^+ added $= 0.030\ L \times 0.10\ mol\ L^{-1} = 0.0030\ mol$

This exceeds the initial amount of NH_3 ($= 0.0025\ mol$). Therefore, the NH_3 is converted entirely to NH_4^+ and

$0.0030 - 0.0025\ mol = 0.0005\ mol$ of H_3O^+ remain.

The pH at this point is

$-\log_{10}(0.0005) = 3.30$

We have the following data with which to plot the titration curve:

pH	$n(NH_3)$ / mol	$n(NH_4^+)$ / mol
11.11	$0.10\ mol\ L^{-1} \times 0.025\ L$ $= 0.0025$	$1.3 \times 10^{-3} \times 0.025\ L$ $= 3.3 \times 10^{-5}$
9.85	0.0020	0.0005
9.25	$0.0025 / 2$ $= 0.00125$	$0.0025 / 2$ $= 0.00125$

	here, concentrations of ammonia and ammonium are equal— i.e., they equal half the total amount	
9.07	0.0010	0.0015
8.65	0.0005	0.0020
8.38	0.0003	0.0022
5.28	5.3×10^{-6} mol L^{-1} × 0.050 L = 2.7×10^{-7}	0. 050 mol L^{-1} × 0.050 L = 0.0025
3.30	0	0.0025

Note that we use amounts here for consistency. The volumes are different, so concentrations cannot be compared directly.

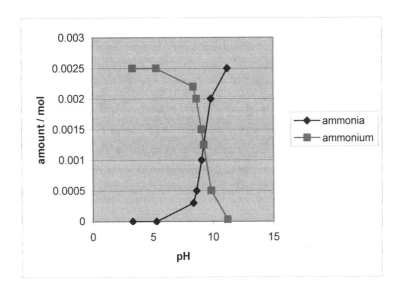

14.111

For an indicator to be useful in a titration, the acid pK_a value should be around the pH value at the equivalence point. The equivalence point is the pH of a solution of the base or acid conjugate to the acid or base. The H^+ or OH^- concentration is the solution to the quadratic equation,

$$x^2 + K x - c K = 0$$

i.e.,

$$x = (-K + (K^2 + 4 c K)^{1/2}) / 2$$

where $K = K_b$ or K_a of the conjugate base or acid, respectively. c is the initial concentration of the conjugate base or acid. It is given by the initial concentration of acid or base being titrated, diluted by the added volume of titrant required to reach equivalence. The latter is determined by the stoichiometry of the neutralization reaction and the concentration of the titrant solution.

(a) The equivalence point of the titration of a 0.1 mol L^{-1} solution (for example) of pyridine ($K_b = 1.7 \times 10^{-9}$) with a 0.1 mol L^{-1} HCl solution is determined as follows:
K_a of pyridine hydrochloride $= 10^{-14} / 1.7 \times 10^{-9} = 5.9 \times 10^{-6}$

$c = 0.05$ mol L^{-1} —the concentration of pyridine hydrochloride is half the original pyridine concentration due to dilution by the titrant

At equivalence, $[H_3O^+] = x = 0.00054$

pH = 3.27

A good indicator would be congo red, methyl yellow, or bromophenol blue.

(b) The equivalence point of the titration of a 0.1 mol L^{-1} solution (for example) of formic acid ($K_a = 1.8 \times 10^{-4}$) with a 0.1 mol L^{-1} NaOH solution is determined as follows:

K_b of formate $= 10^{-14} / 1.8 \times 10^{-4} = 5.6 \times 10^{-11}$

$c = 0.05$ mol L^{-1} is the sodium formate concentration

At equivalence, $[OH^-] = x = 1.7 \times 10^{-6}$

pH $= 14.00 - $ pOH $= 14.00 - 5.77 = 8.23$

A good indicator would be naphtholphthalein or cresol red.

(c) The equivalence point of the titration of a 0.1 mol L^{-1} solution (for example) of ethylenediamine ($pK_a = 9.98$ and 7.52 for ethylenediamine hydrochloride and di(hydrochloride), respectively) with a 0.1 mol L^{-1} HCl solution is determined as follows:

K_a of ethylenediamine hydrochloride $= 1.0 \times 10^{-10}$

$c = 0.05$ mol L^{-1} is the ethylenediamine hydrochloride concentration

At equivalence, $[H_3O^+] = x = 2.2 \times 10^{-6}$

pH = 5.66

A good indicator for this first equivalence point might be bromocresol purple.

K_a of ethylenediamine di(hydrochloride) $= 3.0 \times 10^{-8}$

$c = 0.033$ mol L^{-1} is the ethylenediamine di(hydrochloride) concentration—note the 3-fold dilution (two equivalents of acid are required to neutralize the diprotic base ethylenediamine—this amounts to a 3-fold dilution in this example because the acid titrant has the same concentration as the base).

At equivalence, $[H_3O^+] = x = 3.1 \times 10^{-5}$

pH = 4.51

A good indicator for this second equivalence point might be methyl orange.

SUMMARY AND CONCEPTUAL PROBLEMS

14.113

(a) HB is the stronger acid. For a "strong acid," the equivalence point is at pH = 7.0. The weaker the acid, the stronger is its conjugate base, and the higher the pH of the equivalence point. This can be verified by examining the many calculations of equivalence point pH throughout this chapter. Alternatively, you can simply note that, at the equivalence point, we have a solution of the conjugate base, A^-(aq) or B^-(aq). The stronger the conjugate base, the higher the pH. Since the A^- solution has the highest pH, it is the stronger conjugate base. Consequently, HA is the weaker acid.

(b) A^-(aq) is the stronger conjugate base.

(c) Since the solutions have the same initial amount of acid (same volume, same concentration), then regardless of the fraction ionized in solution before addition of base, they have the same number of ionizable protons, and so they require the same amount of NaOH solution to reach exact neutralization.

14.115

(a) If we use a conjugate acid-base pair with a larger ionization constant of the weak acid, then the pH of the buffer is lower—i.e., it has a higher hydronium ion concentration. The buffer equation,

$$[H_3O^+] = \frac{[HA]}{[A^-]} \times K_a$$

shows that the hydronium ion concentration is proportional to the acid ionization constant, K_a. The K_a of the weak acid of a conjugate acid-base buffer pair determines the pH range in which the solution is effective as a buffer solution.

(b) If the buffer solution has a lower [acid]/[base] ratio, the buffer equation shows that the hydronium concentration decreases—i.e., the pH goes up. Naturally, the higher the relative amount of acid, the more acidic the solution. For a 1:10 change in the ratio, the pH would decrease by 1 unit.

14.117

(a) At very low pH, aquated carbonate species are essentially fully protonated, existing almost entirely as $H_2CO_3(aq)$ molecules. As the pH is increased by addition of, say, $OH^-(aq)$ ions, there is increasing extent of the reaction:

$\quad H_2CO_3(aq) + OH^-(aq) \rightarrow HCO_3^-(aq) + H_2O(\ell)$

So $[H_2CO_3]$ decreases and $[HCO_3^-]$ increases.

$[HCO_3^-]$ increases further and further until $[OH^-]$ is so high that $OH^-(aq)$ ions can begin to compete to "grab" the second proton from the $HCO_3^-(aq)$ ions. Then, as pH is gradually increased, we get more and more of the reaction

$\quad HCO_3^-(aq) + OH^-(aq) \rightarrow CO_3^{2-}(aq) + H_2O(\ell)$

So $[HCO_3^-]$ decreases.

The pH range at which the $[HCO_3^-]$ begins to decrease depends on the relative values of K_{a1} and K_{a2}.

(b) When pH = 6.0, we can see from the graph that $[H_2CO_3] > [HCO_3^-] \gg [CO_3^{2-}]$. This is the case at pH $<$ pK_{a1} = 6.38, that is, when $[H_3O^+] > K_{a1} = 4.2 \times 10^{-7}$ mol L^{-1}. This is clear from the speciation equation

$$\frac{[H_2CO_3]}{[HCO_3^-]} = \frac{[H_3O^+]}{K_{a1}}$$

When pH = 10, the graph shows us that the dominant species is $HCO_3^-(aq)$. This is because at pH $>$ pK_{a1} = 6.38, more than 50% of $H_2CO_3(aq)$ is deprotonated to $HCO_3^-(aq)$ ions, while at pH $<$ pK_{a2} = 10.32, less than 50% of $HCO_3^-(aq)$ ions are deprotonated to $CO_3^{2-}(aq)$ ions. This follows from the speciation equations

$$\frac{[H_2CO_3]}{[HCO_3^-]} = \frac{[H_3O^+]}{K_{a1}} \quad \text{and} \quad \frac{[HCO_3^-]}{[CO_3^{2-}]} = \frac{[H_3O^+]}{K_{a2}}$$

(c) The speciation plot tells us that a solution at pH 11 has $[HCO_3^-] / [CO_3^{2-}] \approx 18\%/82\% = 0.22/1$.

We can check the consistency with buffer equation 14.1:

$$[H_3O^+] = 1 \times 10^{-11} = \frac{[HCO_3^-]}{[CO_3^{2-}]} \times (4.8 \times 10^{-11})$$

$$\frac{[HCO_3^-]}{[CO_3^{2-}]} = \frac{1 \times 10^{-11}}{4.8 \times 10^{-11}} = \frac{0.21}{1}$$

Speciation equations such as the one used here must apply to either (i) very low concentrations of species in solution, the extent of protonation of which is governed by high concentrations of other acids or bases or buffers, (ii) high concentrations of buffer acid-base pairs, which govern the extent of protonation of other species.

Chapter 15
Solubility, Precipitation, and Complexation

IN-CHAPTER EXERCISES

Exercise 15.1—Reaction Quotients for Slightly Soluble Salts in Aqueous Solution

(a) $AgI(s) \rightleftharpoons Ag^+(aq) + I^-(aq)$

$Q = [Ag^+][I^-] = K_{sp} = 1.8 \times 10^{-10}$ at equilibrium

(b) $BaF_2(s) \rightleftharpoons Ba^{2+}(aq) + 2\,F^-(aq)$

$Q = [Ba^{2+}][F^-]^2 = K_{sp} = 1.8 \times 10^{-7}$ at equilibrium

(c) $Ag_2CO_3(s) \rightleftharpoons 2\,Ag^+(aq) + CO_3^{2-}(aq)$

$Q = [Ag^+]^2[CO_3^{2-}] = K_{sp} = 8.5 \times 10^{-12}$ at equilibrium

Exercise 15.3—K_{sp} from Ion Concentrations in Saturated Solution

$BaF_2(s) \rightleftharpoons Ba^{2+}(aq) + 2\,F^-(aq)$
If $[Ba^{2+}] = 3.6 \times 10^{-3}$ mol L^{-1},
$[F^-] = 2 \times [Ba^{2+}] = 7.2 \times 10^{-3}$ mol L^{-1}
and

$Q = [Ba^{2+}][F^-]^2 = (3.6 \times 10^{-3})(7.2 \times 10^{-3})^2 = 1.9 \times 10^{-7} = K_{sp}$

at the temperature of the measurement, since the solid is in equilibrium with the solution containing Ba^{2+} (aq) and F^-(aq) ions.

Exercise 15.5—Solubility from K_{sp}

$Ca(OH)_2(s) \rightleftharpoons Ca^{2+}(aq) + 2\,OH^-(aq)$
At equilibrium, if the concentration of Ca^{2+}(aq) ions is x mol L^{-1}, then the concentration of OH^-(aq) ions is $2\,x$ mol L^{-1}, and

$Q = [Ca^{2+}][OH^-]^2 = x\,(2x)^2 = 4x^3$

$= K_{sp} = 5.5 \times 10^{-5}$

(a) Therefore,

$$\text{Solubility of } Ca(OH)_2 = x = \sqrt[3]{\frac{K_{sp}}{4}} = \sqrt[3]{\frac{5.5 \times 10^{-5}}{4}} = 2.4 \times 10^{-2} \text{ mol L}^{-1}$$

(b) This corresponds to
2.4×10^{-2} mol $L^{-1} \times 74.10$ g mol$^{-1} = 1.8$ g L^{-1}

Exercise 15.9—Participation of Ions in Other Reactions

$$Ag_3PO_4(s) \rightleftharpoons 3\,Ag^+(aq) + PO_4^{3-}(aq) \qquad K_{sp} = [Ag^+]^3[PO_4^{3-}]$$

The solubility of Ag_3PO_4(s) is larger than that predicted using K_{sp} of the salt because the aquated phosphate ion, PO_4^{3-}(aq), is a weak base and reacts with water to form HPO_4^{2-}(aq) ions (as well as smaller concentrations of $H_2PO_4^-$(aq) ions and H_3PO_4(aq) upon further reaction with water). If we ignore the reaction of PO_4^{3-}(aq) ions as a base, we would presume that $[Ag^+] = 3 \times [PO_4^{3-}]$. In fact, $[PO_4^{3-}]$ is less than this assumption implies, so $[Ag^+]$ must be larger to maintain the $Q = K_{sp}$ condition.

Exercise 15.11—The Common Ion Effect and Salt Solubility

(a) In pure water

	$[Ba^{2+}]$ / (mol L^{-1})	$[SO_4^{2-}]$ / (mol L^{-1})
Initial	0	0
Change in concentration	$+x$	$+x$
Equilibrium concentration	x	x

So $x = \{K_{sp}(BaSO_4)\}^{1/2} = \{1.1 \times 10^{-10}\}^{1/2} = 1.0 \times 10^{-5}$ mol L^{-1}

(b) In the presence of 0.010 mol L^{-1} Ba(NO$_3$)$_2$.

	$[Ba^{2+}]$ / (mol L^{-1})	$[SO_4^{2-}]$ / (mol L^{-1})
Initial (before adding BaSO$_4$)	0.010	0
Change in concentration	$+x$	$+x$
Equilibrium concentration	$0.010 + x$	x

Now,

$$Q = [Ba^{2+}][SO_4^{2-}] = K_{sp} = 1.1 \times 10^{-10}$$

$$= (0.010 + x)x$$

which gives the following quadratic equation

$$x^2 + 0.010\,x - 1.1 \times 10^{-10} = 0$$

with (sensible) solution,

$$x = 1.1 \times 10^{-8} \text{ mol L}^{-1}$$

Exercise 15.15—pH-Dependence of Solubility of Salts

(a) PbS(s) is more soluble in nitric acid solution than in pure water because acidic conditions increase the extent of the reactions of S^{2-}(aq) ions as a moderately weak base

$$S^{2-}(aq) + H_3O^+(aq) \rightleftharpoons HS^-(aq) + H_2O(\ell)$$

and $HS^-(aq) + H_3O^+(aq) \rightleftharpoons H_2S(aq) + H_2O(\ell)$

decreasing the concentration of S^{2-}(aq) ions , so that more PbS(s) must dissolve to achieve the condition $Q = K_{sp}$(PbS). Cl$^-$(aq) ions undergo negligible reaction as a base, because HCl(aq) is a strong acid. Their concentration is not reduced by acidifying the solution. So, the solubility of PbS(s) increases more than the solubility of PbCl$_2$(s) does.

(b) The solubility of $Ag_2CO_3(s)$ is greater in acidic solution than in pure water because, like sulfide in part (a), $CO_3^{2-}(aq)$ ions are a weak base, so their concentration is reduced as a result of reaction with $H_3O^+(aq)$ ions. $HI(aq)$ is a strong acid, so $I^-(aq)$ ions undergo negligible reaction as a base, and the solubility of $AgI(s)$ is not changed by acidifying of the solution.

(c) The solubility of $Al(OH)_3(s)$—as measured by the concentration of $Al^{3+}(aq)$ ions in solution—is greater in acidic solutions than in pure water because aquated hydroxide ions, $OH^-(aq)$, are a strong base neutralized by $H_3O^+(aq)$ ions.

Exercise 15.17—Mixing Solutions: Is the Mixture Saturated?

$$Q = [Sr^{2+}][SO_4^{2-}] = (2.5 \times 10^{-4})(2.5 \times 10^{-4}) = 6.3 \times 10^{-8}$$

$$< K_{sp} = 3.4 \times 10^{-7}$$

The mixture is not saturated. $SrSO_4(s)$ will not precipitate.

Exercise 15.21—Separation of Metals by Selective Precipitation

Because K_{sp} for copper hydroxide is much smaller than that for magnesium hydroxide, it is possible to separate the metals via precipitation of $Cu(OH)_2(s)$ before any significant precipitation of $Mg(OH)_2(s)$. The maximum separation is achieved by adjusting the concentration of $OH^-(aq)$ ions that just makes the solution a saturated solution of $Mg(OH)_2$. If $[OH^-]$ is not increased beyond this, no $Mg(OH)_2(s)$ actually precipitates. At the same time, this concentration of hydroxide produces the lowest concentration of $Cu^{2+}(aq)$ ions remaining in solution.

To achieve a solution saturated with respect to $Mg(OH)_2$:

$$Q = [Mg^{2+}][OH^-]^2 = (0.200)\,[OH^-]^2 = K_{sp} = 5.6 \times 10^{-12}$$

i.e., $[OH^-] = (5.6 \times 10^{-12} / 0.200)^{1/2} = 5.3 \times 10^{-6}$ mol L^{-1}

Exercise 15.23—Competition: Precipitation vs. Complexation

$$Cu(OH)_2(s) \rightleftharpoons Cu^{2+}(aq) + 2\,OH^-(aq) \qquad K_{sp} = 2.2 \times 10^{-20}$$

$$\underline{Cu^{2+}(aq) + 4\,NH_3(aq) \rightleftharpoons Cu(NH_3)_4^{2+}(aq) \qquad K_f = 1 \times 10^{13}}$$

$$Cu(OH)_2(s) + 4\,NH_3(aq) \rightleftharpoons Cu(NH_3)_4^{2+}(aq) + 2\,OH^-(aq)$$

$$K_{net} = K_{sp} \times K_f = 2.2 \times 10^{-20} \times 1 \times 10^{13} = 2 \times 10^{-7}$$

Exercise 15.25—Complexation vs. Lewis Base Protonation

At pH = 1, the cyanide is entirely present as aquated hydrogen cyanide, $HCN(aq)$. Silver appears only as $Ag^+(aq)$ or $[Ag(OH_2)_n]^+(aq)$. As pH of solution is increased, an increasing fraction of $HCN(aq)$ molecules are de-protonated, and at pH = pK_a, $[HCN] = [CN^-]$. As pH increases past $pK_a = -\log_{10}(3.5 \times 10^{-4}) = 3.5$, $[CN^-] > [HCN]$.

Depending on the relative amounts of cyanide and silver, some or all of the cyanide ions form complexes with silver ions, $Ag^+(aq)$ – the $Ag^+(aq)$ concentration decreases and the $[Ag(CN)_2]^-$ increases.

The concentrations remain constant from pH about 6 (i.e., when very little HCN(aq) remains) all the way to 13.

REVIEW QUESTIONS

Section 15.1: Ocean Acidification: Ocean Ecology at Risk

15.27

When aragonite "dissolves" in seawater, the equation is:

$$CaCO_3(s) \xrightleftharpoons{H_2O(\ell)} Ca^{2+}(aq) + CO_3^{2-}(aq)$$

When this happens, the $[CO_3^{2-}]$ decreases as this anion is a weak base and reacts further with residual H^+ (from the ionization of H_2CO_3 in water) to produce HCO_3^- :

$$CO_3^{2-}(aq) + H^+(aq) \rightarrow HCO_3^-(aq)$$

Therefore, as H^+ is consumed in this manner, the $[CO_3^{2-}]$ decreases and the solubility of the slightly soluble aragonite increases.

15.29

Aquatic organisms, such as coral reefs and normal marine life, are currently being harmed with the modern day pH of 8.07, which is a change of 0.11 from preindustrial times. As we learned in question 15.28, pH is a logarithmic measure and therefore the actual change in H^+ is much more pronounced. Already, we are observing aquatic life struggle which snowballs into issues for tourism, food supply chains and endangering plant life. The pH does not need to be < 7 in order for the water to be acidic, it simply needs to continue this trend and be more acidic than before.

15.31

(a) (i) 9 **(ii)** 1 **(iii)** 0.11
(b) (i) 9 **(ii)** 1 **(iii)** 0.11
As the pH of the solution increases, the protonated species disappear.

15.33

The carbonic acid/hydrogencarbonate buffer system is important in blood plasma. Four equilibria are important in determining the relative concentrations of the various carbonate species in blood plasma as well as for CO_2 in the atmosphere and seawater. While increasing concentration of $CO_2(g)$ in the atmosphere causes a redistribution to produce more $H_3O^+(aq)$ ions in seawater, so changing the concentration of $CO_2(g)$ in the lungs can change blood pH.

$$CO_2(g) \xrightleftharpoons[]{H_2O(\ell)} CO_2(aq)$$

$$CO_2(aq) + H_2O(\ell) \rightleftharpoons H_2CO_3(aq)$$

$$H_2CO_3(aq) + H_2O(\ell) \rightleftharpoons HCO_3^-(aq) + H_3O^+(aq)$$

$$HCO_3^-(aq) + H_2O(\ell) \rightleftharpoons CO_3^{2-}(aq) + H_3O^+(aq)$$

Section 15.2: Solubility and Precipitation of Ionic Salts

15.35

$$K_{sp} = [Tl^+][Br^-] = (1.9 \times 10^{-3})(1.9 \times 10^{-3}) = 3.6 \times 10^{-6}$$

15.37

Amount of radium sulfate $= 0.025\ g / 322.09\ g\ mol^{-1} = 7.8 \times 10^{-5}\ mol$
If all the radium sulfate dissolved, its concentration would be $7.8 \times 10^{-5}\ mol / 0.100\ L = 7.8 \times 10^{-4}\ mol\ L^{-1}$.
The appropriate reaction quotient is

$$Q = [Ra^{2+}][SO_4^{2-}] = (7.8 \times 10^{-4})(7.8 \times 10^{-4}) = 6.1 \times 10^{-7}$$
$$> K_{sp} = 4.2 \times 10^{-11}$$

Therefore, not all of the radium sulfate dissolves. The concentration of a solution saturated with respect to $RaSO_4$ is

$$(K_{sp})^{1/2} = 6.5 \times 10^{-6}\ mol\ L^{-1}$$

This corresponds to $6.5 \times 10^{-6}\ mol\ L^{-1} \times 0.100\ L = 6.5 \times 10^{-7}\ mol$
and $6.5 \times 10^{-7}\ mol \times 322.09\ g\ mol^{-1} = 2.1 \times 10^{-4}\ g = 0.21\ mg$
i.e., most of the radium sulfate $(25 - 0.21 = 24.8\ mg)$ remains undissolved.

15.41

(a) In pure water

	$[Zn^{2+}] / (mol\ L^{-1})$	$[CN^-] / (mol\ L^{-1})$
Initial	0	0
Change in concentration	$+x$	$+2x$
Equilibrium concentration	x	$2x$

$$Q = [Zn^{2+}][CN^-]^2 = K_{sp} = 8.0 \times 10^{-12}$$

$$= x(2x)^2 = 4x^3$$
So $x = \{K_{sp}(\,Zn(CN)_2\,) / 4\}^{1/3} = \{8.0 \times 10^{-12} / 4\}^{1/3} = 1.3 \times 10^{-4}\ mol\ L^{-1}$

(b) In the presence of $0.10\ \text{mol L}^{-1}\ Zn^{2+}(aq)$ ions:

	$[Zn^{2+}] / (\text{mol L}^{-1})$	$[CN^-] / (\text{mol L}^{-1})$
Initial (before adding $Zn(CN)_2$)	0.10	0
Change in concentration	$+x$	$+2x$
Equilibrium concentration	$0.10 + x$	$2x$

Now,

$$Q = [Zn^{2+}][CN^-]^2 = K_{sp} = 8.0 \times 10^{-12}$$

$$= (0.10 + x)(2x)^2 = 4x^3 + 0.40x^2$$

which gives the following cubic equation

$$4x^3 + 0.40\,x^2 - 8.0 \times 10^{-12} = 0$$

Rather than solve this equation numerically, we can simplify it by noting that the solution, x, must be very small—the very small K_{sp} value is equal to $4x^3 + 0.40\,x^2$. The factor, $0.10 + x$ is consequently essentially equal to 0.10 (it certainly is so to within the accuracy of the available data). With this substitution, we get a much simpler equation; namely,

$$0.40\,x^2 - 8.0 \times 10^{-12} = 0$$

from which we get

$$x = \{8.0 \times 10^{-12}/0.40\}^{1/2} = 4.5 \times 10^{-6}\ \text{mol L}^{-1}$$

Now we can see that the assumption that $0.10 + x = 0.10$ was an accurate approximation to make.

Section 15.3: Precipitation Reactions

15.43

$$Q = [Pb^{2+}][Cl^-]^2 = (0.0012)(0.010)^2 = 1.2 \times 10^{-7}$$

$$< K_{sp} = 1.7 \times 10^{-5}$$

$PbCl_2$ will not precipitate.

15.45

(a) Because $BaSO_4(s)$ or $SrSO_4(s)$ are both 1:1 salts, the solid with the smaller K_{sp} will precipitate first. The K_{sp} values are 1.1×10^{-10} and 3.4×10^{-7}, respectively. $BaSO_4(s)$ will precipitate first.

(b) The second, more soluble salt (i.e., $SrSO_4$) begins to precipitate when $Q = K_{sp}(SrSO_4)$. Neglecting the possible dilution of $Sr^{2+}(aq)$ ions upon adding sulfate solution, we have

$$Q = [Sr^{2+}][SO_4^{2-}] = (0.10)[SO_4^{2-}] = K_{sp} = 3.4 \times 10^{-7}$$

from which we get

$$[SO_4^{2-}] = 3.4 \times 10^{-7} / 0.10 = 3.4 \times 10^{-6}\ \text{mol L}^{-1}$$

The concentration of $Ba^{2+}(aq)$ ions can now be determined from the K_{sp} of $BaSO_4(s)$.

$$Q = [Ba^{2+}][SO_4^{2-}] = [Ba^{2+}](3.4 \times 10^{-6}) = K_{sp} = 1.1 \times 10^{-10}$$

from which we get

$$[Ba^{2+}] = 1.1 \times 10^{-10} / 3.4 \times 10^{-6} = 3.2 \times 10^{-5} \text{ mol L}^{-1}$$

This is the concentration of aquated barium ions remaining in solution at the sulfate concentration computed above.

15.47

(a) To precipitate the maximum amount of Ca^{2+} ions without precipitating $BaF_2(s)$, the $F^-(aq)$ ion concentration must be such that the reaction quotient of $BaF_2(s)$ dissolution is just equal to its solubility product.

$$Q = [Ba^{2+}][F^-]^2 = (0.10)[F^-]^2 = K_{sp} = 1.8 \times 10^{-7}$$

from which we get

$$[F^-] = \{1.8 \times 10^{-7} / 0.10\}^{1/2} = 1.3 \times 10^{-3} \text{ mol L}^{-1}$$

(b) The concentration of $Ca^{2+}(aq)$ ions remaining in solution is determined from the K_{sp} of $CaF_2(s)$.

$$Q = [Ca^{2+}][F^-]^2 = [Ca^{2+}](1.3 \times 10^{-3})^2 = K_{sp} = 5.3 \times 10^{-11}$$

from which we get

$$[Ca^{2+}] = 5.3 \times 10^{-11} / (1.3 \times 10^{-3})^2 = 3.1 \times 10^{-5} \text{ mol L}^{-1}$$

We see that most of the calcium ions have been removed in the $CaF_2(s)$ precipitate.

Section 15.5: Complexation vs. Lewis Base Protonation

15.49

At pH = 1, ethylamine is fully protonated. The concentrations of aquated ethylammonium ions, $CH_3CH_2NH_3^+(aq)$ and $Cu^{2+}(aq)$ ions remain very small until the pH approaches the pK_a of ethylammonium ions. As the pH approaches 10, the ethylamine concentration increases from zero. This causes the $Cu^{2+}(aq)$ ion concentration to decrease, while the $[Cu(CH_3CH_2NH_2)]^{2+}(aq)$ ion concentration increases. These two concentrations decrease and increase, respectively, until the pH approaches 13, whereupon the $Cu^{2+}(aq)$ settles into its minimum value—near zero, if there is sufficient ethylamine to complex all of the copper—and $[Cu(CH_3CH_2NH_2)]^{2+}(aq)$ approaches its maximum value.

SUMMARY AND CONCEPTUAL QUESTIONS

15.51

$Ba(OH)_2(s)$ and $BaCO_3(s)$ will dissolve in HCl solution. The hydroxide and carbonate ions are bases that are protonated in acidic solution. Removing these products of the dissolution reactions allows more of the solid to dissolve. For example, a lower $[CO_3^{2-}]$ means that $[Ba^{2+}]$ must be higher to satisfy the $Q = K$ condition for a saturated solution.

15.53

$K_{sp}(CuS) = 6 \times 10^{-37}$
$K_{sp}(Cu(OH)_2) = 2.2 \times 10^{-20}$

$CuS(s)$ is much more insoluble than $Cu(OH)_2(s)$. In addition, OH^- ions are a strong base, while S^{2-} ions are a relatively weak base. If we add an acidic solution, gradually and with stirring, at some level of $[H^+]$ which we could calculate, all of the $Cu(OH)_2(s)$ will have dissolved, while essentially all of the $CuS(s)$ remains undissolved.

CHAPTER 16
Electron Transfer Reactions and Electrochemistry

IN-CHAPTER EXERCISES

Exercise 16.1—Calculating Oxidation States

(a) $\underline{Fe}_2O_3$
The sum of oxidation states equals zero for a neutral compound.
$2 \times$ Oxidation State(Fe) + $3 \times$ Oxidation State(O) = 0
$2 \times$ Oxidation State(Fe) + $3 \times (-2)$ = 0
Oxidation State(Fe) = $3 \times 2 / 2$ = +3

(b) $H_2\underline{S}O_4$
The sum of oxidation states equals zero for a neutral compound.
$2 \times$ Oxidation State(H) + Oxidation State(S) + $4 \times$ Oxidation State(O) = 0
$2 \times 1 +$ Oxidation State(S) + $4 \times (-2)$ = 0
Oxidation State(S) = $4 \times 2 - 2 \times 1$ = +6

(c) $\underline{C}O_3^{2-}$
The sum of oxidation states equals the charge on an ionic species.
Oxidation State(C) + $3 \times$ Oxidation State(O) = -2
Oxidation State(C) + $3 \times (-2)$ = -2
Oxidation State(C) = $-2 + 3 \times 2$ = +4

(d) $\underline{N}O_2^{+}$
The sum of oxidation states equals the charge on an ionic species.
Oxidation State(N) + $2 \times$ Oxidation State(O) = +1
Oxidation State(N) + $2 \times (-2)$ = +1
Oxidation State(C) = $1 + 2 \times 2$ = +5

Exercise 16.3—Recognizing Oxidation and Reduction

$$3C_2H_5OH(aq) + 2Cr_2O_7^{2-}(aq) + 16H^+(aq) \longrightarrow 3CH_3COOH(aq) + 4Cr^{3+}(aq) + 11H_2O(\ell)$$

Left Side: Right Side:

$OS(C) = (0 - 6 \times 1 - (-2))/2 = -2$ $OS(C) = (0 - 4 \times 1 - 2 \times (-2))/2 = 0$

-2 is the average $OS(C)$ in ethanol. It is possible to assign separate oxidation states to the two carbon atoms in ethanol, -3 and -1 (the C atom bonded to the O atom).

0 is the average $OS(C)$ in acetic acid. It is possible to assign separate oxidation states to the two carbon atoms in acetic acid, -3 and $+3$ (the C atom bonded to two O atoms).

$OS(Cr) = (-2 - 7 \times (-2))/2 = +6$ $OS(Cr) = +3$

The reaction increases the oxidation state of carbon, $OS(C)$. CH₃CH₂OH(aq) is oxidized. CH₃CH₂OH(aq) is the reducing agent. The reaction decreases the oxidation state of chromium, $OS(Cr)$. $Cr_2O_7^{2-}$(aq) ions are reduced. $Cr_2O_7^{2-}$(aq) ions are the oxidizing agent.

Exercise 16.5—Voltaic Cells

Reduction occurs at the cathode. In accord with the reduction half-reaction, a silver electrode immersed in an aqueous solution containing Ag^+(aq) ions (such as an $AgNO_3$ solution) provides the cathode. A nickel electrode immersed in a solution containing Ni^{2+}(aq) ions (such as a $Ni(NO_3)_2$ solution) provides the anode—where oxidation occurs.

The equation for the overall cell reaction is obtained by adding the half-reactions so as the number of electrons removed is the same as the number of electrons gained:

$$2 \times \left(Ag^+ (aq) + e^- \longrightarrow Ag(s) \right)$$
$$\underline{Ni(s) \longrightarrow Ni^{2+}(aq) + 2\,e^-}$$
$$2\,Ag^+(aq) + Ni(s) \longrightarrow 2\,Ag(s) + Ni^{2+}(aq)$$

Anions in the salt bridge flow from cathode to anode within the salt bridge, while cations in the salt bridge flow in the opposite direction (toward the cathode).

Exercise 16.7—Electrochemical Cell Conventions

(a) $Zn(s)|Zn^{2+}(aq)||Ni^{2+}(aq)|Ni(s)$

(b) $C(s)|Fe^{2+}(aq)|Fe^{3+}(aq)||Ag^+(aq)|Ag(s)$

(c) $Mg(s)|Mg^{2+}(aq)||Br_2(l)|Br^-(aq)|Pt(s)$

Cell diagrams:

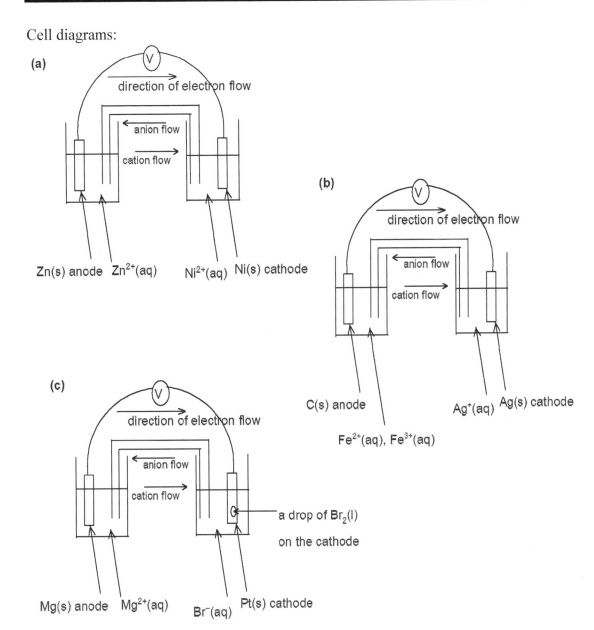

(a)

direction of electron flow

anion flow

cation flow

Zn(s) anode Zn^{2+}(aq) Ni^{2+}(aq)   Ni(s) cathode

(b)

direction of electron flow

anion flow

cation flow

C(s) anode Ag^+(aq)   Ag(s) cathode

Fe^{2+}(aq), Fe^{3+}(aq)

(c)

direction of electron flow

anion flow

cation flow

a drop of Br_2(l)

on the cathode

Mg(s) anode Mg^{2+}(aq) Br^-(aq)   Pt(s) cathode

Exercise 16.9—Standard Half-Cell Reduction Potentials

This is a similar exercise to Exercise 16.8(b), except that the conditions here specify that the $Cl_2(g)|Cl^-$(aq) half-cell is a standard cell, so we use the symbol $E°$ for this half-cell and for the cell.

$$E°_{cell} = 1.36 \text{ V} = E°[Cl_2(g)|Cl^-(aq)|Pt(s)] - 0 \text{ V}$$

i.e., $E°[Cl_2(g)|Cl^-(aq)|Pt(s)] = +1.36$ V

Exercise 16.11—$E°_{cell}$ from Standard Half-Cell Reduction Potentials

(a) $2 I^-(aq) + Zn^{2+}(aq) \longrightarrow I_2(s) + Zn(s)$

$E°_{cell} = E°[Zn^{2+}(aq)|Zn(s)] - E°[I_2(s)|I^-(aq)]$

$\quad\quad = -0.763 \text{ V} - 0.535 \text{ V} = -1.298 \text{ V}$

Because we get a negative value, we know that the spontaneous cell reaction is in the reverse direction to that written—iodine and iodide ions are at the cathode and zinc and $Zn^{2+}(aq)$ ions are at the anode.

(b) $Zn^{2+}(aq) + Ni(s) \longrightarrow Zn(s) + Ni^{2+}(aq)$

$E°_{cell} = E°[Zn^{2+}(aq)|Zn(s)] - E°[Ni^{2+}(aq)|Ni(s)]$
$= -0.763 \text{ V} - (-0.25 \text{ V}) = -0.51 \text{ V}$

Because we get a negative value, we know that the spontaneous cell reaction is in the reverse direction to that written—nickel is the cathode and zinc is the anode.

(c) $2 Cl^-(aq) + Cu^{2+}(aq) \longrightarrow Cu(s) + Cl_2(g)$

$E°_{cell} = E°[Cu^{2+}(aq)|Cu(s)] - E°[Cl_2(g)|Cl^-(aq)]$
$= 0.337 \text{ V} - (1.36 \text{ V}) = -1.023 \text{ V}$

Because we get a negative value, we know that the spontaneous cell reaction is in the reverse direction to that written—chlorine and aquated chloride ions are at the cathode and copper and aquated copper(II) ions are at the anode.

(d) $Fe^{2+}(aq) + Ag^+(aq) \longrightarrow Fe^{3+}(aq) + Ag(s)$

$E°_{cell} = E°[Ag^+(aq)|Ag(s)] - E°[Fe^{3+}(aq),Fe^{2+}(aq)|Pt]$
$= 0.7994 \text{ V} - (0.771 \text{ V}) = 0.028 \text{ V}$

Because we get a positive value, we know that the spontaneous cell reaction is in this direction—silver is the cathode and $Fe^{3+}(aq)$ and $Fe^{2+}(aq)$ ions are at the anode.

Exercise 16.15— Relative Oxidizing and Reducing Abilities

$E°[Cu^{2+}(aq)|Cu(s)] = 0.337 \text{ V}$ $E°[Sn^{2+}(aq)|Sn(s)] = -0.14 \text{ V}$ $E°[Fe^{2+}(aq)|Fe(s)] = -0.44 \text{ V}$
$E°[Zn^{2+}(aq)|Zn(s)] = -0.763 \text{ V}$ $E°[Al^{3+}(aq)|Al(s)] = -1.66 \text{ V}$

(a) Aluminum is the most easily oxidized of these metals.

(b) Aluminum and zinc are capable of reducing $Fe^{2+}(aq)$ ions to $Fe(s)$.

(c) $Fe^{2+}(aq) + Sn(s) \rightarrow Fe(s) + Sn^{2+}(aq)$
Because the reduction potential for $Sn^{2+}(aq)|Sn(s)$ is greater than that for $Fe^{2+}(aq)|Fe(s)$, this is NOT the direction of spontaneous reaction. E_{cell} is negative.

(d) $Zn^{2+}(aq) + Sn(s) \rightarrow Zn(s) + Sn^{2+}(aq)$
Because the reduction potential for $Sn^{2+}(aq)|Sn(s)$ is greater than that for $Zn^{2+}(aq)|Zn(s)$, this is NOT the direction of spontaneous reaction.

Exercise 16.21—Ion Concentration from Cell emf

The $Zn^{2+}(aq)|Zn(s)$ half-cell is a standard half-cell, but it is most unlikely that $Ag^+(aq)|Ag(s)$ is a standard half-cell, so we use the symbol E (not $E°$) for this half-cell and for the cell. First, we do need to know the value of an imagined standard cell comprised of the same components.

$E°_{cell} = E°[Ag^+(aq)|Ag(s)] - E°[Zn^{2+}(aq)|Zn(s)] = 0.7994 \text{ V} - (-0.763 \text{ V}) = 1.562 \text{ V}$

$2 \times (Ag^+(aq) + e^- \rightarrow Ag(s))$
$\underline{Zn(s) \rightarrow Zn^{2+}(aq) + 2 e^-}$
$2 Ag^+(aq) + Zn(s) \rightarrow 2 Ag(s) + Zn^{2+}(aq)$

$$Q = \frac{[Zn^{2+}]}{[Ag^+]^2} = \frac{1.0}{[Ag^+]^2}$$

$$E_{cell} = E^\circ_{cell} - \frac{0.0257\,\text{V}}{n} \times \ln Q = +1.562\,\text{V} - \frac{0.0257\,\text{V}}{2} \times \ln\left(\frac{1.0}{[Ag^+]^2}\right) = +1.48\,\text{V}$$

i.e.,

$$\frac{0.0257\,\text{V}}{2} \times \ln\left(\frac{1.0}{[Ag^+]^2}\right) = 1.562 - 1.48\,\text{V} = 0.082\,\text{V}$$

$$\ln\left(\frac{1.0}{[Ag^+]^2}\right) = \frac{2}{0.0257\,\text{V}} \times 0.082\,\text{V} = 6.4$$

$$\frac{1.0}{[Ag^+]^2} = e^{6.4} = 602$$

$$[Ag^+] = 1.0/\sqrt{602} = 0.041\,\text{mol L}^{-1}$$

Alternatively, you can use $E_{cell} = E[Ag^+(aq)|Ag(s)] - E^\circ[Zn^{2+}(aq)|Zn(s)]$ to calculate $E[Ag^+(aq)|Ag(s)]$. Then you can apply the Nernst equation to this half-cell, using the appropriate reduction half-equation and the tabulated value of $E^\circ[Ag^+(aq)|Ag(s)]$. Check that you get the same answer.

Exercise 16.23—pH-Dependence of Oxidizing Power of Oxoanions

$E^\circ[MnO_4^-(aq)|Mn^{2+}(aq)] = 1.51\,\text{V}$
$MnO_4^-(aq) + 8\,H^+ + 5\,e^- \rightarrow Mn^{2+}(aq) + 4\,H_2O(\ell)$
(a) At pH = 0.0, $[H^+] = 1.0\,\text{mol L}^{-1}$

$$Q = \frac{[Mn^{2+}]}{[MnO_4^-][H^+]^8} = \frac{1.00}{(1.00)[H^+]^8} = \frac{1.00}{(1.00)(1.00\times10^{-0})^8} = 1.00$$

and

$$E = E^\circ - \frac{0.0257\,\text{V}}{n} \times \ln Q = +1.51\,\text{V} - \frac{0.0257\,\text{V}}{5} \times \ln(1.00)$$

$$= +1.51\,\text{V} - \frac{0.0257\,\text{V}}{5} \times (0) = +1.51\,\text{V}$$

(b) At pH = 5.0, $[H^+] = 1 \times 10^{-5.0}\,\text{mol L}^{-1}$

$$Q = \frac{[Mn^{2+}]}{[MnO_4^-][H^+]^8} = \frac{1.00}{(1.00)[H^+]^8} = \frac{1.00}{(1.00)(1.00\times10^{-5.0})^8} = 1.00\times10^{40}$$

and

$$E = E^\circ - \frac{0.0257\,\text{V}}{n} \times \ln Q = +1.51\,\text{V} - \frac{0.0257\,\text{V}}{5} \times \ln(1.00\times10^{40})$$

$$= +1.04\,\text{V}$$

The half-cell reduction potential is greater at pH = 0 than at pH 5.0. This means permanganate is a more powerful oxidizing agent at pH = 0 than at pH 5.0.

Exercise 16.29—Corrosion of Iron

The main reason for corrosion at the water line is differential aeration—the different concentrations of O_2 in the air and dissolved in the water. This creates a difference in reduction potentials, giving a driving force for electrons (from where the concentration is least, to where it is highest). Parts of the pier out of the water have almost uniform O_2 concentrations so a cell is not set up. The deeper the water, the greater is the difference between O_2 concentrations in the air and in the water. But also the deeper the water, the greater is the resistance to movement of ions through the water. The optimum "trade-off" between increasing potential (due to increasing difference of O_2 concentrations) and increasing resistance, occurs not far below the water line.

REVIEW QUESTIONS

Section 16.1: Artificial Leaves: Personal Energy Sources for Everybody by Mimicking Nature

16.31

(a) 174 petawatts (174×10^{15} Watts) fall on the surface of the earth daily.
(b) 143.851 petawatthours yearly or 0.394 petawatthours daily for world energy consumption
  If my estimates are correct, then the time during which the amount of energy falling on the surface of the earth is equivalent to that used for human activities each day is

$$\text{time} = \frac{0.394\,\cancel{PW} \times 24\text{h}}{174\,\cancel{PW}} = 0.0543\,\cancel{h} \times \frac{60\,\text{min}}{\cancel{h}} = 3.26\,\text{min}$$

16.33

The oxidation, or "splitting" of water to oxygen gas has a high activation energy (requires a lot of energy) giving rise to a high overvoltage for oxygen production. As in nature, a mechanism must be put in place to counter this, such as a catalyst. As well, the light-absorbing semiconductor materials are expensive, and the hydrogen produced as a by-product cannot be stored or used in homes.

16.35

When compared to carbohydrates, ethanol is a liquid fuel that is portable, easily stored, and an excellent, clean fuel source for use in homes.

16.37

adenosine triphosphate (ATP)

nicotinamide adenine dinucleotide phosphate (NADPH)

The overall reaction for the Calvin cycle is

$3\ CO_2 + 6\ NADPH + 5\ H_2O + 9\ ATP \rightarrow G3P + 2\ H^+ + 6\ NADP^+ + 9\ ADP + 8\ P_i$

where G3P = glyceraldehyde-3-phosphate and P_i = inorganic phosphate

Unlike photosynthesis, which produces 6C carbohydrates, the Calvin cycle only produces 3C carbohydrates through the following mechanism:

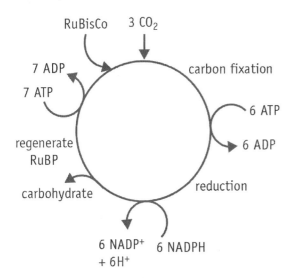

Section 16.2: Oxidation-Reduction Reactions

16.39

(a) PF_6^-
$OS(F) = -1$
$-1 = OS(P) + 6 \times (-1)$
$OS(P) = -1 - 6 \times (-1) = +5$

(b) $H_2AsO_4^-$
$OS(H) = +1$ & $OS(O) = -2$
$-1 = 2 \times (+1) + OS(As) + 4 \times (-2)$
$OS(As) = -1 - 2 \times (+1) - 4 \times (-2) = +5$

(c) UO^{2+}
$OS(O) = -2$
$+2 = OS(U) + (-2)$
$OS(U) = +4$

(d) N_2O_5
$OS(O) = -2$
$0 = 2 \times OS(N) + 5 \times (-2)$
$OS(N) = (0 - 5 \times (-2)) / 2 = +5$

(e) $POCl_3$
$OS(Cl) = -1$ & $OS(O) = -2$
$0 = OS(P) + 3 \times (-1) + (-2)$
$OS(P) = +5$

(f) XeO_4^{2-}
$OS(O) = -2$
$-2 = OS(Xe) + 4 \times (-2)$
$OS(Xe) = +6$

16.41

(a) $OH^-(aq) + H^+(aq) \longrightarrow H_2O(\ell)$
is NOT an oxidation-reduction reaction. Oxidation states are the same on both sides: +1 and −2 for H and O, respectively.

(b) $Cu(s) + Cl_2(g) \longrightarrow CuCl_2(s)$
is an oxidation-reduction reaction. The oxidation state of Cu increases from 0 to +2, while that of Cl decreases from 0 to −1.

(c) $CO_3^{2-}(aq) + 2H^+(aq) \longrightarrow CO_2(g) + H_2O(\ell)$
is NOT an oxidation-reduction reaction. The oxidation state of C equals +4 on both sides. Also, we have +1 and −2 for H and O on both sides.

(d) $2S_2O_3^{2-}(aq) + I_2(s) \longrightarrow S_4O_6^-(aq) + 2I^-(aq)$
is an oxidation-reduction reaction. The oxidation state of S increases from +2 to +11/4, while that of I decreases from 0 to −1.

16.43

(a) $Sn(s) + H^+(aq) \longrightarrow Sn^{2+}(aq) + H_2(g)$
First, we balance the electrons. 2 electrons are taken from each Sn atom, while two $H^+(aq)$ ions each gain one electron.
$Sn(s) + 2H^+(aq) \longrightarrow Sn^{2+}(aq) + H_2(g)$
The reaction is now balanced.

(b) $Cr_2O_7^{2-}(aq) + Fe^{2+}(aq) \longrightarrow Cr^{3+}(aq) + Fe^{3+}(aq)$
First, we balance the electrons. Each $Cr_2O_7^{2-}(aq)$ ion gains 6 electrons (write the half-equation), while one electron is taken from each $Fe^{2+}(aq)$ ion.
To balance the electrons, the 6 $Fe^{2+}(aq)$ ions must be oxidized for every $Cr_2O_7^{2-}(aq)$ ion that is reduced. This gives

$$Cr_2O_7^{2-}(aq) + 6\,Fe^{2+}(aq) \longrightarrow 2\,Cr^{3+}(aq) + 6\,Fe^{3+}(aq)$$

What remains is to balance the O atoms via participation of the water solvent. Balancing in acid, we balance the O atoms by adding 7 H_2O to the O-deficient side—the right-hand side. This introduces 14 H atoms on the right, which must be balanced by adding 14 $H^+(aq)$ ions to the left. The final balanced equation is

$$Cr_2O_7^{2-}(aq) + 6\,Fe^{2+}(aq) + 14\,H^+(aq) \longrightarrow 2\,Cr^{3+}(aq) + 6\,Fe^{3+}(aq) + 7\,H_2O(\ell)$$

(c) $MnO_2(s) + Cl^-(aq) \longrightarrow Mn^{2+}(aq) + Cl_2(g)$

First, we balance the electrons. Each mol of $MnO_2(s)$ gains 2 mol of electrons, while 1 mol of electrons is removed from each mol of $Cl^-(aq)$ ions.

To balance the electrons, 2 mol of $Cl^-(aq)$ ions must be oxidized for every 1 mol of $MnO_2(s)$ reduced. This gives

$$MnO_2(s) + 2\,Cl^-(aq) \longrightarrow Mn^{2+}(aq) + Cl_2(g)$$

What remains is to balance the O atoms via participation of the aqueous solvent. Balancing in acid, we balance the O atoms by adding 2 H_2O molecules to the O-deficient side—the right. This introduces 4 H atoms on the right, which must be balanced by adding 4 $H^+(aq)$ ions to the left. The final balanced equation is

$$MnO_2(s) + 2\,Cl^-(aq) + 4\,H^+(aq) \longrightarrow Mn^{2+}(aq) + Cl_2(g) + 2\,H_2O(l)$$

(d) $HCHO(aq) + Ag^+(aq) \longrightarrow HCOOH(aq) + Ag(s)$

First, we balance the electrons. Each $Ag^+(aq)$ ion gains one electron, while two electrons are removed from each HCHO(aq) molecule.

To balance the electrons, 2 mol of $Ag^+(aq)$ ions must be reduced for every 1 mol of HCHO(aq) oxidized. This gives

$$HCHO(aq) + 2\,Ag^+(aq) \longrightarrow HCOOH(aq) + 2\,Ag(s)$$

What remains is to balance the O atoms via participation of the aqueous solvent. Balancing in acid, we balance the O atoms by adding one H_2O molecule to the O-deficient side—the left. Now we add 2 $H^+(aq)$ ions to the H-deficient side, the right. The final balanced equation is

$$HCHO(aq) + 2\,Ag^+(aq) + H_2O(l) \longrightarrow HCOOH(aq) + 2\,Ag(s) + 2\,H^+(aq)$$

Section 16.3: Voltaic Cells: Electricity from Chemical Change

16.45

A voltaic cell is constructed using the reaction of chromium metal and aquated iron(II) ions
$$2\,Cr(s) + 3\,Fe^{2+}(aq) \longrightarrow 2\,Cr^{3+}(aq) + 3\,Fe(s)$$
Electrons in the external circuit flow from the $\underline{Cr\,|Cr^{3+}}$ electrode (the <u>anode</u>) to the $\underline{Fe^{2+}|Fe}$ electrode (the <u>cathode</u>). Negative ions move in the salt bridge from the $\underline{Fe^{2+}|Fe}$ half-cell (the <u>reduction</u> half-cell) to the $\underline{Cr\,|Cr^{3+}}$ half-cell (the <u>oxidation</u> half-cell). The half-reaction at the anode is $\underline{Cr(s) \to Cr^{3+}(aq) + 3\,e^-}$ and that at the cathode is $\underline{Fe^{2+}(aq) + 2\,e^- \to Fe(s)}$.

16.47

(a) and (c)

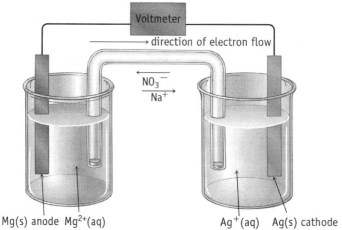

Mg(s) anode Mg^{2+}(aq) Ag^+(aq)   Ag(s) cathode

(b) $Mg(s) \longrightarrow Mg^{2+}(aq) + 2\,e^-$ is the oxidation half-reaction that occurs at the anode.

$Ag^+(aq) + e^- \longrightarrow Ag(s)$ is the reduction half-reaction that occurs at the cathode.

The salt bridge is needed to complete the circuit. Current flows only when the circuit is completed.

Section 16.4: Cell emf, and Half-Cell Reduction Potentials

16.49

(a) $E°_{Fe\ scale}[Cu^{2+}(aq)|Cu(s)] = E°[Cu^{2+}(aq)|Cu(s)] - E°[Fe^{2+}(aq)|Fe(s)]$
$= 0.337\ V - (-0.44\ V) = 0.78\ V$

(b) $E°_{Fe\ scale}[Zn^{2+}(aq)|Zn(s)] = E°[Zn^{2+}(aq)|Zn(s)] - E°[Fe^{2+}(aq)|Fe(s)]$
$= -0.763\ V - (-0.44\ V) = -0.32\ V$

16.51

$$NO_3^-(aq) + 4\,H^+(aq) + 3e^- \longrightarrow NO(g) + 2\,H_2O(\ell)$$

(a) $E[\,NO_3^-(aq)|NO(g)\,] = E°[\,NO_3^-(aq)|NO(g)\,]$
when $[NO_3^-] = 1.00\ mol\ L^{-1}$, $[H^+] = 1.00\ mol\ L^{-1}$ and $p(NO) = 1$ bar,
or, more generally, whenever

$$Q = \frac{p(NO)}{[NO_3^-][H^+]^4} = 1.00$$

(b) $E°_{cell} = E°[\,NO_3^-(aq)|NO(g)\,] - E°[\,H^+(aq)|H_2(g)\,] = 0.96\ V$
So, $E°[\,NO_3^-(aq)|NO(g)\,] = 0.96\ V$

16.53

(a) To make a cell with potential close to 1.1 V, using the $Zn^{2+}(aq)\,|\,Zn(s)$ half-cell, we could use the $Cu^{2+}(aq)\,|\,Cu(s)$ half-cell as the cathode and the $Zn^{2+}(aq)\,|\,Zn(s)$ half-cell as the anode.

$E°_{cell} = E°[Cu^{2+}(aq)|Cu(s)] - E°[Zn^{2+}(aq)|Zn(s)] = 0.337\ V - (-0.763\ V) = 1.10\ V$

Alternatively we could use any cell with $E°_{cell}$ approximately 1.1 V and calculate, using the Nernst equation, concentrations of species that would result in $E_{cell} = 1.1$ V.

For example, a cell that we could use, in which the $Zn^{2+}(aq)\,|\,Zn(s)$ half-cell is the anode is

$Zn(s)|Zn^{2+}(aq)||I_2(s)|I^-(aq)$ \qquad $E°_{cell} = 0.535\ V - (-0.763\ V) = 1.30\ V$

A cell in which the $Zn^{2+}(aq)\,|\,Zn(s)$ half-cell is the cathode is

$Al(s)|Al^{3+}(aq)||Zn^{2+}(aq)\,|\,Zn(s)$ \qquad $E°_{cell} = -0.763\ V - (-1.66\ V) = 0.90\ V$

(b) To make a cell with potential close to 0.5 V, using the $Zn^{2+}(aq)\,|\,Zn(s)$ half-cell, we could use the $Ni^{2+}(aq)\,|\,Ni(s)$ half-cell as the cathode and the $Zn^{2+}(aq)\,|\,Zn(s)$ half-cell as the anode.

$E°_{cell} = E°[\,Ni^{2+}(aq)\,|\,Ni(s)\,] - E°[Zn^{2+}(aq)|Zn(s)] = -0.25\ V - (-0.763\ V) = 0.51\ V$

We can then use the Nernst equation to calculate concentrations of $Ni^{2+}(aq)$ and $Zn^{2+}(aq)$ ions that result in $E_{cell} = 0.50$ V.

A cell with the $Zn^{2+}(aq)\,|\,Zn(s)$ half-cell as the cathode and which has $E°_{cell} \approx 0.5$ V is

$Cd(s),\ S^{2-}(aq)|CdS(s)||\ Zn^{2+}(aq)\,|\,Zn(s)$ \qquad $E°_{cell} = -0.763\ V - (-1.21\ V) = 0.45\ V$

16.55

Ranked in order of half-cell reduction potentials:
$E°[F_2(g)|F^-(aq)] = 2.87\ V$
$E°[Cl_2(g)|Cl^-(aq)] = 1.36V$
$E°[O_2(g)|H^+(aq)|H_2O(\ell)] = 1.229\ V$
$E°[Br_2(\ell)|Br^-(aq)] = 1.08\ V$
$E°[I_2(s)|I^-(aq)] = 0.535\ V$
$E°[S(s)|H^+(aq)|H_2S(aq)] = 0.14\ V$
$E°[Se(s)|H^+(aq)|H_2Se(\ell)] = -0.40\ V$

(a) Se(s) is the least powerful oxidizing agent.
(b) $F^-(aq)$ ions are the least powerful reducing agent.
(c) $F_2(g)$ and $Cl_2(g)$ are capable of oxidizing $H_2O(\ell)$ to $O_2(g)$.
(d) $F_2(g)$, $Cl_2(g)$, $Br_2(\ell)$, $I_2(s)$ and $O_2(g)$ are capable of oxidizing $H_2S(g)$ to S(s).
(e) $O_2(g)$ is capable of oxidizing $I^-(aq)$ ions to $I_2(s)$ in acidic solution.
(f) S(s) is NOT capable of oxidizing $I^-(aq)$ ions to $I_2(s)$ in acidic solution.
(g) $H_2S(aq) + Se(s) \rightarrow H_2Se(aq) + S(s)$ is NOT the direction of spontaneous reaction when all species are at standard concentrations at 25°C.
(h) $H_2S(aq) + I_2(s) \rightarrow 2\ H^+(aq) + 2\ I^-(aq) + S(s)$ is the direction of spontaneous reaction when all species are at standard concentrations at 25 °C.

16.57

If $E°_{cell}$, computed for the direction of reaction as written has a positive value, then at standard conditions the equation indicates the direction of spontaneous reaction.

(a) $Ni^{2+}(aq) + H_2(g) \longrightarrow Ni(s) + 2H^+(aq)$
   is NOT the direction of spontaneous reaction.
(b) $Fe^{3+}(aq) + 2I^-(aq) \longrightarrow Fe^{2+}(aq) + I_2(s)$
   is the direction of spontaneous reaction.
(c) $Br_2(\ell) + 2Cl^-(aq) \longrightarrow 2\,Br^-(aq) + Cl_2(g)$
   is NOT the direction of spontaneous reaction.
(d) $Cr_2O_7^{2-}(aq) + 6\,Fe^{2+}(aq) + 14\,H^+(aq) \longrightarrow 2\,Cr^{3+}(aq) + 6\,Fe^{3+}(aq) + 7\,H_2O(\ell)$
   is the direction of spontaneous reaction.

16.61

If $E°_{cell}$, computed for the direction of reaction as written has a positive value, then at standard conditions the equation indicates the direction of spontaneous reaction.

(a) $Zn(s) + I_2(s) \longrightarrow Zn^{2+}(aq) + 2\,I^-(aq)$
   is the direction of spontaneous reaction.
(b) $2\,Cl^-(aq) + I_2(s) \longrightarrow Cl_2(g) + 2\,I^-(aq)$
   is NOT the direction of spontaneous reaction.
(c) $2\,Na^+(aq) + 2\,Cl^-(aq) \longrightarrow 2\,Na(s) + Cl_2(g)$
   is NOT the direction of spontaneous reaction.
(d) $2\,K(s) + H_2O(l) \longrightarrow 2\,K^+(aq) + H_2(g) + 2\,OH^-(aq)$
   is the direction of spontaneous reaction.

16.63

Adding a KI solution to a standard acidic $Cu(NO_3)_2$ solution causes a brown colour and a precipitate to form because the nitrate ions oxidize iodide ions under these conditions. The $I_2(s)$ that is formed first reacts with $I^-(aq)$ ions to form brown $I_3^-(aq)$ ions. When all of the $I^-(aq)$ ions have reacted, further iodine formed remains as the insoluble solid precipitate. A similar response is not observed when KCl or KBr solutions are added because nitrate cannot oxidize chloride or bromide under standard conditions—there is no reaction.
$6\,I^-(aq) + 2\,NO_3^-(aq) + 8\,H^+(aq) \rightarrow 3\,I_2(s) + 2\,NO(g) + 4\,H_2O(\ell)$
is the balanced equation for reaction with $I^-(aq)$ ions.

Section 16.5: Electrochemical Cells under Non-standard Conditions

16.65

$E°_{cell} = E°[Ag^+(aq)|Ag(s)] - E°[Zn^{2+}(aq)|Zn(s)] = 0.7994\ V - (-0.763\ V) = +1.562\ V$
$\qquad Zn(s) \rightarrow Zn^{2+}(aq) + 2\,e^-$
$\qquad \underline{2\times(\ Ag^+(aq) + e^- \rightarrow Ag(s)\ )}$
$Zn(s) + 2\,Ag^+(aq) \rightarrow Zn^{2+}(aq) + 2\,Ag(s)$
$$Q = \frac{[Zn^{2+}]}{[Ag^+]^2} = \frac{(0.010)}{(0.25)^2} = 0.16$$

$$E_{cell} = E°_{cell} - \frac{0.0257\,V}{n} \times \ln Q = +1.562\,V - \frac{0.0257\,V}{2} \times \ln(0.16)$$

$$= +1.562\,V - \frac{0.0257\,V}{2} \times (-1.8) = +1.585\,V$$

16.67

(a)

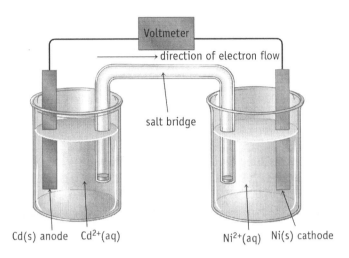

(b)

$$Ni^{2+}(aq) + 2e^- \longrightarrow Ni(s)$$

$$\underline{Cd(s) \longrightarrow Cd^{2+}(aq) + 2e^-}$$

$$Ni^{2+}(aq) + Cd(s) \longrightarrow Ni(s) + Cd^{2+}(aq)$$

(c) Since $Ni^{2+}(aq)|Ni(s)$ has the higher reduction potential, the Ni(s) electrode is the cathode—it is labelled (+). The $Cd^{2+}(aq)|Cd(s)$ half-cell is the anode compartment, labelled (−).

(d) $E°_{cell} = E°[Ni^{2+}(aq)|Ni(s)] - E°[Cd^{2+}(aq)|Cd(s)] = -0.25\,V - (-0.403\,V) = +0.15\,V$

(e) Electrons flow from the anode to cathode in the external circuit.

(f) The $Na^+(aq)$ ions (cations) in the salt bridge move from the anode compartment toward the cathode compartment. The $NO_3^-(aq)$ ions (anions) move from the cathode compartment toward the anode compartment.

(g) The cell is at equilibrium when $E_{cell} = 0$.

$$E_{cell} = E°_{cell} - \frac{0.0257\,V}{n} \times \ln Q = 0\,V$$

when $Q = K$. So

$$\ln K = \frac{2}{0.0257\,V} \times E°_{cell} = \frac{2}{0.0257\,V} \times (+0.15\,V)$$

$$= 11.7$$

and

$$K = 1.2 \times 10^5$$

(h) If the concentration of $Cd^{2+}(aq)$ ions is reduced to 0.010 mol L^{-1}, keeping $[Ni^{2+}] = 1.0$ mol L^{-1}, then

$$E_{cell} = E^\circ_{cell} - \frac{0.0257\,\text{V}}{n} \times \ln Q = +0.15\,\text{V} - \frac{0.0257\,\text{V}}{2} \times \ln\left(\frac{[\text{Cd}^{2+}]}{[\text{Ni}^{2+}]}\right)$$

$$= +0.15\,\text{V} - \frac{0.0257\,\text{V}}{2} \times \ln\left(\frac{0.010}{1.0}\right) = +0.21\,\text{V}$$

Since E_{cell} is still positive, the net reaction is in the same direction given in part (b). In fact, the cell emf is increased, as we might have expected as a result of decreasing the concentration of a product species.

Section 16.6: Standard Cell emf and Equilibrium Constant

16.69

(a) $2\,\text{Fe}^{3+}(aq) + 2\,\text{I}^-(aq) \longrightarrow 2\,\text{Fe}^{2+}(aq) + \text{I}_2(s)$

$E^\circ_{cell} = E^\circ[\text{Fe}^{3+}(aq)|\text{Fe}^{2+}(aq)] - E^\circ[\text{I}_2(s)|\text{I}^-(aq)] = 0.771\,\text{V} - (0.535\,\text{V}) = +0.236\,\text{V}$

$$\ln K = \frac{2}{0.0257\,\text{V}} \times E^\circ_{cell} = \frac{2}{0.0257\,\text{V}} \times (+0.236\,\text{V})$$

$$= 18.4$$

and

$$K = 9.5\times10^7$$

(b) $\text{I}_2(s) + 2\,\text{Br}^-(aq) \longrightarrow 2\,\text{I}^-(aq) + \text{Br}_2(\ell)$

$E^\circ_{cell} = E^\circ[\text{I}_2(s)|\text{I}^-(aq)] - E^\circ[\text{Br}_2(\ell)|\text{Br}^-(aq)] = 0.535\,\text{V} - (1.08\,\text{V}) = -0.545\,\text{V}$

$$\ln K = \frac{2}{0.0257\,\text{V}} \times E^\circ_{cell} = \frac{2}{0.0257\,\text{V}} \times (-0.545\,\text{V})$$

$$= -42.4$$

and

$$K = 3.9\times10^{-19}$$

16.73

$$\text{Au}^{3+}(aq) + 3\,e^- \longrightarrow \text{Au}(s)$$

$$\underline{\text{Au}(s) + 4\,\text{Cl}^-(aq) \longrightarrow [\text{AuCl}_4]^-(aq) + 3\,e^-}$$

$$\text{Au}^{3+}(aq) + 4\,\text{Cl}^-(aq) \longrightarrow [\text{AuCl}_4]^-(aq)$$

The formation reaction is expressed here as the sum of reduction and oxidation half-reactions.

$E^\circ_{cell} = E^\circ[\text{Au}^{3+}(aq)|\text{Au}(s)] - E^\circ\{[\text{AuCl}_4]^-(aq)|\text{Au}(s)\} = 1.50\,\text{V} - (1.00\,\text{V}) = +0.50\,\text{V}$

$$\ln K_{f} = \frac{3}{0.0257\,\text{V}} \times E^{\circ}_{\text{cell}} = \frac{3}{0.0257\,\text{V}} \times (0.50\,\text{V})$$

$$= 58.4$$

and

$$K_{f} = 2.3 \times 10^{25}$$

Section 16.7: Electrolysis: Chemical Change Using Electrical Energy

16.75

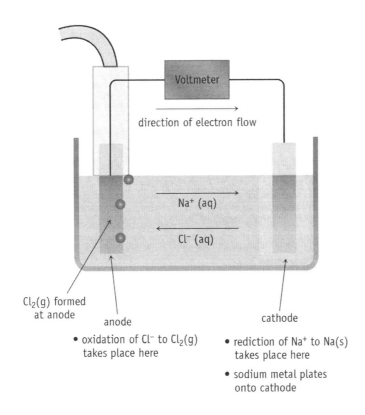

Cl₂(g) formed
at anode
         anode
         cathode

• oxidation of Cl⁻ to Cl₂(g)   • rediction of Na⁺ to Na(s)
  takes place here   takes place here

                                             • sodium metal plates
                                              onto cathode

16.77

Fluorine, $F_2(g)$, has the largest reduction potential of all species—it is easily reduced to $F^-(aq)$ ions. This means that $F^-(aq)$ ions are difficult to oxidize to $F_2(g)$. Oxygen gas forms at the anode in the electrolysis of an aqueous KF solution.

16.79

The bumper on an off-road vehicle is likely to get scratched. The chrome plated steel will corrode at the scratch faster than it would without the plating. This is because chromium is a (slightly) more noble metal than iron, and a voltaic cell is set up with the iron as the anode area. The galvanized steel stays protected, even with scratches, until all of the zinc plating is oxidized by corrosion. Note that corrosion of zinc is slower than that for iron because zinc oxide coats the zinc protecting it from atmospheric oxygen. When iron oxidizes, its oxide does not coat the metal, so it is vulnerable to corrosion.

Section 16.8: Corrosion of Iron

16.81

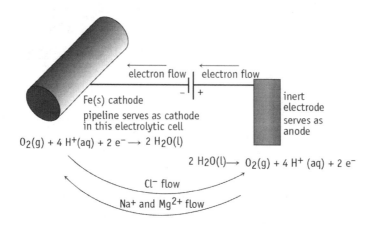

$O_2(g) + 4 H^+(aq) + 2 e^- \longrightarrow 2 H_2O(l)$

$2 H_2O(l) \longrightarrow O_2(g) + 4 H^+(aq) + 2 e^-$

By keeping the pipeline at a higher potential than the inert electrode, the pipeline behaves like a metal even more noble - it does not corrode.

SUMMARY AND CONCEPTUAL QUESTIONS

16.83

Both of the problems encountered are examples of a more noble metal polarizing a less noble metal, making it more susceptible to corrosion.
(a) In this case, the iron screws in contact with the more noble copper alloy corrode rapidly. They crumble, leaving holes in the copper alloy sheets where water leaks in.
(b) In this case, the iron wrench is the more noble metal. It accelerates the corrosion of the aluminum in contact with it by creating a voltaic cell in which the aluminium is the anode area. The aluminum corrodes rapidly leaving a wrench-shaped hole in the hull.

16.85

If the emf is applied in the wrong direction, the corrosion of the pipeline will be accelerated. Electrons would be taken away from the pier, causing oxidation of the iron—i.e., rendering the pier the anode of a cell. The electrons would be directed to the platinum-coated titanium, where the only possible reaction would be reduction of $O_2(g)$ at its surface.

CHAPTER 17
Spontaneous Change: How Far?

IN-CHAPTER EXERCISES

Exercise 17.1—Entropy Change

The change in entropy for a reversible process (i.e., add the heat very slowly) is just heat flow into the system divided by temperature. Therefore, the entropy of the water and hexane change by the same amount—it depends only on the temperature of the system and the amount of heat added.

$$\Delta S = q / T = 1 \times 10^6 \, J / 323.15 \, K = 3095 \, J \, K^{-1}$$

Exercise 17.3—Entropy Change

$$\Delta S = q / T = 29.1 \times 10^3 \, J / 239.7 \, K = 121 \, J \, K^{-1}$$

i.e., the entropy of 1 mol of ammonia vapour is 121 J K^{-1} greater than the entropy of 1 mol of the liquid.

Exercise 17.7—Standard Entropy Change of Reaction

(a) $CaCO_3(s) \longrightarrow CaO(s) + CO_2(g)$

$\Delta_r S° = [1 \times S°(CO_2, g) + 1 \times S°(CaO, s)] - 1 \times S°(CaCO_3, s)$

$\quad\quad = [213.74 \, J \, K^{-1} \, mol^{-1} + 38.2 \, J \, K^{-1} \, mol^{-1})] - 91.7 \, J \, K^{-1} \, mol^{-1}$

$\quad\quad = +160.2 \, J \, K^{-1} \, mol^{-1}$

We predict an increase in entropy as the reaction produces a gas, which is consistent with the calculation.

(b) $N_2(g) + 3H_2(g) \longrightarrow 2NH_3(g)$

$\Delta_r S° = 2 \times S°(NH_3, g) - [1 \times S°(N_2, g) + 3 \times S°(H_2, g)]$

$\quad\quad = 2 \times 192.77 \, J \, K^{-1} \, mol^{-1} - [191.56 \, J \, K^{-1} \, mol^{-1} + 3 \times 130.7 \, J \, K^{-1} \, mol^{-1}]$

$\quad\quad = -198.1 \, J \, K^{-1} \, mol^{-1}$

We predict a decrease in entropy as the reaction converts a total of 4 mol of gas to 2 mol of gas, which is consistent with the calculation.

Exercise 17.9—Predicting Spontaneity from $\Delta S°_{univ}$

$Si(s) + 2Cl_2(g) \longrightarrow SiCl_4(g)$

$\Delta S°_{sys} = 1 \times S°(SiCl_4, g) - [1 \times S°(Si, s) + 2 \times S°(Cl_2, g)]$

$\quad\quad = 330.86 \, J \, K^{-1} \, mol^{-1} - [18.82 \, J \, K^{-1} \, mol^{-1} + 2 \times 223.08 \, J \, K^{-1} \, mol^{-1}]$

$\quad\quad = -134.12 \, J \, K^{-1} \, mol^{-1}$

$$\Delta H^{\circ}{}_{sys} = 1 \times \Delta_f H^{\circ}(SiCl_4, g) - [1 \times \Delta_f H^{\circ}(Si, s) + 2 \times \Delta_f H^{\circ}(Cl_2, g)]$$
$$= -662.75 \, kJ \, mol^{-1} - [0+0]$$
$$= -662.75 \, kJ \, mol^{-1}$$

i.e., this is just the formation reaction for $SiCl_4(g)$.

$$\Delta S^{\circ}{}_{surr} = -\frac{\Delta H^{\circ}{}_{sys}}{T} = -\frac{-662.75 \times 10^3 \, J \, mol^{-1}}{298 \, K} = +2224 \, J \, K^{-1} \, mol^{-1}$$

$$\Delta S^{\circ}{}_{univ} = \Delta S^{\circ}{}_{sys} + \Delta S^{\circ}{}_{surr} = -134.12 \, J \, K^{-1} \, mol^{-1} + 2224 \, J \, K^{-1} \, mol^{-1}$$
$$= +2090 \, J \, K^{-1} \, mol^{-1}$$

Since the change in entropy for the universe upon reaction in the direction indicated by the equation is positive, the reaction is spontaneous.

Exercise 17.13—Calculating $\Delta_r G^{\circ}$ from $\Delta_r H^{\circ}$ and $\Delta_r S^{\circ}$

$$N_2(g) + 3 H_2(g) \longrightarrow 2 NH_3(g)$$
$$\Delta_r S^{\circ} = 2 \times S^{\circ}(NH_3, g) - [1 \times S^{\circ}(N_2, g) + 3 \times S^{\circ}(H_2, g)]$$
$$= 2 \times 192.77 \, J \, K^{-1} \, mol^{-1} - [191.56 \, J \, K^{-1} \, mol^{-1} + 3 \times 130.7 \, J \, K^{-1} \, mol^{-1}]$$
$$= -198.1 \, J \, K^{-1} \, mol^{-1}$$
$$\Delta_r H^{\circ} = 2 \times \Delta_f H^{\circ}(NH_3, g) - [\Delta_f H^{\circ}(N_2, g) + 3 \times \Delta_f H^{\circ}(H_2, g)]$$
$$= -2 \times (-45.90 \, kJ \, mol^{-1}) = -91.80 \, kJ \, mol^{-1}$$
$$\Delta_r G^{\circ} = \Delta_r H^{\circ} - T \Delta_r S^{\circ}$$
$$= -91.80 \, kJ \, mol^{-1} - 298 \, K \times (-198.1 \, J \, K^{-1} mol^{-1}) / 1000 \, J \, / kJ$$
$$= -32.8 \, kJ \, mol^{-1}$$

Exercise 17.15—Standard Molar Free Energy Change of Formation, $\Delta_f G^{\circ}$

(a) $\frac{1}{2} N_2(g) + \frac{3}{2} H_2(g) \rightarrow NH_3(g)$

(b) $2 Fe(s) + \frac{3}{2} O_2(g) \rightarrow Fe_2O_3(s)$

(c) $3 C(s) + 3 H_2(g) + O_2(g) \rightarrow CH_3CH_2COOH(\ell)$

(d) $Ni(s) \rightarrow Ni(s)$ ($Ni(s)$ is the element in its standard state. Its $\Delta_f G^{\circ} = 0$)

Exercise 17.19—Finding the Temperature at Which $\Delta_r G^{\circ} = 0$

$$2 HgO(s) \longrightarrow 2 Hg(\ell) + O_2(g)$$
$$\Delta_r S^{\circ} = 2 \times S^{\circ}(Hg, l) + 1 \times S^{\circ}(O_2, g) - 2 \times S^{\circ}(HgO, s)$$
$$= 2 \times 76.02 \, J \, K^{-1} \, mol^{-1} + 205.07 \, J \, K^{-1} \, mol^{-1} - 2 \times 70.29 \, J \, K^{-1} \, mol^{-1}$$
$$= 216.53 \, J \, K^{-1} \, mol^{-1}$$

$$\Delta_r H^{\circ} = 2 \times \Delta_f H^{\circ}(Hg, l) + 1 \times \Delta_f H^{\circ}(O_2, g) - 2 \times \Delta_f H^{\circ}(HgO, s)$$
$$= 0 + 0 - 2 \times (-90.83 \, kJ \, mol^{-1}) = +181.66 \, kJ \, mol^{-1}$$

$$\Delta_r G° = \Delta_r H° - T\Delta_r S° = 0$$

when

$$T = \frac{\Delta_r H°}{\Delta_r S°} = \frac{181.66 \times 1000 \text{ J mol}^{-1}}{216.53 \text{ J K}^{-1} \text{mol}^{-1}} = 839.0 \text{ K}$$

Exercise 17.23—Calculation of $\Delta_r G$ in a Non-standard Reaction Mixture

(a) $N_2(g) + 3H_2(g) \longrightarrow 2NH_3(g)$

$$\Delta_r G° = 2 \times \Delta_f G°(NH_3,g) - \left[\Delta_f G°(N_2,g) + 3 \times \Delta_f G°(H_2,g)\right]$$
$$= 2 \times \left(-16.37 \text{ kJ mol}^{-1}\right) - \left[0 + 3 \times 0\right]$$
$$= -32.74 \text{ kJ mol}^{-1}$$

(b) The negative value for $\Delta_r G°$ indicates that the reaction proceeds spontaneously in the forward direction under standard conditions.

(c) The conditions provided (25 °C, $p(N_2) = p(H_2) = p(NH_3) = 1$ bar) are standard conditions. Therefore, as described in (b), the reaction will proceed in the forward direction under these conditions.

(d) To obtain $\Delta_r G$, we must first calculate Q:

$$Q = \frac{(pNH_3)^2}{(pH_2)^3(pN_2)}$$
$$= \frac{(1 \text{ bar})^2}{(1.0 \times 10^{-1} \text{ bar})^3(1.0 \times 10^{-2} \text{ bar})}$$
$$= 1 \times 10^5$$

$$\Delta_r G = \Delta_r G° - RT \ln Q$$
$$= -32\,740 \text{ J mol}^{-1} + (8.314 \text{ J mol}^{-1} \text{ K}^{-1})(298 \text{ K})\ln(1 \times 10^5)$$
$$= -4 \text{ kJ mol}^{-1}$$

As the value for $\Delta_r G$ is negative, the reaction would proceed in the forward direction under these conditions.

(e) Calculation of Q:

$$Q = \frac{(pNH_3)^2}{(pH_2)^3(pN_2)}$$

$$= \frac{(10 \text{ bar})^2}{(1.0 \times 10^{-2} \text{ bar})^3 (1.0 \times 10^{-3} \text{ bar})}$$

$$= 1 \times 10^{11}$$

$$\Delta_r G = \Delta_r G^\circ - RT \ln Q$$

$$= -32740 \text{ J mol}^{-1} + (8.314 \text{ J mol}^{-1} \text{ K}^{-1})(298 \text{ K})\ln(1 \times 10^{11})$$

$$= +30 \text{ kJ mol}^{-1}$$

As the value for $\Delta_r G$ is positive, the reaction would proceed in the forward direction under these conditions.

(f) The negative values for $\Delta_r G$ in parts **(c)** and **(d)** indicate that under these conditions, the reaction will proceed spontaneously in the forward direction, with the forward direction being more favourable for **(c)** than **(d)** due to the more negative value for $\Delta_r G$. In contrast, the reaction conditions in **(e)** lead to the reaction proceeding spontaneously in the reverse direction.

(g) For **(c)**:

$$Q = \frac{(pNH_3)^2}{(pH_2)^3(pN_2)}$$

$$= \frac{(1 \text{ bar})^2}{(1 \text{ bar})^3 (1 \text{ bar})}$$

$$= 1$$

Therefore, $Q < K$ and the reaction proceeds spontaneously in the forward direction.

For **(d)**: $Q = 1 \times 10^5$. Therefore, $Q < K$ and the reaction proceeds in the forward direction.

For **(e)**: $Q = 1 \times 10^{11}$. Therefore, $Q > K$ and the reaction proceeds in the reverse direction.

Exercise 17.25—The Relationship between $\Delta_r G^\circ$ and K

$$\Delta_r G^\circ = -RT \ln K = -8.314 \text{ J } K^{-1} \text{ mol}^{-1} \times 298 \text{ } K \times \ln(1.6 \times 10^7)$$

$$= -41 \text{ kJ mol}^{-1}$$

Exercise 17.29—Interrelationships among $\Delta_r G^\circ$, K, and $E^\circ{}_{cell}$

(a) $2 \text{ Fe}^{3+}(\text{aq}) + 2 \text{ I}^-(\text{aq}) \longrightarrow 2 \text{ Fe}^{2+}(\text{aq}) + \text{I}_2(\text{s})$

$n = 2$ here

$$E°_{cell} = E°[Fe^{3+}|Fe^{2+}] - E°[I_2|I^-] = 0.771 \text{ V} - 0.535 \text{ V} = 0.236 \text{ V}$$

$$\Delta_r G° = -nFE°_{cell} = -(2)(96\,450 \text{ C mol}^{-1})(0.236 \text{ V}) = -4.55 \times 10^4 \text{ CV mol}^{-1} = -45.5 \text{ kJ mol}^{-1}$$

$$K = \exp(-\Delta_r G°/RT) = \exp\left(\frac{4.55 \times 10^4 \text{ J mol}^{-1}}{8.314 \text{ J K}^{-1} \text{ mol}^{-1} \times 298 \text{ K}}\right)$$

$$= 9.46 \times 10^7$$

(b) $I_2(aq) + 2\,Br^-(aq) \longrightarrow 2\,I^-(aq) + Br_2(aq)$

$n = 2$ here

$$E° = E°[I_2|I^-] - E°[Br_2|Br^-] = 0.535 \text{ V} - 1.08 \text{ V} = -0.545 \text{ V}$$

$$\Delta_r G° = -nFE°_{cell} = -(2)(96\,450 \text{ C mol}^{-1})(-0.545 \text{ V}) = 1.05 \times 10^5 \text{ CV mol}^{-1} = 105 \text{ kJ mol}^{-1}$$

$$K = \exp(-\Delta_r G°/RT) = \exp\left(\frac{-105 \times 10^3 \text{ J mol}^{-1}}{8.314 \text{ J K}^{-1} \text{ mol}^{-1} \times 298 \text{ K}}\right)$$

$$= 3.93 \times 10^{-19}$$

Exercise 17.31—Dependence of *K* on Temperature

$$2\,SO_3(g) \rightleftharpoons 2\,SO_2(g) + O_2(g)$$

(a) $\Delta_r G°$ at 298 K

$$\Delta_r G° = 2 \times \Delta_f G°(SO_2, g) + 1 \times \Delta_f G°(O_2, g) - 2 \times \Delta_f G°(SO_3, g)$$

$$= 2 \times (-300.13 \text{ kJ mol}^{-1}) + 0 \text{ kJ mol}^{-1} - 2 \times (-371.04 \text{ kJ mol}^{-1})$$

$$= 141.82 \text{ kJ mol}^{-1}$$

(b) K at 298 K

$$\ln K = -\frac{\Delta_r G°}{RT} = -\frac{141.82 \times 10^3 \text{ J mol}^{-1}}{8.314 \text{ J K}^{-1} \text{ mol}^{-1} \times 298 \text{ K}}$$

$$= -57.24$$

$$K = e^{-57.24} = 1.38 \times 10^{-25}$$

(c) K at 1500 °C

$$\ln K = -\frac{\Delta_r G°}{RT} = -\frac{141.82 \times 10^3 \text{ J mol}^{-1}}{8.314 \text{ J K}^{-1} \text{ mol}^{-1} \times 1773 \text{ K}}$$

$$= -9.62$$

$$K = e^{-9.62} = 6.64 \times 10^{-5}$$

REVIEW QUESTIONS

Section 17.1: Photochemical Smog and Chemical Equilibrium

17.35

(a) Anthropomorphism: attribution of human characteristics to non-human entities. The statement implies that hydrogen and oxygen have a desire to react with each other. Of course, atoms and molecules are incapable of desire, and their behaviour is directly attributable to the laws of thermodynamics.

(b) Some non-anthropomorphic alternatives for this statement: The reaction between hydrogen and oxygen to form water is thermodynamically favourable. Hydrogen and oxygen spontaneously react to form water.

(c) This statement deals with thermodynamics as it describes a reaction that occurs spontaneously to produce a product that is more stable than the starting materials. Of course, the kinetics of this process are very strongly influenced by the temperature of the system, as the reaction proceeds negligibly slowly at room temperature but is extremely fast at elevated temperatures.

17.37

It is not appropriate to refer to a system as thermodynamically stable but kinetically reactive as the laws of thermodynamics define the outcome of a process, while kinetics deals with how quickly this outcome will be achieved. If a system is thermodynamically stable, it will not undergo a change and the rate of change is therefore irrelevant.

Section 17.2: Spontaneous Direction of Change and Equilibrium

17.39

Many responses possible. One example: *A chemical reaction will proceed in the spontaneous direction of reaction until chemical equilibrium is achieved.*

17.41

Many responses possible. Some examples: ice melting at T > 0°C; dissolution of ammonium nitrate in water; evaporation of water

Section 17.4—Entropy and the Second Law of Thermodynamics

17.43

$\Delta S = q / T = 6.01 \times 10^3 \text{ J mol}^{-1} / 273.15 \text{ K} = 22.0 \text{ J K}^{-1} \text{ mol}^{-1}$

17.45

(a) NaCl(g) has more entropy than NaCl(aq), which has more entropy than NaCl(s).
The greatest disorder is in the gas state. NaCl(s) is a highly constrained arrangement of the ions.
Check: $S°[NaCl(g)] = 229.79 > S°[NaCl(aq)] = 115.5 > S°[NaCl(s)] = 72.11$ (all in J K^{-1} mol^{-1})

(b) $H_2S(g)$ has more entropy than $H_2O(g)$ because S is heavier than O. Heavier molecules have more entropy (everything else equal). This is because energy distributed among heavier gas molecules corresponds to a larger range of momenta—which means more ways of distributing the energy. At the same temperature, molecules with less mass travel faster (on average).
Check: $S°[H_2S(g)] = 205.79 > S°[H_2O(g)] = 188.84$ (all in J K^{-1} mol^{-1})

(c) $C_2H_4(g)$ has more entropy than $N_2(g)$ (with the same molar mass).
More complex molecules have more entropy than less complex molecules with the same mass. They have more internal motions which mean more ways of distributing energy.
Check: $S°[C_2H_4(g)] = 219.36 > S°[N_2(g)] = 191.56$ (all in J K^{-1} mol^{-1})

(d) $H_2SO_4(\ell)$ has more entropy than $H_2SO_4(aq)$.
On the surface, it would appear that since $H_2SO_4(aq)$ explores a larger volume—the volume of its aqueous environment—it should have a greater entropy than the same amount of $H_2SO_4(\ell)$. (Note that standard entropies correspond to 1 M concentration in case of aqueous species.) However, the aqueous sulfuric acid is strongly associated with water via the acid-base reaction. The stronger forces in the ionized solution reduce its entropy, more than compensating for the increase in entropy due to the greater freedom of the aqueous species.
Check: $S°[H_2SO_4(\ell)] = 156.9 > S°[H_2SO_4(aq)] = 20.1$ (all in J K^{-1} mol^{-1})

17.47

(a) KOH(s) $\longrightarrow$ KOH(aq)

$$\Delta_r S° = S°(KOH, aq) - S°(KOH, s)$$
$$= 91.6\,J\,K^{-1}\,mol^{-1} - 78.9\,J\,K^{-1}\,mol^{-1}$$
$$= +12.7\,J\,K^{-1}\,mol^{-1}$$

In solution, the aquated ions have more freedom—i.e., greater entropy.

(b) Na(g) $\longrightarrow$ Na(s)

$$\Delta_r S° = S°(Na, s) - S°(Na, g)$$
$$= 51.21\,J\,K^{-1}\,mol^{-1} - 153.765\,J\,K^{-1}\,mol^{-1}$$
$$= -102.55\,J\,K^{-1}\,mol^{-1}$$

The solid is a more ordered/constrained arrangement of the ions—i.e., lower entropy.

(c) $Br_2(\ell) \longrightarrow Br_2(g)$

$$\Delta_r S° = S°(Br_2, g) - S°(Br_2, \ell)$$
$$= 245.47\,J\,K^{-1}\,mol^{-1} - 152.2\,J\,K^{-1}\,mol^{-1}$$
$$= +93.3\,J\,K^{-1}\,mol^{-1}$$

In the gaseous phase, the molecules are less constrained—i.e., greater entropy.

(d) HCl(g) $\longrightarrow$ HCl(aq)

$$\Delta_r S^\circ = S^\circ(\text{HCl,aq}) - S^\circ(\text{HCl,g})$$
$$= 56.5\,\text{J}\,\text{K}^{-1}\,\text{mol}^{-1} - 186.2\,\text{J}\,\text{K}^{-1}\,\text{mol}^{-1}$$
$$= -129.7\,\text{J}\,\text{K}^{-1}\,\text{mol}^{-1}$$

The aqueous species has less freedom than the gas phase species—i.e., lower entropy. Note that under standard conditions, the concentration of a gas is

$$\frac{n}{V} = \frac{p}{RT} = \frac{100\,\text{kPa}}{8.314\,\text{J}\,\text{K}^{-1}\text{mol}^{-1} \times 298\,\text{K}} = 0.04\,\text{mol}\,\text{L}^{-1}$$

17.51

(a) $\tfrac{1}{2}\,H_2(g) + \tfrac{1}{2}\,Cl_2(g) \rightarrow HCl(g)$
$$\Delta_f S^\circ[\text{HCl(g)}] = S^\circ[\text{HCl(g)}] - (\tfrac{1}{2}\,S^\circ[H_2(g)] + \tfrac{1}{2}\,S^\circ[Cl_2(g)])$$
$$= 186.2\,\text{J}\,\text{K}^{-1}\,\text{mol}^{-1} - (\tfrac{1}{2} \times 130.7\,\text{J}\,\text{K}^{-1}\,\text{mol}^{-1} + \tfrac{1}{2} \times 223.08\,\text{J}\,\text{K}^{-1}\,\text{mol}^{-1}) =$$
$9.3\,\text{J}\,\text{K}^{-1}\,\text{mol}^{-1}$

(b) $Ca(s) + H_2(g) + O_2(g) \rightarrow Ca(OH)_2(s)$
$$\Delta_f S^\circ[\text{Ca(OH)}_2(s)] = S^\circ[\text{Ca(OH)}_2(s)] - (S^\circ[\text{Ca(s)}] + S^\circ[H_2(g)] + S^\circ[O_2(g)])$$
$$= 83.39\,\text{J}\,\text{K}^{-1}\,\text{mol}^{-1} - (41.59\,\text{J}\,\text{K}^{-1}\,\text{mol}^{-1} + 130.7\,\text{J}\,\text{K}^{-1}\,\text{mol}^{-1} + 205.07\,\text{J}\,\text{K}^{-1}$$
$\text{mol}^{-1}) = -294.0\,\text{J}\,\text{K}^{-1}\,\text{mol}^{-1}$

Section 17.5: Entropy Changes and Spontaneity: The Second Law

17.55

At 25°C,

$$\Delta S^\circ_{surr} = -\frac{\Delta H^\circ_{sys}}{T} = -\frac{467.9 \times 10^3\,\text{J}\,\text{mol}^{-1}}{298\,\text{K}} = -1570\,\text{J}\,\text{K}^{-1}\,\text{mol}^{-1}$$

and

$$\Delta S^\circ_{univ} = \Delta S^\circ_{sys} + \Delta S^\circ_{surr} = 560.7\,\text{J}\,\text{K}^{-1}\,\text{mol}^{-1} - 1570\,\text{J}\,\text{K}^{-1}\,\text{mol}^{-1}$$
$$= -1009.3\,\text{J}\,\text{K}^{-1}\,\text{mol}^{-1}$$

The reaction is NOT spontaneous at 25°C. However, if we increase the temperature, ΔS°_{surr} would be smaller in magnitude (it is negative). A sufficiently large temperature would make ΔS°_{sys} the dominant contribution, and ΔS°_{univ} would be positive.

17.57

(a)
$Fe_2O_3(s) + 2\,Al(s) \longrightarrow 2\,Fe(s) + Al_2O_3(s)$ $\qquad \Delta_r H^\circ = -851.5\,\text{kJ}\,\text{mol}^{-1},\ \Delta_r S^\circ = -375.2\,\text{J}\,\text{K}^{-1}\,\text{mol}^{-1}$
This reaction is spontaneous at low temperature, but NOT spontaneous at high temperature—i.e., type 2.

(b)
$N_2(g) + 2\,O_2(g) \longrightarrow 2\,NO_2(g)$ $\qquad \Delta_r H^\circ = +66.2\,\text{kJ}\,\text{mol}^{-1},\ \Delta_r S^\circ = -121.6\,\text{J}\,\text{K}^{-1}\,\text{mol}^{-1}$
This reaction is NOT spontaneous at any temperature—i.e., type 4.

Section 17.6—Gibbs Free Energy

17.59

(a)

$2 \, Pb(s) + O_2(g) \longrightarrow 2 PbO(s)$

$\Delta_r S° = 2 \, S°[PbO(s)] - (2 \, S°[Pb(s)] + 1 \times S°[O_2(g)])$

$\quad = 2 \times 66.5 \, J \, K^{-1} \, mol^{-1} - (2 \times 64.81 \, J \, K^{-1} \, mol^{-1} + 205.07 \, J \, K^{-1} \, mol^{-1}) = -201.7 \, J \, K^{-1} \, mol^{-1}$

$\Delta_r H° = 2 \, \Delta_f H°[PbO(s)] - (2 \, \Delta_f H°[Pb(s)] + 1 \times \Delta_f H°[O_2(g)])$

$\quad = 2 \times (-219 \, kJ \, mol^{-1}) - (0 + 0) = -438 \, kJ \, mol^{-1}$

$\Delta_r G° = \Delta_r H° - T \, \Delta_r S° = -438 \, kJ \, mol^{-1} - (298 \, K) \times (-201.7 \, J \, K^{-1} \, mol^{-1}) / (1000 \, J / kJ)$

$\quad = -378 \, kJ \, mol^{-1}$

This reaction is spontaneous. It is enthalpy driven—the negative enthalpy change more than compensates for the negative entropy change at 25°C.

(b)

$CaO(s) + CO_2(g) \longrightarrow CaCO_3(s)$

$\Delta_r S° = S°[CaCO_3(s)] - (S°[CaO(s)] + S°[CO_2(g)])$

$\quad = 91.7 \, J \, K^{-1} \, mol^{-1} - (38.2 \, J \, K^{-1} \, mol^{-1} + 213.74 \, J \, K^{-1} \, mol^{-1}) = -160.2 \, J \, K^{-1} \, mol^{-1}$

$\Delta_r H° = \Delta_f H°[CaCO_3(s)] - (\Delta_f H°[CaO(s)] + \Delta_f H°[CO_2(g)])$

$\quad = -1207.6 \, kJ \, mol^{-1} - [(-635.09 \, kJ \, mol^{-1}) + (-393.509 \, kJ \, mol^{-1})] = -179.0 \, kJ \, mol^{-1}$

$\Delta_r G° = \Delta_r H° - T \, \Delta_r S° = -179.0 \, kJ \, mol^{-1} - (298 \, K) \times (-160.2 \, J \, K^{-1} \, mol^{-1}) / (1000 \, J / kJ)$

$\quad = -131.3 \, kJ \, mol^{-1}$

This reaction is spontaneous. It is enthalpy driven—the negative enthalpy change more than compensates for the negative entropy change at 25°C.

17.61

$C_6H_6(\ell) + 3 H_2(g) \longrightarrow C_6H_{12}(\ell) \qquad \Delta_r H° = -206.7 \, kJ \, mol^{-1}, \, \Delta_r S° = -316.5 \, J \, K^{-1} \, mol^{-1}$

$\Delta_r G° = \Delta_r H° - T \, \Delta_r S° = -206.7 \, kJ \, mol^{-1} - (298 \, K) \times (-316.5 \, J \, K^{-1} \, mol^{-1}) / (1000 \, J / kJ)$

$\quad = -112.4 \, kJ \, mol^{-1}$

This reaction is spontaneous at 25°C in a reaction mixture with all reagent species in their standard states. It is enthalpy driven—the negative enthalpy change more than compensates for the negative entropy change at 25°C.

17.63

(a) $\quad 2 \, K(s) + Cl_2(g) \longrightarrow 2 \, KCl(s)$

$\quad \Delta_r G° = 2 \times \Delta_f G°[KCl(s)] = 2 \times (-408.77 \, kJ \, mol^{-1}) - [0 + 0] = -817.54 \, kJ \, mol^{-1}$

$\quad$ This reaction is spontaneous at 25°C in a reaction mixture with all reagent species in their standard states.

(b) $\quad 2 \, H_2S(g) + 3 O_2(g) \longrightarrow 2 H_2O(g) + 2 SO_2(g)$

$\quad \Delta_r G° = 2 \, \Delta_f G°[SO_2(g)] + 2 \, \Delta_f G°[H_2O(g)] - (2 \, \Delta_f G°[H_2S(g)] + 3 \, \Delta_f G°[O_2(g)])$

$$= 2 \times (-300.13 \text{ kJ mol}^{-1}) + 2 \times (-228.59 \text{ kJ mol}^{-1}) - (2 \times (-33.56 \text{ kJ mol}^{-1}) + 3 \times (0))$$
$$= -990.32 \text{ kJ mol}^{-1}$$

This reaction is spontaneous at 25°C in a reaction mixture with all reagent species in their standard states.

(c) $\quad 4 NH_3(g) + 7 O_2(g) \longrightarrow 4NO_2(g) + 6 H_2O(g)$

$\Delta_r G° = 4 \Delta_f G°[NO_2(g)] + 6 \Delta_f G°[H_2O(g)] - (4 \Delta_f G°[NH_3(g)] + 7 \Delta_f G°[O_2(g)])$
$= 4 \times (51.23 \text{ kJ mol}^{-1}) + 6 \times (-228.59 \text{ kJ mol}^{-1}) - (4 \times (-16.37 \text{ kJ mol}^{-1}) + 7 \times (0)) =$
$-1101.14 \text{ kJ mol}^{-1}$

This reaction is spontaneous at 25°C in a reaction mixture with all reagent species in their standard states.

17.65

$BaCO_3(s) \longrightarrow BaO(s) + CO_2(g)$
$\Delta_r G° = \Delta_f G°[BaO(s)] + \Delta_f G°[CO_2(g)] - \Delta_f G°[BaCO_3(s)] = 219.7 \text{ kJ mol}^{-1}$
$\quad = -520.38 \text{ kJ mol}^{-1} + (-394.359 \text{ kJ mol}^{-1}) - \Delta_f G°[BaCO_3(s)] = 219.7 \text{ kJ mol}^{-1}$
So,
$\Delta_f G°[BaCO_3(s)] = -520.38 \text{ kJ mol}^{-1} + (-394.359 \text{ kJ mol}^{-1}) - 219.7 \text{ kJ mol}^{-1} =$
$-1134.4 \text{ kJ mol}^{-1}$

17.67

$Ni(CO)_4(\ell) \longrightarrow Ni(s) + 4CO(g) \qquad \Delta_r G° \text{ at } 25°C = 40 \text{ kJ mol}^{-1}$
$\Delta_r S° = S°[Ni(s)] + 4 S°[CO(g)] - S°[Ni(CO)_4(\ell)]$
$\quad = 29.87 \text{ J K}^{-1} \text{ mol}^{-1} + 4 \times 197.674 \text{ J K}^{-1} \text{ mol}^{-1} - 320 \text{ J K}^{-1} \text{ mol}^{-1} = 501 \text{ J K}^{-1} \text{ mol}^{-1}$
$\Delta_r H° = \Delta_f H°[Ni(s)] + 4 \Delta_f H°[CO(g)] - \Delta_f H°[Ni(CO)_4(\ell)]$
$\quad = 0 + 4 \times (-110.525 \text{ kJ mol}^{-1}) - (-632 \text{ kJ mol}^{-1}) = 190 \text{ kJ mol}^{-1}$
Assuming $\Delta_r H°$ and $\Delta_r S°$ are temperature independent, we get the temperature where $Ni(CO)_4(\ell)$ is in equilibrium with 1 bar CO(g) via
$\Delta_r G° = \Delta_r H° - T_{eq} \Delta_r S° = 0$
i.e.,
$T_{eq} = \Delta_r H° / \Delta_r S° = 190 \times 1000 \text{ J mol}^{-1} / (501 \text{ J K}^{-1} \text{ mol}^{-1}) = 379 \text{ K} = 106°C.$

17.69

$C_2H_6(g) + \frac{7}{2}O_2(g) \longrightarrow 2CO_2(g) + 3H_2O(g)$
$\Delta_r G° = 2 \Delta_f G°[CO_2(g)] + 3 \Delta_f G°[H_2O(g)] - (\Delta_f G°[C_2H_6(g)] + \frac{7}{2} \Delta_f G°[O_2(g)])$
$\quad = 2 \times (-394.359 \text{ kJ mol}^{-1}) + 3 \times (-228.59 \text{ kJ mol}^{-1}) - (-31.89 \text{ kJ mol}^{-1} + \frac{7}{2} \times 0)$
$\quad = -1442.60 \text{ kJ mol}^{-1}$
This is in accord with experience. Ethane is a component of natural gas—it is a fuel. Its combustion is highly favoured—it releases a lot of free energy.

17.71

For the reaction described in the figure, $\Delta_r G° > 0$ and therefore $K < 1$. A standard reaction mixture would proceed toward the reactants (i.e., in the reverse direction) until equilibrium is achieved.

Section 17.7—The Relationship between $\Delta_r G°$ and K

17.73

$$C_2H_4(g) + H_2(g) \longrightarrow C_2H_6(g)$$
$$\Delta_r G° = \Delta_f G°[C_2H_6(g)] - (\Delta_f G°[C_2H_4(g)] + \Delta_f G°[H_2(g)])$$
$$= -31.89 \text{ kJ mol}^{-1} - (68.35 \text{ kJ mol}^{-1} + 0) = -100.24 \text{ kJ mol}^{-1}$$
$$\ln K = -\frac{\Delta_r G°}{RT} = -\frac{-100.24 \times 10^3 \text{ J mol}^{-1}}{8.314 \text{ J K}^{-1} \text{ mol}^{-1} \times 298 \text{ K}}$$
$$= 40.46$$

$$K = e^{40.46} = 3.73 \times 10^{17}$$

This reaction is product favoured in a reaction mixture at 25°C and all reactants and products in their standard states (i.e., the conditions that apply for specification of the <u>standard</u> free energy change of reaction). The negative standard free energy change is in accord with the very large equilibrium constant.

Section 17.8—$\Delta_r G°$ and $E°_{cell}$ for voltaic cell reactions

17.75

(a) $Zn^{2+}(aq) + Ni(s) \longrightarrow Zn(s) + Ni^{2+}(aq)$
$$E° = E°[Zn^{2+}(aq)|Zn(s)] - E°[Ni^{2+}(aq)|Ni(s)] = -0.763 \text{ V} - (-0.25 \text{ V}) = -0.51 \text{ V}$$
$$\Delta_r G° = -nF E° = -2 \times 96450 \text{ C mol}^{-1} \times (-0.51 \text{ V}) = 98379 \text{ CV mol}^{-1} = 98.4 \text{ kJ mol}^{-1}$$
$$\ln K = -\frac{\Delta_r G°}{RT} = -\frac{98.4 \times 10^3 \text{ J mol}^{-1}}{8.314 \text{ J K}^{-1} \text{ mol}^{-1} \times 298 \text{ K}}$$
$$= -39.72$$

$$K = e^{-39.72} = 5.62 \times 10^{-18}$$

(b) $Cu(s) + 2 Ag^+(aq) \longrightarrow Cu^{2+}(aq) + 2 Ag(s)$
$$E° = E°[Ag^+(aq)|Ag(s)] - E°[Cu^{2+}(aq)|Cu(s)] = 0.7994 \text{ V} - (0.337 \text{ V}) = 0.462 \text{ V}$$
$$\Delta_r G° = -nF E° = -2 \times 96450 \text{ C mol}^{-1} \times (0.462 \text{ V}) = -89120 \text{ CV mol}^{-1} = -89.12 \text{ kJ mol}^{-1}$$

$$\ln K = -\frac{\Delta_r G^\circ}{RT} = -\frac{-89.12 \times 10^3 \text{ J mol}^{-1}}{8.314 \text{ J K}^{-1} \text{ mol}^{-1} \times 298 \text{ K}}$$

$$= 35.97$$

$$K = e^{35.97} = 4.18 \times 10^{15}$$

Section 17.9—Dependence of Equilibrium Constants on Temperature

17.77

$CH_3CH_2OH(\ell) \rightarrow CH_3CH_2OH(g)$
(a) $\Delta_{vap}G^\circ = \Delta_f G^\circ[CH_3CH_2OH(g)] - \Delta_f G^\circ[CH_3CH_2OH(\ell)]$
$= -168.49 \text{ kJ mol}^{-1} - (-174.7 \text{ kJ mol}^{-1}) = 6.21 \text{ kJ mol}^{-1}$

$$\ln K = -\frac{\Delta_r G^\circ}{RT} = -\frac{6.21 \times 10^3 \text{ J mol}^{-1}}{8.314 \text{ J K}^{-1} \text{ mol}^{-1} \times 298 \text{ K}}$$

$$= -2.51$$

$$K = e^{-2.51} = 0.0813$$

$$p(\text{ethanol}) = 0.0813 \text{ bar}$$

The equilibrium constant for vaporization of ethanol is just the ethanol vapour pressure.
(b) At 298 K,
$\Delta_{vap}G^\circ = \Delta_{vap}H^\circ - 298 \text{ K} \times \Delta_{vap}S^\circ = 6.21 \text{ kJ mol}^{-1}$
So, $\Delta_{vap}S^\circ = (39.33 \text{ kJ mol}^{-1} - 6.21 \text{ kJ mol}^{-1}) / (298 \text{ K}) = 0.111 \text{ kJ K}^{-1} \text{ mol}^{-1} = 111 \text{ J K}^{-1} \text{ mol}^{-1}$
At the boiling point,
$\Delta_{vap}G^\circ = \Delta_{vap}H^\circ - T_{bp} \Delta_{vap}S^\circ = 0$
i.e.,
$T_{bp} = \Delta_{vap}H^\circ / \Delta_{vap}S^\circ = 39.33 \text{ kJ mol}^{-1}/ (0.111 \text{ kJ K}^{-1} \text{ mol}^{-1}) = 355 \text{ K} = 82 \text{ °C}$

Section 17.10: Photochemical Smogs and the Dependence of *K* on *T*

17.79

Thermodynamic: the equilibrium constant for the formation of NO from N_2 and O_2 becomes sufficiently large at high temperatures to lead to significant production of NO.
Kinetic: During peak morning traffic, NO is produced more quickly than it is consumed, resulting in an increase in atmospheric levels. Following the morning peak, NO is consumed through reaction with reactive oxygen species such as ozone and peroxy radicals.

17.81

(a) There is no build-up of NO concentration during the 5 p.m. peak. This is partly attributable to the abundance of ozone later in the day, which rapidly oxidizes NO as it is produced.

(b) UV light leads to the formation of radicals, which in turn react with hydrocarbons and other substances present in the atmosphere to produce secondary pollutants. Because there is less sunlight during the afternoon peak, the formation of radicals is reduced relative to the morning peak and the production of secondary pollutants is significantly reduced.

SUMMARY AND CONCEPTUAL QUESTIONS

17.83

Many responses possible. Some examples:
- Human activity: production of atmospheric pollutants in vehicle exhaust; production of oxygen gas by Joseph Priestley; generating electrical current with voltaic cells
- Chemical reactivity: generation of NO from NO_2; formation of ozone; combustion of methane

17.85

$$CaCO_3(s) \longrightarrow CaO(s) + CO_2(g)$$
$$\Delta_r G^\circ = \Delta_r H^\circ - T_{eq} \Delta_r S^\circ = 0$$
if calcium carbonate is in equilibrium with 1.00 bar $CO_2(g)$, at any temperature.
So,
$$\Delta_r S^\circ = \Delta_r H^\circ / T_{eq} = 179.0 \times 10^3 \text{ J mol}^{-1}/ 1170 \text{ K} = 153 \text{ J K}^{-1} \text{ mol}^{-1}$$
We ignore the temperature dependence of the reaction enthalpy in making this estimate.

17.87

(a) $\Delta_r G^\circ = \{\Delta_f G^\circ[CO(g)] + \Delta_f G^\circ[H_2(g)]\} - \{\Delta_f G^\circ[C(s)] + \Delta_f G^\circ[H_2O(g)]\}$
$= -137.168 \text{ kJ mol}^{-1} + 0 - 0 - (-228.59 \text{ kJ mol}^{-1}) = 91.42 \text{ kJ mol}^{-1}$

(b)

$$\ln K = -\frac{\Delta_r G^\circ}{RT} = -\frac{91.42 \times 10^3 \text{ J mol}^{-1}}{8.314 \text{ J K}^{-1} \text{ mol}^{-1} \times 298 \text{ K}}$$
$$= -36.90$$

$$K = e^{-36.90} = 9.43 \times 10^{-17}$$

(c) This reaction is NOT spontaneous at 25°C in a reaction mixture with all reagents in their standard states. To determine a temperature at which the reaction is spontaneous, we need $\Delta_r H^\circ$ and $\Delta_r S^\circ$.
$\Delta_r H^\circ = \Delta_f H^\circ[CO(g)] + \Delta_f H^\circ[H_2(g)] - \{\Delta_f H^\circ[C(s)] + \Delta_f H^\circ[H_2O(g)]\}$
$= -110.525 \text{ kJ mol}^{-1} + 0 - 0 - (-241.83 \text{ kJ mol}^{-1}) = 131.30 \text{ kJ mol}^{-1}$
$\Delta_r G^\circ = \Delta_r H^\circ - T \Delta_r S^\circ$

$\Delta_r S° = (\Delta_r H° - \Delta_r G°) / T$
  $= (131.30 \text{ kJ mol}^{-1} - 91.42 \text{ kJ mol}^{-1}) / 298 \text{ K} = 0.1338 \text{ kJ K}^{-1} \text{ mol}^{-1}$

The minimum temperature at which the reaction becomes spontaneous is such that

$\Delta_r G° = \Delta_r H° - T_{spon} \Delta_r S° = 0$ (above T_{spon}, $\Delta_r G° < 0$)

$T_{spon} = \Delta_r H° / \Delta_r S° = 131.30 \text{ kJ mol}^{-1} / 0.1338 \text{ kJ K}^{-1} \text{ mol}^{-1} = 981.3 \text{ K}$

17.89

Reaction 1: $CaCO_3(s) + SO_2(g) + \frac{1}{2}H_2O(l) \rightleftharpoons CaSO_3 \cdot \frac{1}{2}H_2O(s) + CO_2(g)$

$\Delta_r H° = \Delta_f H°[CaSO_3 \cdot \frac{1}{2}H_2O(s)] + \Delta_f H°[CO_2(g)] - (\Delta_f H°[CaCO_3(s)] + \Delta_f H°[SO_2(g)] + \frac{1}{2} \Delta_f H°[H_2O(\ell)])$

  $= -1574.65 \text{ kJ mol}^{-1} + (-393.509 \text{ kJ mol}^{-1}) - (-1207.6 \text{ kJ mol}^{-1} + (-296.84 \text{ kJ mol}^{-1})$
  $+ \frac{1}{2}(-285.83 \text{ kJ mol}^{-1}))$

  $= -320.8 \text{ kJ mol}^{-1}$

$\Delta_r S° = S°[CaSO_3 \cdot \frac{1}{2}H_2O(s)] + S°[CO_2(g)] - (S°[CaCO_3(s)] + S°[SO_2(g)] + \frac{1}{2} S°[H_2O(\ell)])$

  $= 121.3 \text{ J K}^{-1} \text{ mol}^{-1} + (213.74 \text{ J K}^{-1} \text{ mol}^{-1}) - (91.7 \text{ J K}^{-1} \text{ mol}^{-1} + 248.21 \text{ J K}^{-1} \text{ mol}^{-1} +$
  $\frac{1}{2}(69.95 \text{ J K}^{-1} \text{ mol}^{-1})$

  $= -39.8 \text{ J K}^{-1} \text{ mol}^{-1}$

$\Delta_r G° = \Delta_r H° - T \Delta_r S° = -320.8 \text{ kJ mol}^{-1} - 298 \text{ K} \times (-39.8 \text{ J K}^{-1} \text{ mol}^{-1}) / (1000 \text{ J /kJ}) =$
  $-308.9 \text{ kJ mol}^{-1}$

Reaction 2: $CaCO_3(s) + SO_2(g) + \frac{1}{2}H_2O(\ell) + \frac{1}{2}O_2(g) \rightleftharpoons CaSO_4 \cdot \frac{1}{2}H_2O(s) + CO_2(g)$

$\Delta_r H° = \Delta_f H°[CaSO_4 \cdot \frac{1}{2}H_2O(s)] + \Delta_f H°[CO_2(g)] - (\Delta_f H°[CaCO_3(s)] + \Delta_f H°[SO_2(g)] + \frac{1}{2} \Delta_f H°[H_2O(\ell)])$

  $= -1311.7 \text{ kJ mol}^{-1} + (-393.509 \text{ kJ mol}^{-1}) - (-1207.6 \text{ kJ mol}^{-1} + (-296.84 \text{ kJ mol}^{-1}) +$
  $\frac{1}{2}(-285.83 \text{ kJ mol}^{-1}))$

  $= -57.9 \text{ kJ mol}^{-1}$

$\Delta_r S° = S°[CaSO_4 \cdot \frac{1}{2}H_2O(s)] + S°[CO_2(g)] - (S°[CaCO_3(s)] + S°[SO_2(g)] + \frac{1}{2} S°[H_2O(\ell)] +$
  $\frac{1}{2} S°[O_2(g)])$

  $= 134.8 \text{ J K}^{-1} \text{ mol}^{-1} + (213.74 \text{ J K}^{-1} \text{ mol}^{-1}) - (91.7 \text{ J K}^{-1} \text{ mol}^{-1} + (248.21 \text{ J K}^{-1} \text{ mol}^{-1})$
  $+ \frac{1}{2}(69.95 \text{ J K}^{-1} \text{ mol}^{-1}) + \frac{1}{2}(205.07 \text{ J K}^{-1} \text{ mol}^{-1}))$

$= -128.9 \text{ J K}^{-1} \text{ mol}^{-1}$

$\Delta_r G° = \Delta_r H° - T \Delta_r S° = -57.9 \text{ kJ mol}^{-1} - 298 \text{ K} \times (-128.9 \text{ J K}^{-1} \text{ mol}^{-1}) / (1000 \text{ J /kJ}) =$
  $-19.5 \text{ kJ mol}^{-1}$

Reaction 1, with larger negative value of $\Delta_r G°$, is more product-favoured than reaction 2.

17.91

$CuO(s) + H_2(g) \longrightarrow Cu(s) + H_2O(g)$

$\Delta_r G° = \Delta_f G°[H_2O(g)] - \Delta_f G°[CuO(s)]$
  $= -228.59 \text{ kJ mol}^{-1} - (-128.3 \text{ kJ mol}^{-1}) = -100.3 \text{ kJ mol}^{-1}$

This reaction is product-favoured at 298 K and with all reactants and products at standard conditions.

17.93

$$\tfrac{1}{2}H_2(g) + \tfrac{1}{2}I_2(g) \rightleftharpoons HI(g)$$

$$K = \frac{p(HI)}{p(H_2)^{1/2}\, p(I_2)^{1/2}} = \frac{1.61}{(0.132)^{1/2}(0.295)^{1/2}} = 8.16$$

$\Delta_r G° = -RT \ln(K) = -8.314\ \text{J K}^{-1}\ \text{mol}^{-1} \times 623.15\ \text{K} \times \ln(8.16) = -10880\ \text{J mol}^{-1} = -10.88\ \text{kJ mol}^{-1}$

17.95

$\Delta_r H° = 0 + 0 - 2 \times \Delta_f H°[\text{Ag}_2\text{O}(s)] = -2 \times (-31.1\ \text{kJ mol}^{-1}) = 62.2\ \text{kJ mol}^{-1}$
$\Delta_r S° = 4\, S°[\text{Ag}(s)] + S°[\text{O}_2\,(g)] - (2\, S°[\,\text{Ag}_2\text{O}(s)])$
   $= 4 \times 42.55\ \text{J K}^{-1}\ \text{mol}^{-1} + 205.07\ \text{J K}^{-1}\ \text{mol}^{-1} - (2 \times 121.3\ \text{J K}^{-1}\ \text{mol}^{-1})$
   $= 132.7\ \text{J K}^{-1}\ \text{mol}^{-1}$
$\Delta_r G° = \Delta_r H° - T\Delta_r S° = 62.2\ \text{kJ mol}^{-1} - 298.15\ \text{K} \times (132.7\ \text{J K}^{-1}\ \text{mol}^{-1}) / (1000\ \text{J /kJ}) = 22.6\ \text{kJ mol}^{-1}$
This reaction is NOT product favoured. This is why silver tarnishes when exposed to the atmosphere.
However, silver oxide will decompose at temperatures above T such that
$\Delta_r G° = \Delta_r H° - T\Delta_r S° = 0$
i.e., $T = \Delta_r H° / \Delta_r S° = 62.2 \times 10^3\ \text{J mol}^{-1} / (132.7\ \text{J K}^{-1}\ \text{mol}^{-1}) = 469\ \text{K}$
The reaction is product-favoured above this temperature.

17.97

(a) The entropy of a substance increases on going from the liquid to the vapour state at any temperature. **TRUE**
The gas always has more entropy than the corresponding liquid.
(b) An exothermic reaction will always be spontaneous. **FALSE**
Some exothermic processes, namely those with negative entropy change, are not spontaneous at sufficiently high temperature.
(c) Reactions with a positive $\Delta_r H°$ and a positive $\Delta_r S°$ can never be product-favoured. **FALSE**
Such reactions are spontaneous at sufficiently high temperature (if the process is possible at such temperatures). They are entropy driven processes at these temperatures.
(d) If $\Delta_r G°$ for a reaction is negative, the reaction has $K > 1$. **TRUE**
$K = \exp(-\Delta_r G° / RT) > 1$ if $\Delta_r G° < 0$

17.99

$$C_2H_6(g) + \tfrac{7}{2}\,O_2(g) \rightarrow 2\,CO_2(g) + 3\,H_2O(g)$$

(a) + (b)
We can deduce that $\Delta S°_{sys} > 0$, because 4 ½ moles of gas are transformed into 5 moles of gas. We can also predict that $\Delta H°_{sys} < 0$. Ethane is a fuel—a component of natural gas. It is combusted to release energy.
Energy conservation gives us $\Delta H°_{surr} > 0$ and $\Delta S°_{surr} = -\Delta H°_{sys} / T > 0$.

Therefore, $\Delta S^\circ_{univ} = \Delta S^\circ_{sys} + \Delta S^\circ_{surr} > 0$.

We also have $\Delta G^\circ_{sys} = \Delta H^\circ_{sys} - T\Delta S^\circ_{sys} < 0$.

The reaction is spontaneous at all temperatures.

(c)

$K = \exp(-\Delta G^\circ_{sys}/RT) > 1$

because $\Delta G^\circ_{sys} < 0$. In fact, K is very large because of the large exothermicity of the reaction—the magnitude of ΔG°_{sys} is large.

17.101

(a) The decomposition of liquid water to give gaseous oxygen and hydrogen.

$\Delta_r H^\circ > 0$

a very endothermic process

$\Delta_r S^\circ > 0$

1 ½ moles of gas are produced from 1 mole of liquid

$\Delta_r G^\circ > 0$

NOT a spontaneous process

(b) The explosive decomposition of nitroglycerin.

$\Delta_r H^\circ < 0$

a very exothermic process

$\Delta_r S^\circ > 0$

gases are produced from a solid

$\Delta_r G^\circ < 0$

a spontaneous process

(c) The combustion of octane.

$$2\,C_8H_{18}(g) + 25\,O_2(g) \longrightarrow 16\,CO_2(g) + 18\,H_2O(g)$$

$\Delta_r H^\circ < 0$

a very exothermic process

$\Delta_r S^\circ > 0$

34 mol gas are produced from 27 mol gas

$\Delta_r G^\circ < 0$

a spontaneous process

17.103

$C(graphite) \rightarrow C(diamond)$

(a)

$\Delta_r H^\circ = \Delta_f H^\circ[C(diamond)] - 0 = 1.8 \text{ kJ mol}^{-1}$

$\Delta_r S^\circ = S^\circ[C(diamond)] - S^\circ[C(graphite)] = 2.377 \text{ K}^{-1} \text{ mol}^{-1} - 5.6 \text{ J K}^{-1} \text{ mol}^{-1} = -3.2$
$\text{J K}^{-1} \text{ mol}^{-1}$

$\Delta_r G^\circ = \Delta_f G^\circ[C(diamond)] = 2.900 \text{ kJ mol}^{-1}$

(b) These calculations are for 1.00 bar and 298 K. At other temperatures and pressures (much higher), $\Delta_r G^\circ$ is negative. Such conditions can be used to manufacture diamonds. However, the most economically viable means of making diamonds is chemical vapour deposition which relies on a small already formed diamond to "grow" a larger diamond by carefully allowing carbon atoms (from the vapour) to deposit onto the crystal faces. The process is spontaneous because the carbon vapour has a higher Gibbs free energy than diamond. The carbon vapour is made by heating graphite—here we put in energy making the subsequent production of diamonds favourable.

17.105

(a) $\Delta_r G° = 2 \Delta_f G° (NO_2, g) - \Delta_f G (N_2O_4, g) = 2(51.23 \text{ kJ mol}^{-1}) - 97.73 \text{ kJ mol}^{-1} = 4.73 \text{ kJ mol}^{-1}$

(b) Because the value for $\Delta_r G°$ is positive, this means that under standard conditions, the reaction proceeds spontaneously in the reverse direction

(c) We can use $\Delta_r G$ obtained in (a) to determine the equilibrium constant, K for this reaction at 25 °C:

$\Delta_r G° = -RT \ln K$

$$K = e^{-\left(\frac{\Delta_r G°}{RT}\right)} = e^{-\left(\frac{4730 \text{ J mol}^{-1}}{(8.314 \text{ J mol}^{-1} \text{ K}^{-1})(298 \text{ K})}\right)} = 0.148$$

Given the conditions provided, we can calculate the reaction quotient, Q:

$$Q = \frac{P_{NO_2}^2}{P_{N_2O_4}} = \frac{(1 \text{ bar})^2}{(1 \text{ bar})} = 1$$

Because the value for Q is greater than the value for K, the reaction proceeds in the reverse direction.

(d)

$$\Delta_r G = \Delta_r G° + RT \ln Q = 4730 \text{ J mol}^{-1} + (8.31 \text{ J mol}^{-1}\text{K}^{-1})(298 \text{ K}) \ln \left[\frac{(10 \text{ bar})^2}{1.0 \times 10^{-2} \text{bar}} \right] = 28 \text{ kJ mol}^{-1}$$

The reaction therefore proceeds in the reverse direction.

(e)

$$\Delta_r G = \Delta_r G° + RT \ln Q = 4730 \text{ J mol}^{-1} + (8.314 \text{ J mol}^{-1}\text{K}^{-1})(298 \text{ K}) \ln \left[\frac{(100 \text{ bar})^2}{1.0 \times 10^{-4} \text{bar}} \right] = 50 \text{ kJ mol}^{-1}$$

The reaction therefore proceeds in the reverse direction.

(f) In each case, the reaction proceeds in reverse direction. As [NO_2] increases and [N_2O_4] decreases, the reverse reaction becomes more spontaneous

(g) $K = 0.148$ (refer to (c))

(h) For (c): $Q = 1$ (refer to (c)); Reaction proceeds in the reverse direction

For (d):

$$Q = \frac{P_{NO_2}^2}{P_{N_2O_4}} = \frac{(10 \text{ bar})^2}{(1 \times 10^{-2} \text{ bar})} = 10^4$$

Reaction proceeds in reverse direction

For (e):

$$Q = \frac{P_{NO_2}{}^2}{P_{N_2O_4}} = \frac{(100 \text{ bar})^2}{(1 \times 10^{-4} \text{ bar})} = 10^8$$

Reaction proceeds in reverse direction

(i) Yes, the answers are consistent. In each case, it is predicted that the reaction will proceed in the reverse direction.

CHAPTER 18
Spontaneous Change: How Fast?

IN-CHAPTER EXERCISES

Exercise 18.1—Relative Rates of [Reactant] Decrease and [Product] Increase

$$2\,NOCl(g) \longrightarrow 2\,NO(g) + Cl_2(g)$$

For every 2 mol of $NOCl(g)$ that react, 2 mol of $NO(g)$ and 1 mol of $Cl_2(g)$ are produced. The relative rate of appearance of Cl_2 is ½ the rate of appearance of NO which equals the rate of disappearance of $NOCl$.

$$-\frac{\Delta[NOCl]}{\Delta t} = \frac{\Delta[NO]}{\Delta t} = 2 \times \frac{\Delta[Cl_2]}{\Delta t}$$

Exercise 18.5—Determining a Rate Equation from Initial Rates

Experiment	Initial concentrations (mol L^{-1})		Initial rate
	[NO] (mol L^{-1})	[O$_2$] (mol L^{-1})	(mol L^{-1} s^{-1})
1	0.020	0.010	0.028
2	0.020	0.020	0.057
3	0.020	0.040	0.114
4	0.040	0.020	0.227
5	0.010	0.020	0.014

In Experiments 1, 2, and 3, with [NO] the same, the relative [O$_2$] values are 1: 2: 4. The corresponding relative rates are 0.028: 0.057: 0.114 = 1: 2: 4, showing that the reaction is first order with respect to the $O_2(g)$ concentration.
In Experiments 5, 2, and 4, with [O$_2$] the same, the relative [NO] values are 1, 2, and 4. The corresponding relative rates are 0.014: 0.057: 0.227 $\approx$ 1: 4: 16 $= (1)^2: (2)^2: (4)^2$, showing that the reaction is second order with respect to the $NO(g)$ concentration.

$$Rate = k\,[NO]^2[O_2]$$

Using the data from Experiment 4, we have

$$k = \frac{Rate}{[NO]^2[O_2]} = \frac{0.227\,mol\,L^{-1}\,s^{-1}}{(0.040\,mol\,L^{-1})(0.040\,mol\,L^{-1})(0.020\,mol\,L^{-1})} = 7.1 \times 10^3\,L^2\,mol^{-2}\,s^{-1}$$

Exercise 18.21—Elementary Steps and Reaction Mechanisms

(a)

Step 1 (Fast): $NH_3(aq) + OCl^-(aq) \longrightarrow \cancel{NH_2Cl(aq)} + \cancel{OH^-(aq)}$

Step 2 (Slow): $\cancel{NH_2Cl(aq)} + NH_3(aq) \longrightarrow \cancel{N_2H_5^+(aq)} + Cl^-(aq)$

Step 3 (Fast): $\cancel{N_2H_5^+(aq)} + \cancel{OH^-(aq)} \longrightarrow N_2H_4(aq) + H_2O(l)$

$$2\,NH_3(aq) + OCl^-(aq) \longrightarrow N_2H_4(aq) + H_2O(l) + Cl^-(aq)$$

(b) Step 2, the slow step is the rate-determining step.
(c) The rate equation for elementary step 2 is
$$\text{Rate} = k\,[NH_2Cl]\,[NH_3]$$
(d) NH_2Cl, $N_2H_5^+$ and OH^- are intermediates in this reaction mechanism.

Exercise 18.23—Nucleophilic Substitution Reactions

(a) The rate of substitution will double if the concentration of *tert*-butyl chloride is doubled, keeping the concentration of iodide ions constant.
(b) The rate of substitution will NOT change if the concentration of iodide ions is doubled, keeping the concentration of *tert*-butyl chloride constant. The rate is independent of $[I^-]$.
(c) Carbocation intermediate.

(d) Whether or not inversion takes place is irrelevant here because the product is the same in both cases—due to symmetry.
(e) The proportion of this reaction that proceeds by the S_N2 mechanism is negligible because the rate of the S_N1 mechanism is enhanced by the relatively stable carbocation intermediate, and because the rate of the S_N2 mechanism is reduced by the presence of the three methyl groups about the carbon center. The portion of collisions between iodide and *tert*-butyl chloride which can lead to reaction is much smaller than the value for methyl chloride (for example). The three methyl groups present an obstacle limiting the rate of the S_N2 reaction. The S_N1 mechanism presents no such obstacle—the iodide reacts with a reactive planar intermediate.

REVIEW QUESTIONS

Section 18.1: Winds of Change

18.25

Many responses possible, some examples are included after each definition. A **monosaccharide** is a basic unit of a carbohydrate with the general formula $C_xH_{2x}O_x$ ($x \geq 3$) that cannot be further hydrolyzed. They can be linear or cyclic. Some examples include glucose, fructose, galactose, and ribose. A **disaccharide** is comprised of two monosaccharide molecules that link together via a condensation reaction (loss of H_2O). Some examples include lactose, sucrose, and maltose. **Oligosaccharides** are a chain of 3–10 simple sugar molecules linked together (small polymer). Some examples include fructo-oligosaccharides (vegetables), glycoproteins (cell recognition) and mannan oligosaccharides (animal feed). **Polysaccharides** are long carbohydrates composed of monosaccharides linked together through covalent (glycosidic) bonds. They can be branched or linear. Some examples include cellulose (plants) and chitin (shells in turtles and crabs).

18.27

The Intergovernmental Panel on Climate Change has said that methane has 72 times the global warming potential of an equal mass of CO_2 over 20-year period and 25 times that of CO_2 over a 100-year period. Micro-organisms in some animals can produce 100–500 L of methane per day (vs. < 1 L for humans). To counter this, there are some changes that could reduce emissions: changing their diet to include more grains or legumes that are easier to break down, enzymes, controlling methanogens in stomachs, and introducing new micro-organisms to aid in oligosaccharide digestion.

Section 18.2: The Concept of Reaction Rate

18.29

$$N_2(g) + 3H_2(g) \longrightarrow 2NH_3(g)$$
If $-\Delta[H_2]/\Delta t = 4.5 \times 10^{-4}$ mol L^{-1} min^{-1}, $\Delta[NH_3]/\Delta t = \frac{2}{3} \times 4.5 \times 10^{-4}$ mol L^{-1} min^{-1}
$$= 3.0 \times 10^{-4} \text{ mol L}^{-1} \text{ min}^{-1}$$

Section 18.3: Conditions That Affect the Rate of a Reaction

18.31

$$\text{Rate} = k[A]^2[B]$$
The reaction is second order in [A] and first order in [B]. The total order of the reaction is three.

Section 18.4: Dependence of Rate on Reactant Concentration

18.33

There is not enough information provided to determine the actual order of a reaction (it can only be determined experimentally).

18.37

$\text{Rate} = k[A]^2$
The rate increases by a factor of $9 = 3^2$ if [A] is tripled.
The rate decreases by a factor of $\frac{1}{4} = \frac{1}{2}^2$ if [A] is halved.

18.39

$$CO(g) + NO_2(g) \longrightarrow CO_2(g) + NO(g)$$

(a) When [NO₂] is doubled with [CO] fixed, the rate increases by the factor $3.4 \times 10^{-8} / 1.7 \times 10^{-8} = 2$.
The reaction is first order in [NO₂].
When [CO] is doubled, with [NO₂] fixed, the rate increases by the factor $6.8 \times 10^{-8} / 3.4 \times 10^{-8} = 2$.
The reaction is first order in [CO].
$\text{Rate} = k \, [CO][NO_2]$

(b) Using the third data set,
$k = 6.8 \times 10^{-8} \text{ mol L}^{-1} \text{ h}^{-1} / (1.0 \times 10^{-3} \text{ mol L}^{-1})(0.36 \times 10^{-4} \text{ mol L}^{-1}) = 1.9 \text{ L mol}^{-1} \text{ h}^{-1}$

(c) The initial rate in experiment 4 is
$\text{Rate} = 1.9 \text{ L mol}^{-1} \text{ h}^{-1} \times (1.5 \times 10^{-3} \text{ mol L}^{-1})(0.72 \times 10^{-4} \text{ mol L}^{-1}) = 2.0 \times 10^{-7} \text{ mol L}^{-1} \text{ h}^{-1}$

18.41

(a) Doubling [NO], keeping [O₂] fixed (Exp 2 vs. 1), increases the rate of reaction by the factor, $4 = 2^2$. The reaction is second order with respect to NO(g) concentration.
Doubling [O₂], keeping [NO] fixed, increases the rate of reaction by the factor, $2 = 2^1$. The reaction is first order with respect to O₂(g) concentration.
$\text{Rate} = k \, [NO]^2[O_2]$

(b) Using the third data set,
$k = 1.7 \times 10^{-8} \text{ mol L}^{-1} \text{ h}^{-1} / (1.8 \times 10^{-4} \text{ mol L}^{-1})^2 (1.04 \times 10^{-2} \text{ mol L}^{-1}) = 50 \text{ L}^2 \text{ mol}^{-2} \text{ h}^{-1}$

(c) For the fourth experiment,
$\text{Rate} = 50 \text{ L}^2 \text{ mol}^{-2} \text{ h}^{-1} \times (1.8 \times 10^{-4} \text{ mol L}^{-1})^2 (5.2 \times 10^{-2} \text{ mol L}^{-1}) = 8.4 \times 10^{-8} \text{ mol L}^{-1} \text{ h}^{-1}$
We could get this answer directly by noting that experiment 4 is like experiment 3 except that the oxygen concentration is increased by a factor of 5. So, we just have to scale the experiment 3 rate by 5, since the reaction is first order with respect to [O₂].

Section 18.5: Concentration-Time Relationships: Integrated Rate Equations

18.43

$$\ln \frac{[\text{sucrose}]_{2.57\,h}}{[\text{sucrose}]_0} = -kt = -k\,(2.57\text{ h})$$

$$k = -\frac{1}{2.57\text{ h}}\ln \frac{0.0132}{0.0146} = 0.0392\text{ h}^{-1}$$

18.47

$$\ln \frac{[\text{HCOOH}]_t}{[\text{HCOOH}]_0} = -kt = -(2.4 \times 10^{-3}\text{ s}^{-1})\,t$$

$$t = -\frac{1}{2.4 \times 10^{-3}\text{ s}^{-1}}\ln \frac{3}{4} = 120\text{ s}$$

18.49

(a)

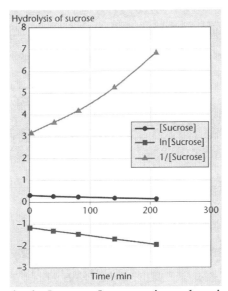

Close examination reveals that the ln [sucrose] versus time plot gives the straightest line. The reaction is first order.

(b) Rate $= k\,[\text{sucrose}]$

$$\ln \frac{[\text{sucrose}]_{210\,\text{min}}}{[\text{sucrose}]_0} = -kt = -k\,(210\text{ min})$$

$$k = -\frac{1}{210\text{ min}}\ln \frac{0.146}{0.316} = 0.00368\text{ min}^{-1}$$

(c)

$$\ln \frac{[\text{sucrose}]_{175\ \text{min}}}{[\text{sucrose}]_0} = -kt = -(0.00368\ \text{min}^{-1})\,(175\ \text{min}) = -0.644$$

$$\frac{[\text{sucrose}]_{175\ \text{min}}}{0.316\ \text{mol L}^{-1}} = 0.525$$

$$[\text{sucrose}]_{175\ \text{min}} = 0.525 \times (0.316\ \text{mol L}^{-1}) = 0.166\ \text{mol L}^{-1}$$

18.51

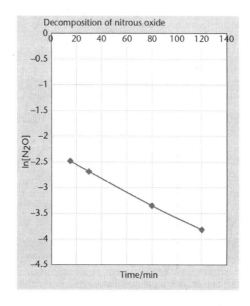

The ln[N$_2$O] plot is well fitted by a straight line. The slope is
$$(-4.2 - (-2.3)) / (105\ \text{min}) = -0.018\ \text{min}^{-1}$$
$$k = 0.018\ \text{min}^{-1}$$
The decomposition rate at 900 °C when [N$_2$O] = 0.035 mol L^{-1} is
$$\text{Rate} = -\Delta[\text{N}_2\text{O}] / \Delta t = k\,[\text{N}_2\text{O}] = 0.018\ \text{min}^{-1} \times 0.035\ \text{mol L}^{-1} = 0.00063\ \text{mol L}^{-1}\ \text{min}^{-1}$$

18.55

$$\ln \frac{m(\text{azomethane})_{0.0500\ \text{min}}}{m(\text{azomethane})_0} = -kt = -(40.8\ \text{min}^{-1})\,(0.0500\ \text{min}) = -2.04$$

$$\frac{m(\text{azomethane})_{0.0500\ \text{min}}}{2.00\ \text{g}} = 0.130$$

$$m(\text{azomethane})_{0.0500\ \text{min}} = 0.130 \times (2.00\ \text{g}) = 0.260\ \text{g}$$

Note that for first order reactions we can substitute any measure of the amount of a substance in place of the concentration. Here, we use mass.
Mass of azomethane decomposed = (2.00 − 0.260) g = 1.74 g.
Amount of azomethane decomposed = 1.74 g / 58.09 g mol^{-1} = 0.0300 mol
Amount of N$_2$ formed = amount of azomethane decomposed = 0.0300 mol (i.e., 0.84 g)

Section 18.6: A Microscopic View of Reaction Rates: Collision Theory

18.57

$$\ln \frac{1.5 \times 10^{-3} \text{ s}^{-1}}{3.46 \times 10^{-5} \text{ s}^{-1}} = \frac{E_a}{8.314 \text{ J K}^{-1} \text{ mol}^{-1}} \left(\frac{1}{298 \text{ K}} - \frac{1}{328 \text{ K}} \right)$$

$$E_a = 8.314 \text{ J K}^{-1} \text{ mol}^{-1} \times 3.77 \times \left(3.07 \times 10^{-4} \right)^{-1}$$

$$= 1.02 \times 10^5 \text{ J mol}^{-1} = 102 \text{ kJ mol}^{-1}$$

18.59

$$\ln \frac{k_{850K}}{k_{800K}} = \frac{E_a}{8.314 \text{ J K}^{-1} \text{ mol}^{-1}} \left(\frac{1}{800 \text{ K}} - \frac{1}{850 \text{ K}} \right) = \frac{260 \times 10^3 \text{ J mol}^{-1}}{8.314 \text{ J K}^{-1} \text{ mol}^{-1}} \left(7.35 \times 10^{-5} \right)$$

$$\ln \frac{k_{850K}}{0.0315 \text{ s}^{-1}} = 2.30$$

$$k_{850K} = 0.0315 \text{ s}^{-1} \times e^{2.30} = 0.314 \text{ s}^{-1}$$

Section 18.7: Reaction Mechanisms

18.61

Elementary reactions
(a) $NO(g) + NO_3(g) \rightarrow 2\, NO_2(g)$
   Rate $= k\,[NO][NO_3]$
(b) $Cl(g) + H_2(g) \rightarrow HCl(g) + H(g)$
   Rate $= k\,[Cl][H_2]$
(c) $(CH_3)_3CBr(aq) \rightarrow (CH_3)_3C^+(aq) + Br^-(aq)$
   Rate $= k\,[(CH_3)_3CBr]$

18.63

(a)
$$NO_2(g) + NO_2(g) \longrightarrow NO(g) + NO_3(g)$$
$$\underline{NO_3(g) + CO(g) \longrightarrow NO_2(g) + CO_2(g)}$$
$$NO_2(g) + CO(g) \longrightarrow NO(g) + CO_2(g)$$
(b) The molecularity of both steps is 2. These are bimolecular processes.
(c) If the first step is rate limiting, the rate equation is
   Rate $= k\,[NO_2]^2$
(d) $NO_3(g)$ is an intermediate.

18.65

(a)

Step 1   Fast, endothermic $CH_3OH + H^+ \rightleftharpoons CH_3OH_2^+$

Step 2   Slow $CH_3OH_2^+ + Br^- \longrightarrow CH_3Br + H_2O$

$CH_3OH + H^+ + Br^- \longrightarrow CH_3Br + H_2O \text{ (overall)}$

(b)

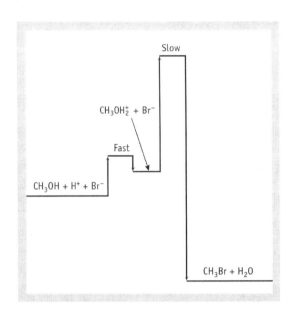

(c)

The rate is determined by the slow reaction.

Rate $= k_2 \, [CH_3OH_2^+] \, [Br^-]$

The $CH_3OH_2^+$ concentration is determined by the equilibrium represented in step 1.

$[CH_3OH_2^+] = K_1 \, [CH_3OH][H^+]$

So,

Rate $= k_2 \, K_1 \, [CH_3OH][H^+][Br^-]$

i.e.,

$-\Delta \, [CH_3OH] / \Delta t = k \, [CH_3OH][H^+][Br^-]$

18.69

(a)

Step 1   Slow $(CH_3)_3 CBr \longrightarrow (CH_3)_3 C^+ + Br^-$

Step 2   Fast $(CH_3)_3 C^+ + H_2O \longrightarrow (CH_3)_3 COH_2^+$

Step 3   Fast $(CH_3)_3 COH_2^+ + Br^- \longrightarrow (CH_3)_3 COH + HBr$

$(CH_3)_3 CBr + H_2O \longrightarrow (CH_3)_3 COH + HBr$

(b) Step 1 is rate-determining.

(c) The expected rate equation is   Rate $= k \, [(CH_3)_3CBr]$

Section 18.8: Nucleophilic Substitution Reactions

18.71

(a) $(CH_3)_3CBr + CH_3OH \rightarrow (CH_3)_3COCH_3 + HBr$

(b)

(c) Rate = $k[(CH_3)_3CBr]$

   (i) If the concentration of $(CH_3)_3CBr$ is doubled and that of CH_3OH is halved, the rate increases 2-fold.

   (ii) If the concentration of $(CH_3)_3CBr$ is halved and that of CH_3OH is doubled, the rate decreases 2-fold—i.e., it is half as big.

   (iii) If the concentrations of both $(CH_3)_3CBr$ and CH_3OH are tripled, the rate increases 3-fold.

   (iv) If the reaction temperature is lowered, the rate of reaction decreases—there is less thermal energy available allowing the alkyl halide to overcome the barrier to dissociation.

   (v) The reaction rate will decrease.

18.73

(a) The overall reaction is exothermic as the products are at a lower energy than the reactants.

(b) There are two intermediates and three transition states.

(c) There are three steps in this reaction. The second step is the rate-determining step as it has the highest energy barrier (activation energy – E_A).

(d) The fastest step in the reaction is the third step.

(e) The activation energy for the overall reaction would be drawn with an arrow from A straight down level with D.

SUMMARY AND CONCEPTUAL QUESTIONS

18.75

There are many possible responses to this question. A Canadian example is the Grunwald-Winstein equation (1948, physical organic chemistry) proposed by Ernest Grunwald and Saul Winstein (Canadian-American born in Montreal, Quebec). No, I cannot find another equation named after a woman.

18.77

Finely divided rhodium is a much more efficient catalyst than a small block of the metal, because the former has a much greater total surface area. The catalytic processes take place on the surface of the metal.

18.79

$$Cl + O_3 \longrightarrow ClO + O_2$$
$$ClO + O \longrightarrow Cl + O_2$$
$$O_3 \longrightarrow O + O_2$$
$$\overline{2\,O_3 \longrightarrow 3\,O_2}$$

Key observations:
Ozone is converted to oxygen.
Cl atoms are regenerated—they act catalytically. Many ozone molecules are converted to oxygen for every chlorine atom. The atom's catalytic ability is limited only by rare termolecular processes wherein chlorine atoms form more stable species—e.g., 2 Cl atoms combining to form Cl_2. A third molecule is required in this process to take away the excess energy. Two chlorine atoms cannot form a stable molecule without releasing energy. ClO is a reaction intermediate—its name is chlorine monoxide.

18.81

(a) Reactions are faster at a higher temperature because activation energies are lower. **FALSE**
Reactions are faster at a higher temperature because there are more frequent, more energetic collisions, more frequently leading to products.

(b) Rates increase with increasing concentration of reactants because there are more collisions between reactant molecules. **TRUE**
However, sometimes rates do not change with increasing concentration of reactant. This is the case of a zero order reaction. Zero order reactions eventually become first (or other) order when reactants are depleted.

(c) At higher temperatures a larger fraction of molecules have enough energy to get over the activation energy barrier. **TRUE**

(d) Catalyzed and uncatalyzed reactions occur by the same mechanism. **FALSE**
Catalyzed and uncatalyzed reactions occur by different mechanisms. The catalyst provides a different pathway with lower activation energy than the uncatalyzed pathway.

CHAPTER 19
Understanding Structure, Understanding Reactivity: Alkenes, Alkynes, and Aromatics

IN-CHAPTER EXERCISES

Exercise 19.1—Classifying Functional Groups by Level

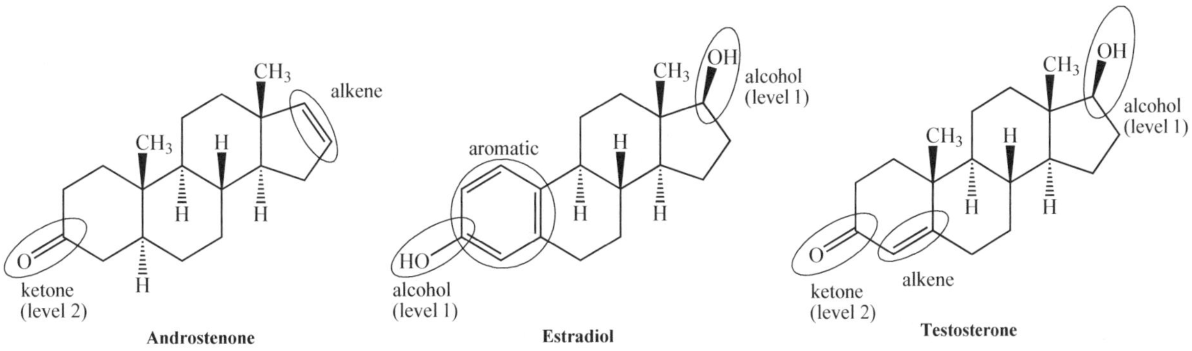

Androstenone   Estradiol   Testosterone

Exercise 19.3—Classifying Organic Reactions

(a) $CH_3Br + {}^-OH \rightarrow CH_3OH + {}^-Br$

Substitution. The bromo- group in the alkyl halide reactant and the hydroxyl group in the alcohol product are both level 1 functional groups.

(b) $CH_3CH_2OH \rightarrow H_2C=CH_2 + H_2O$

Elimination. The hydroxyl group in the alcohol reactant is a level 1 functional group, while the alkene in the product has no assigned level.

(c) $H_2C=CH_2 + H_2 \rightarrow CH_3CH_3$

Addition. The alkene functional group in the reactant has no assigned level, while the product lacks any functional groups.

Exercise 19.5—Recognizing Electrophiles and Nucleophiles

(a) NH_4^+

Ammonium is an electrophile. It is attracted to concentrations of negative charge, and will donate a proton to proton acceptors—i.e., nucleophiles.

(b) CN^-

Cyanide is a nucleophile. It is attracted to concentrations of positive charge (i.e., regions of depleted electron density) such as acidic protons and carbonyl carbon atoms.

(c) Br⁻

Bromide is a nucleophile. It is attracted to concentrations of positive charge (i.e., regions of depleted electron density).

(d) CH₃NH₂

The amine N atom is a nucleophile. It is attracted to concentrations of positive charge (i.e., regions of depleted electron density).

(e) H–C≡C–H

Ethyne is a nucleophile. It has a relatively loosely held cloud of π electrons that electrophiles are attracted to—and vice versa.

Exercise 19.7—Using Electrostatic Potential Maps

Lewis acid
(electron deficient)

Lewis base
(electron rich)

Exercise 19.9— IUPAC Names

(a)   (b)   (c)

(a) 1,3 is meta   **(b)** 1,4 is para   **(c)** 1,2 is ortho

Exercise 19.11—IR Spectra of Alkynes

The peak just above 3300 cm⁻¹ looks like an alkyne C–H stretch, while the peak just above 2100 cm⁻¹ looks like an alkyne C≡C stretch. There also appear to be alkyl C–H stretch peaks just below 3000 cm⁻¹. The compound likely has alkyl H atoms in addition to the alkynyl H atom.

Exercise 19.13—Using IR and NMR to Determine the Structure of a Compound

The ¹³C NMR spectrum shows three C atoms, only one of them a downfield alkene C. Because we know there are 6 C atoms, the molecule must be symmetrical, with three pairs of indistinguishable carbons, including a pair of equivalent alkene carbon atoms. The IR spectrum shows alkyl C–H stretches, a C=C stretch, and C–H stretch just above 3000 cm⁻¹, consistent with an alkene. Note that the formula, C₆H₁₀, corresponds to two degrees of unsaturation—i.e., a total of two double bonds plus rings. The ¹H NMR spectrum shows three types of H atoms in the ratio 1:2:2. This corresponds to 2, 4, and 4 H atoms of each type. The 2 equivalent (one cluster of peaks) downfield H atoms have chemical shift around 5.8 ppm, consistent with two equivalent vinyl H atoms. The equivalence of 4 H atoms corresponds to two pairs of CH_2 type H atoms equivalent by symmetry. These observations are consistent with cyclohexene.

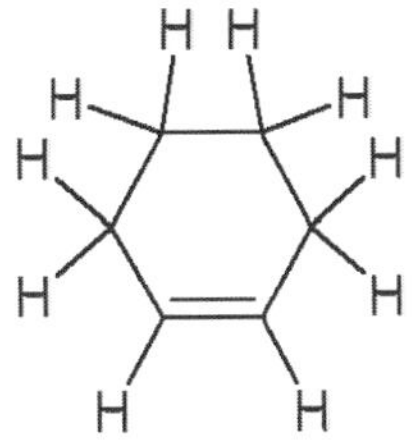

Other structures involving two C=C double bonds are ruled out because they would produce at least two inequivalent alkene carbon atoms.

Exercise 19.15—Assigning *E, Z* Configurations

(a) –H or –Br
–Br has higher priority
(b) –Cl or –Br
–Br has higher priority
(c) –CH_3 or –CH_2CH_3
–CH_2CH_3 has higher priority
(d) –NH_2 or –OH
–OH has higher priority
(e) –CH_2OH or –CH_3
–CH_2OH has higher priority
(f) –CH_2OH or –CH=O
–CH=O has higher priority

Exercise 19.19—Addition of HX to alkenes

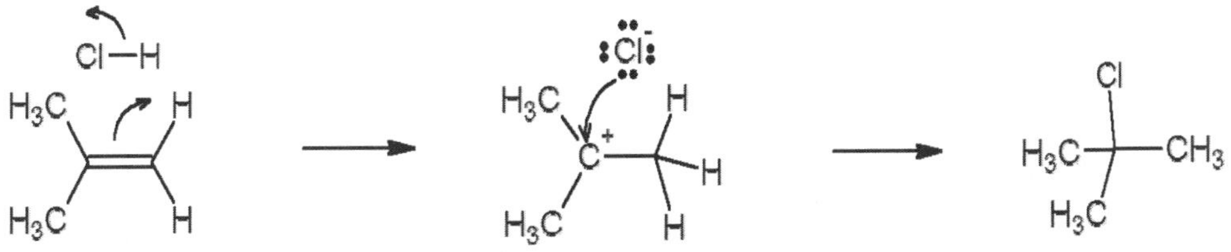

Exercise 19.21—Addition of HX to Alkenes

(a) Bromocyclopentane can be prepared from cyclopentene and HBr.
(b) 3-Bromohexane can be prepared from hex-2-ene and HBr.
(c) 1-Iodo-1-isopropylcyclohexane can be prepared from:

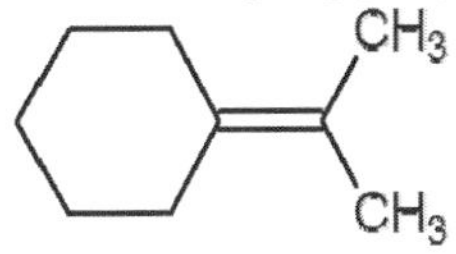

(d)

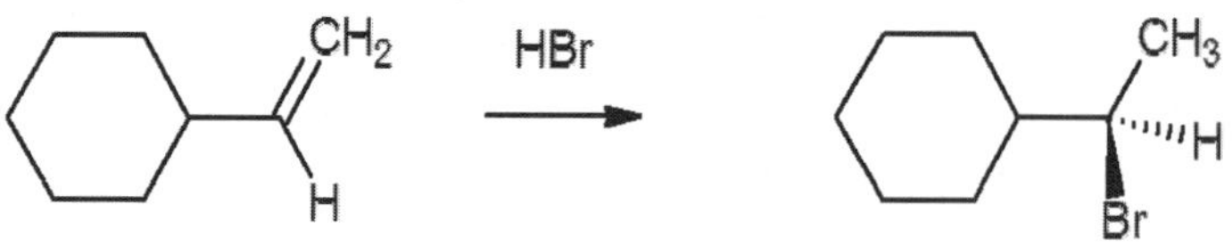

Exercise 19.23—Addition of H$_2$O to Alkenes

(a)

Step 1

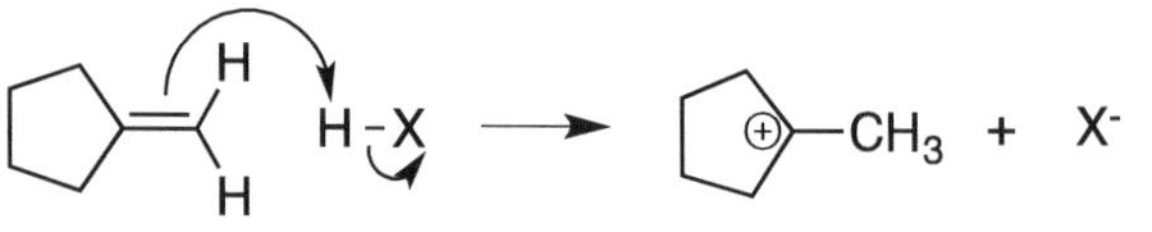

Step 2

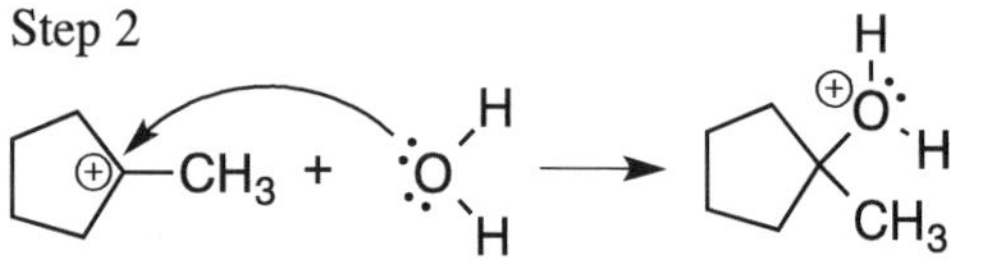

Step 3

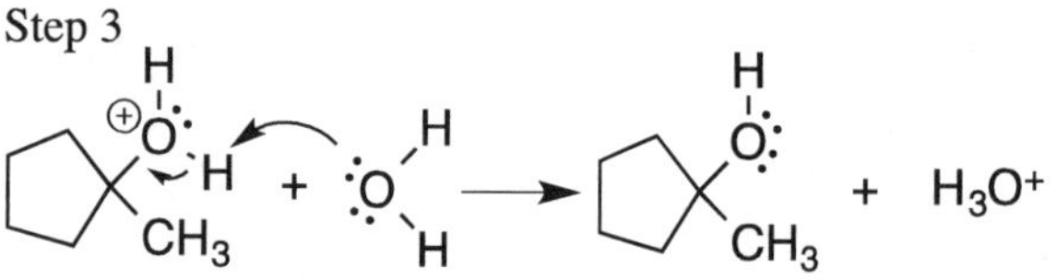

(b) The other possible product would be:

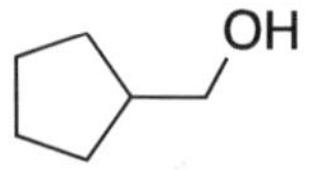

This is the product that results if the carbocation is formed on the exocylic carbon in the first step, rather than the ring carbon. The amount of this product formed, however, would be extremely small because this alternate carbocation is primary. A tertiary carbocation (formed in the mechanism shown in (a)) is much, much more stable than a primary carbocation. This alternate reaction mechanism is therefore much less favourable than the mechanism shown in (a).

Exercise 19.25—Addition of H$_2$O to Alkenes

(a) Butan-2-ol can be made by addition of water to but-1-ene or but-2-ene.
(b) 3-Methylpentan-3-ol can be made by addition of water to 3-methylpent-2-ene or 2-ethylbut-1-ene.
(c) 1,2-Dimethylcyclohexanol can be made by addition of water to 1,2-dimethylcyclohexene or 1,6-dimethylcyclohexene.

Exercise 19.27—Oxidation of Alkenes

(a) $(CH_3)_2C=O + CO_2$ is the product upon treating $(CH_3)_2C=CH_2$ with acidic $KMnO_4$.
(b) 2 mol $CH_3CH_2CO_2H$ is the product upon treating one mol of $CH_3CH_2CH=CHCH_2CH_3$ with acidic $KMnO_4$.

Exercise 19.29—Addition of H₂O to Alkynes

Exercise 19.31—Hückel's Rule

(a) Cyclobutadiene is monocyclic, planar and fully conjugated. However, there are $4 = 4 \times 1$ $\neq 4n + 2$ π electrons. Cyclobutadiene is anti-aromatic.

(b) Cyclohepta-1,3,5-triene is monocyclic and planar, but not fully conjugated—there is a methylene group separating the two outer double bonds.

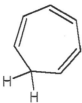

(c) Cyclopenta-1,3-diene is monocyclic and planar, but not fully conjugated—there is a methylene group separating the double bonds on one side.

(d) Cyclooctatetraene is monocyclic, planar and fully conjugated. However, there are $8 = 4 \times 2$ $\neq 4n + 2$ π electrons. Cyclooctatetraene is anti-aromatic.

Exercise 19.33—IR Spectra of Aromatic Compounds

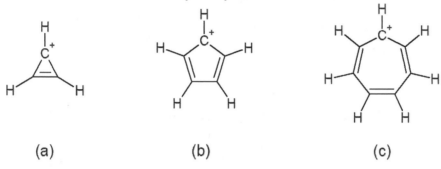

(a) There is an aryl C–H stretching band with three peaks at 3100, 3070, and 3030 cm^{-1}. There is an alkyl C–H stretching band with principal peaks at 2925 and 2850 cm^{-1}.

(b) The out-of-plane C–H bending vibrations give rise to two peaks at 740 and 700 cm^{-1}.

(c) The in-plane C–H bending vibrations produce peaks at 1090 and 1030 cm^{-1}. The aryl C–C stretches produce a band with principal peaks at 1610, 1510, and 1470 cm^{-1}.

Exercise 19.35—Aromaticity of Cyclic Ions

(a)	(b)	(c)

(a) $C_3H_3^+$ has $2 = 4 \times 0 + 2$ π electrons. It is aromatic.

(b) $C_5H_5^+$ has $4 = 4 \times 1$ π electrons. It is anti-aromatic.

(c) $C_7H_7^+$ has $6 = 4 \times 1 + 2$ π electrons. It is aromatic.

Exercise 19.37—Electrophilic Aromatic Bromination

o-bromotoluene, *m*-bromotoluene, and *p*-bromotoluene, respectively

REVIEW QUESTIONS

Section 19.1: Making Scents of the Mountain Pine Beetle

19.39

Forests serve as an important carbon sink that help to sequester CO_2 and thereby counter climate changes caused by the greenhouse effect.

19.41

As trees are killed by the pine beetles, these trees can no longer sequester CO_2 from the atmosphere. Additionally, when the dead trees decay, they release their previously trapped CO_2 back to the atmosphere.

19.43

Possible examples:

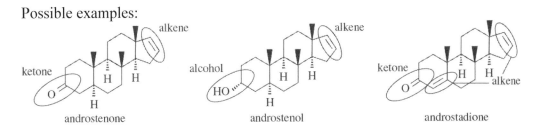

androstenone   androstenol   androstadione

Section 19.2: Overview of Structure and Reactivity of Carbon Compounds

19.45

Biomimicry: imitation of biological structures, processes or phenomena for human benefit. Mountain pine beetle pheromones are exploited by humans to influence the behaviour of this pest.

19.47

Answer depends on examples chosen for 19.42. Some examples:
1-octanol: IR: O–H stretch (~3300–3500 cm^{-1}); ^{13}C NMR: In addition to the aliphatic peaks, there would be a peak at ~40–90 ppm for C–OH.
Ethyl oleate: IR: ester C=O stretch (~1730–1750 cm^{-1}), two C–O bands (1000–1300 cm^{-1}); ^{13}C NMR: In addition to the many aliphatic carbon signals, there would be a peak at ~160–180 ppm for the C=O and two peaks between ~130–160 ppm for the alkene carbons.

19.49

Answer depends on examples chosen for 19.43. The distinguishing features in the IR and the 13C NMR spectra for the three compounds provided in the solution for 19.43 are provided below. Note that because all three compounds contain alkenes, the IR range corresponding to the alkene C–H (~3000–3100 cm^{-1}) would not be especially diagnostic.
Androstenone: IR: ketone C=O stretch (1705–1725 cm^{-1}); ^{13}C NMR: C=O ~190–220 ppm, alkene ~130-160 ppm (two peaks)
Androstenol: IR: O–H stretch (3200–3550 cm^{-1}); ^{13}C NMR: alkene ~130–160 ppm (two peaks), C–OH ~40–90 ppm
Androstadione: IR: conjugated ketone C=O stretch (1680-1700 cm^{-1}); ^{13}C NMR: ketone C=O stretch (1705–1725 cm^{-1}), alkene ~130–160 ppm (four peaks)

19.51

(a) Cyclohexanol to cyclohexanone: Level 1 (alcohol) to Level 2 (ketone)
(b) An oxidizing agent must be used as the oxidation state changes from +1 for the alcohol to +2 for the ketone.
(c) This is an elimination reaction (elimination of H_2).

19.53

A reaction equation gives information about the identities of reactants and products and the stoichiometric mol ratios) involved in the reaction.
A reaction mechanism shows the specific sequence of steps that leads from reactant(s) to product(s), including all intermediates.

19.55

Higher electron density (red) = greater ability to act as a nucleophile; lower electron density (blue) = greater ability to act as an electrophile

19.57

Polar reaction: involves separation of charges, electrons redistributed in pairs. Example: nucleophilic substitution.
Free radical reaction: generally no separation of charges, electrons redistributed individually. Example: radical halogenation of alkanes

19.59

(a)

$CH_3CH_2C \equiv N$ nitrile

(b)

OCH₃ ether

(c)

ketone

$CH_3CCH_2COCH_3$ ester

(d)

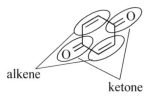

alkene

ketone

(e)

alkene   amide

NH₂

(f)

O

H aldehyde

19.61

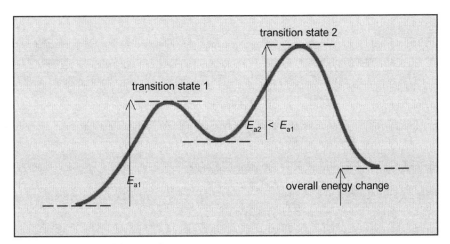

Reaction progress $\longrightarrow$

(a) The overall energy change for the reaction is positive.

(b) There are two steps involved in the reaction.

(c) The second step is slightly faster.

(d) There are two transition states.

Section 19.3: Alkenes, Alkynes, and Aromatic Compounds

19.63

(a) 3-propylhept-2-ene

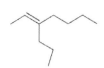

plus Z isomer – above is the E isomer

(b) 2,4-dimethylhex-2-ene

(c) octa-1,5-diene

or

plus *cis* or Z isomer – above is the E isomer

(d) 4-methylpenta-1,3-diene

(e) *cis*-4,4-Dimethylhex-2-ene

(f) (*E*)-3-methylhept-3-ene

19.65

The compound with formula C_9H_{14} has $(20 - 14)/2 = 3$ degrees of unsaturation. Since it has no rings or triple bonds, it must have 3 double bonds.

19.67

(a) 1,2-dimethylcyclohexene
(b) 4,4-dimethylcycloheptene

(c) 3-isopropylcyclopentene or 3-(1-methylethyl)-cyclopentene

19.69

There are four compounds with the formula C_7H_7Cl:

chloromethylbenzene:

1-chloro-3-methylbenzene:

1-chloro-2-methylbenzene:

1-chloro-4-methylbenzene:

19.71

(a) *m*-bromophenol

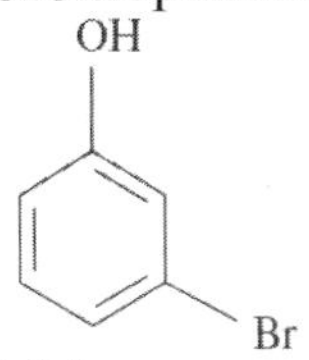

(b) 1,3,5-benzenetriol

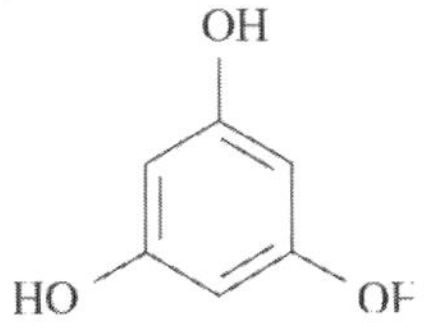

(c) *p*-iodonitrobenzene

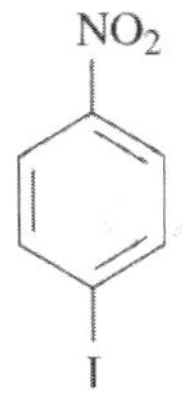

(d) 2,4,6-trinitrotoluene (TNT)

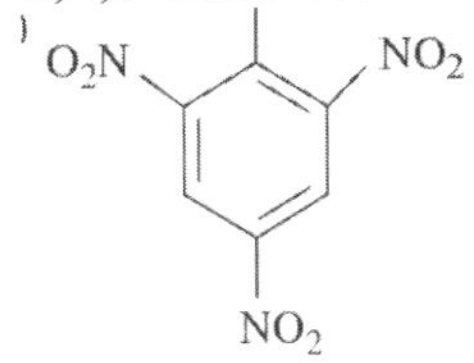

(e) *o*-aminobenzoic acid

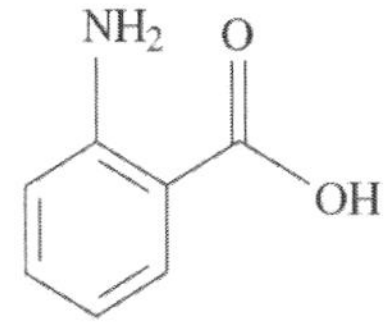

(f) 3-methyl-2-phenylhexane

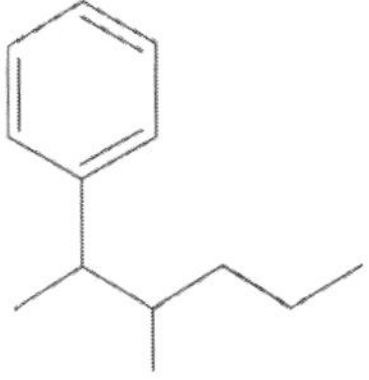

Section 19.4: Structure and Reactivity of Alkenes

19.73

(a) aromatic carbons: sp^2; carbonyl carbon: sp^2; CH_2/CH_3: sp^3
(b) CH_2: sp^3; $CH(OH)$: sp^3; $CH(ring)$: sp^3; $C=C$: sp^2; $C=O$: sp^2

19.75

Methylcarboxylate, $-COOCH_3$, has higher prioroty than carboxyl, $-COOH$.

19.77

(a) *Z*. Although the methyl groups are on opposite sides, the higher priority group on the left and on the right are on the same side.
(b) *Z*.

19.79

$CH_3CH_2CH=CHCH_3 + HCl \rightarrow CH_3CH_2CHClCH_2CH_3$ or $CH_3CH_2CH_2CHClCH_3$

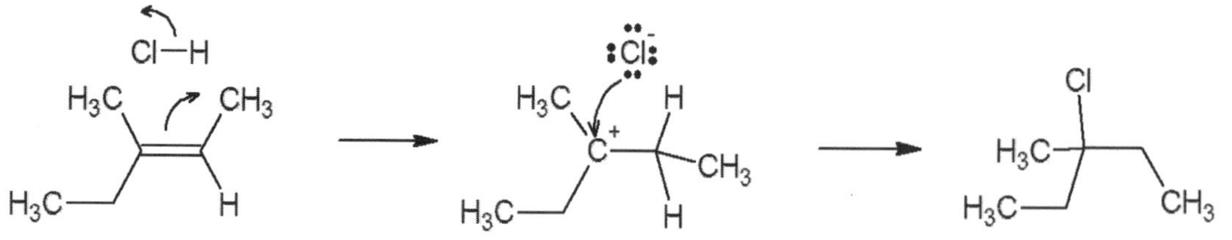

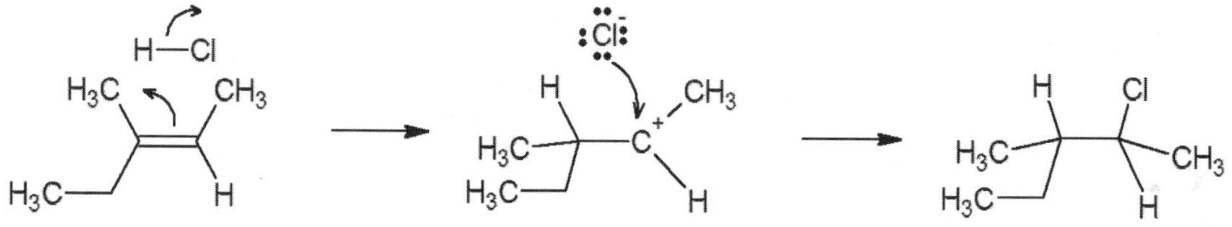

19.81

(a)

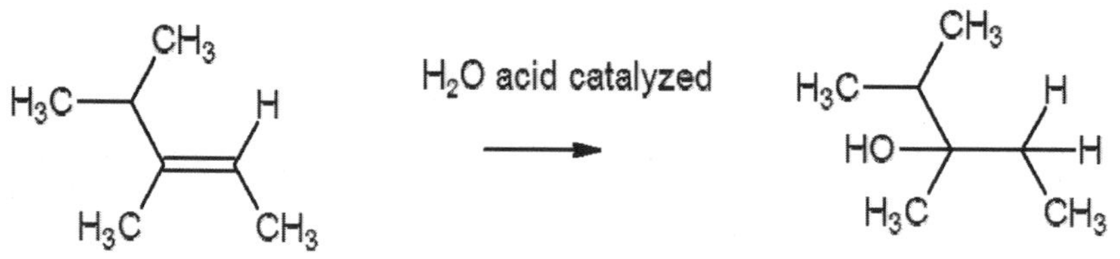

(b)

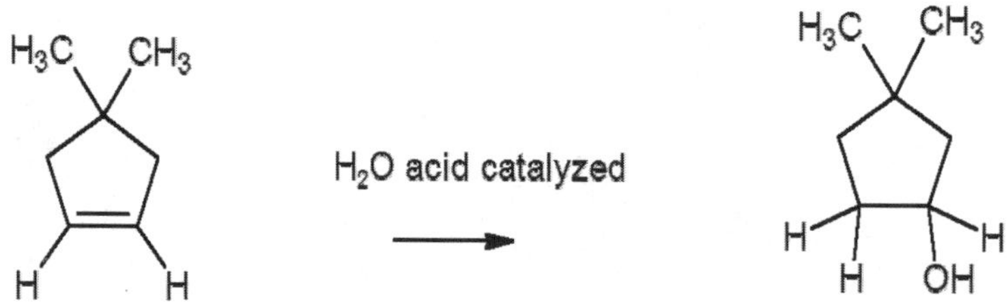

19.83

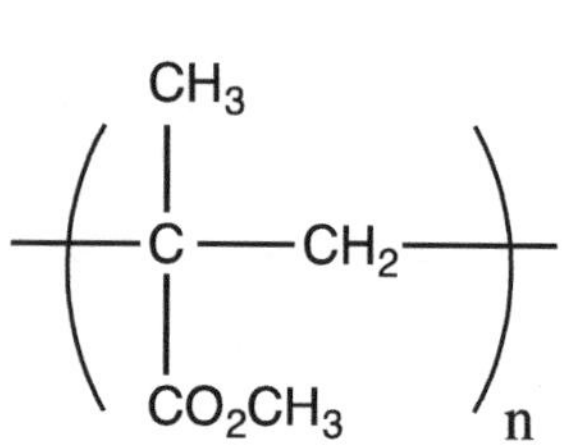

Section 19.5: Structure and Reactivity of Alkynes

19.85

The IR spectra of hex-1-yne and hex-2-yne are distinguished by the alkynyl C–H stretch seen only in the spectrum (a peak around 3300 cm^{-1}) of the terminal alkyne.

19.87

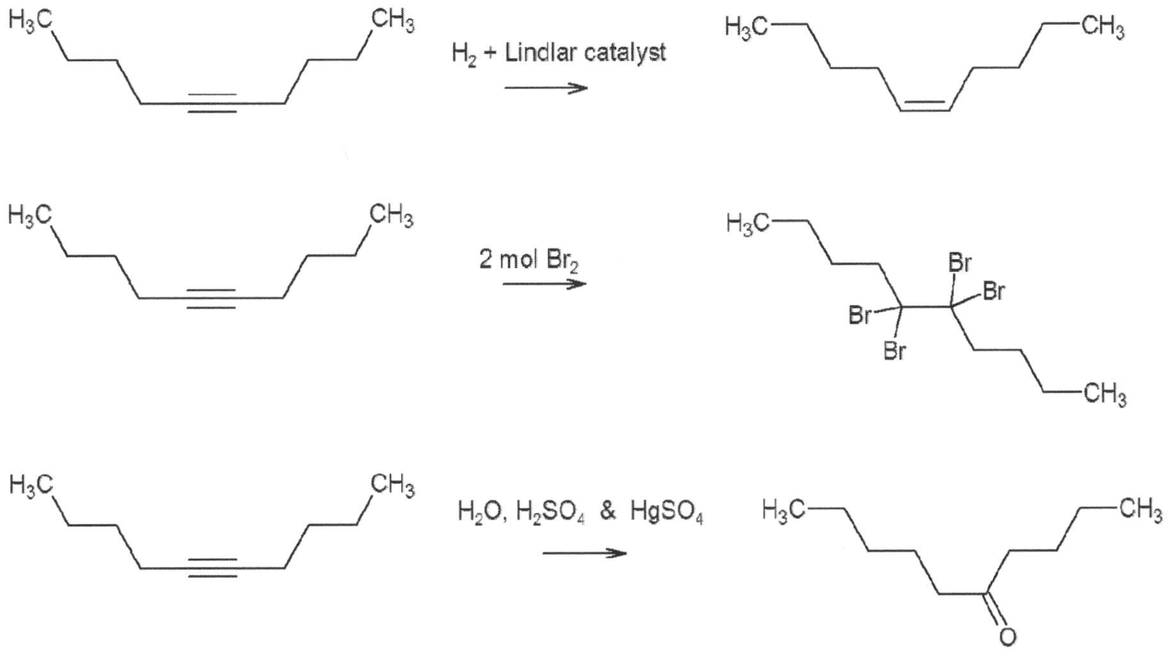

19.89

To get 4-methylpentan-2-one, you would hydrate 4-methylpent-1yne,

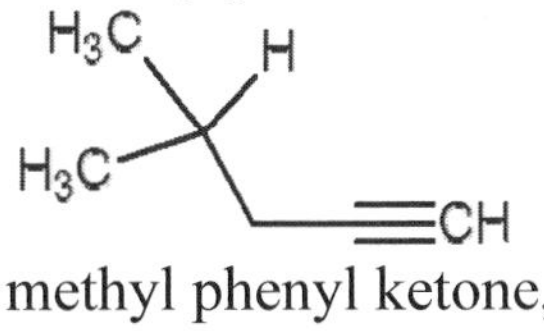

To get methyl phenyl ketone, you would hydrate phenylethyne,

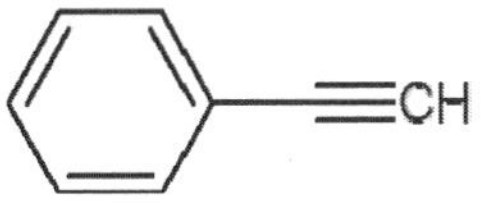

Section 19.6: Structure and Reactivity of Aromatic Compounds

19.91

The four resonance structures of anthracene

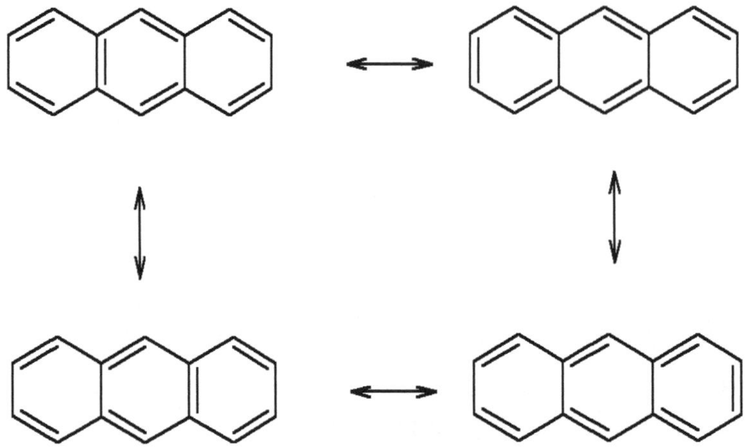

19.93

There are three non-equivalent H's in anthracene—the central vertical pair, and two pairs (related via reflection through the vertical axis) of vertical pairs of H's. The chemical shifts are expected to be in the 6.5 to 8 ppm range typical of aromatics.

19.95

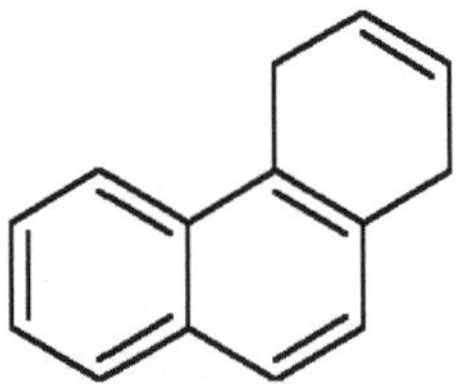

Phenanthrene has $14 = 4 \times 3 + 2$ π electrons. Although it is not monocyclic, it certainly qualifies as aromatic. The outer ring of atoms provides the principal path over which $4 \times 3 + 2$ electrons are delocalized in accord with Hückel's rule. Consequently, the term "aromatic" applies equally to polycyclic systems.

19.97

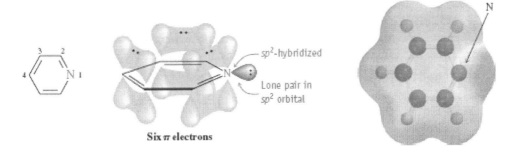

19.99

Pyridine has $6 = 4 \times 1 + 2$ π electrons, just like benzene. Like benzene, it has special stability in accord with Hückel's rule. Note that the lone pair of electrons on the N is not in the π system—they are in an sp^2 orbital. The orbital depiction of this special stability is just that of benzene. In benzene, there is a filled two-orbital shell—i.e., the electrons are optimally packed for interaction with the nuclei. Pyridine is similar except it lacks the six-fold symmetry of benzene and the HOMO and LUMO are not quite degenerate.

anti-bonding orbitals

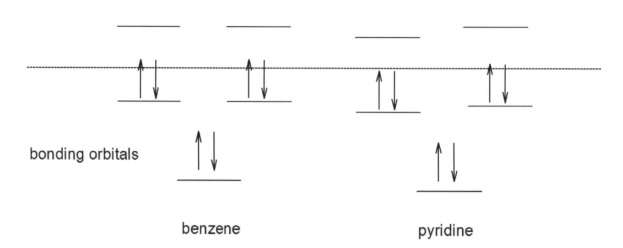

bonding orbitals

benzene   pyridine

There are six π electrons delocalized in pyridine. The lone pair electrons of the nitrogen atom are not part of the π system.

19.101

Cyclononatetraenyl radical has 9 C's each contributing an electron in a *p* orbital to make the π system. It is not aromatic. If we subtract an electron to get the cation, there are $8 = 4 \times 2$ π electrons. The cation is anti-aromatic. If we add an electron to get the anion, there are $10 = 4 \times 2 + 2$ π electrons. The anion is aromatic. It is expected to be most stable.

19.103

1)

$$H-\overset{\overset{\displaystyle H}{|}}{\underset{\underset{\displaystyle H}{|}}{C}}-\ddot{\underset{\cdot\cdot}{C}l}: + AlCl_3 \longrightarrow H-\overset{\overset{\displaystyle H}{|}}{\underset{\underset{\displaystyle H}{|}}{C}}-\overset{+}{\underset{\cdot\cdot}{\ddot{C}l}}-\overset{-}{AlCl_3}$$

2)

$$H-\overset{\overset{\displaystyle H}{|}}{\underset{\underset{\displaystyle H}{|}}{C}}-\overset{+}{\ddot{C}l}-\overset{-}{AlCl_3} + \bigcirc \longrightarrow \quad + \ddot{\underset{\cdot\cdot}{C}l}-\overset{-}{AlCl_3}$$

3)

$$H_3C \quad \bigcirc \quad + \ddot{\underset{\cdot\cdot}{C}l}-\overset{-}{AlCl_3} \longrightarrow \quad H_3C \quad \bigcirc \quad + AlCl_3 +$$

19.105

Energy applications of graphene: electrodes in next-generation lithium ion batteries; supercapacitors for energy storage. Other applications: photodetectors, composite materials and coatings, sensors, drug delivery.

19.107

Some examples: C_{20} (non-aromatic), C_{28} (non-aromatic), C_{50} (aromatic), C_{70} (non-aromatic), C_{72} (aromatic), C_{76} (non-aromatic), C_{84} (non-aromatic).

19.109

Some examples:

Benz[a]anthracene:

Chrysene (found in coal tar and creosote, a wood preservative):

Coronene (a pollutant that is produced as a byproduct of petroleum refining):

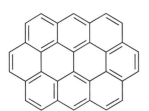

Ovalene (also a pollutant produced in petroleum refining):

SUMMARY AND CONCEPTUAL QUESTIONS

19.111

An O–H bond is formed, a C–H bond is broken, a C–C bond is formed (to make a double bond) and a C–Br bond is broken.

19.113

Because of these three additional resonance structures, the + charge of the carbocation is spread out—delocalized—over four carbon atoms. This makes the benzyl carbocation especially stable—for a carbocation.

19.115

19.117

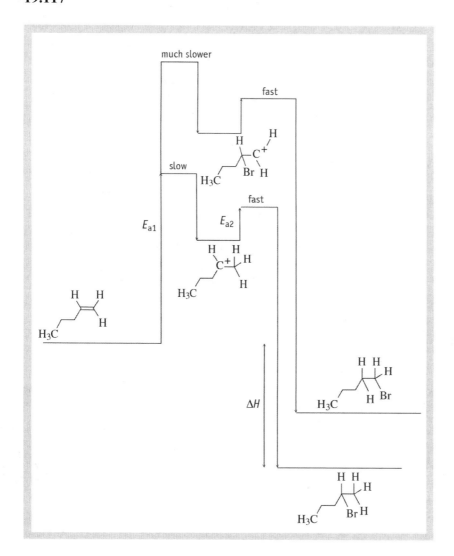

19.119

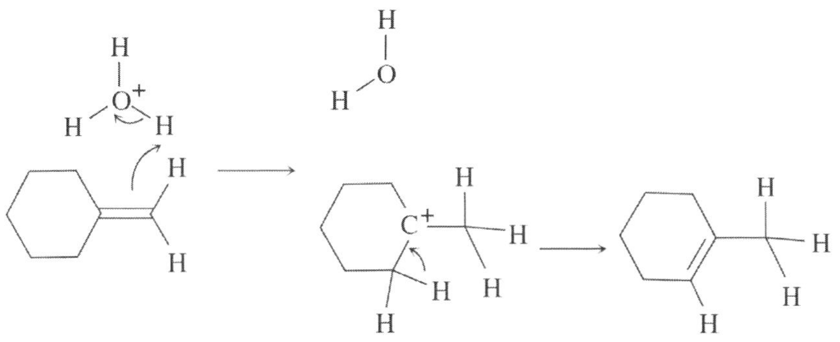

19.121

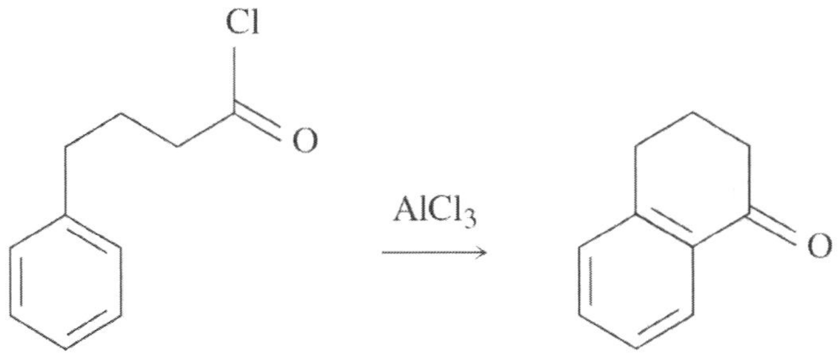

19.123

CHAPTER 20
Understanding Structure, Understanding Reactivity: Alcohols, Amines, and Alkyl Halides

IN-CHAPTER EXERCISES

Exercise 20.1—Naming Alcohols

(a) 2-Methylhexan-2-ol

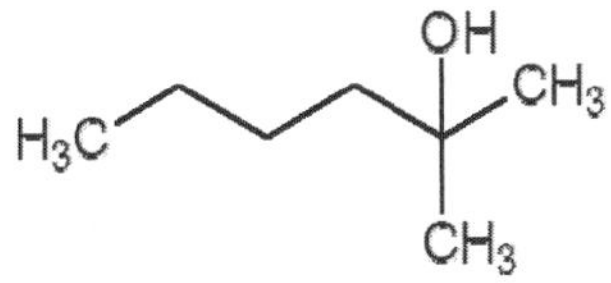

(b) Hexane-1,5-diol

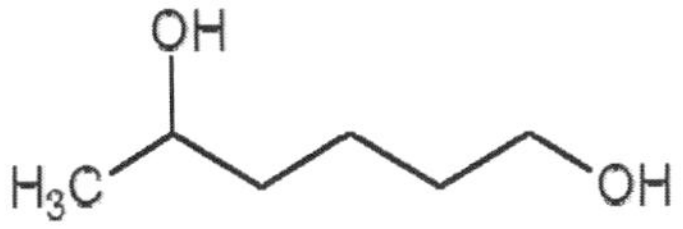

(c) 2-Ethylbut-2-en-1-ol

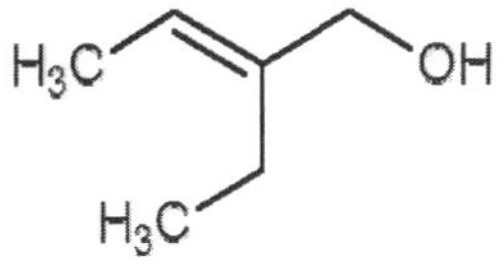

(d) Cyclohex-3-en-1-ol

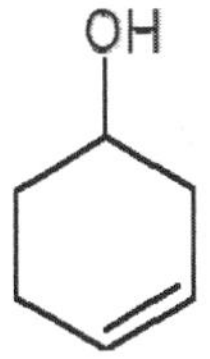

(e) 2-Bromophenol

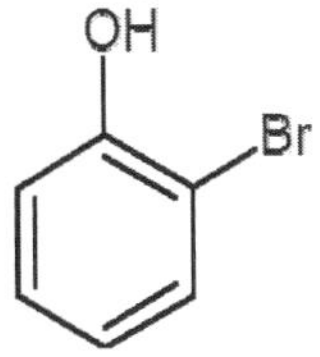

(f) 2,4,6-Trinitrophenol

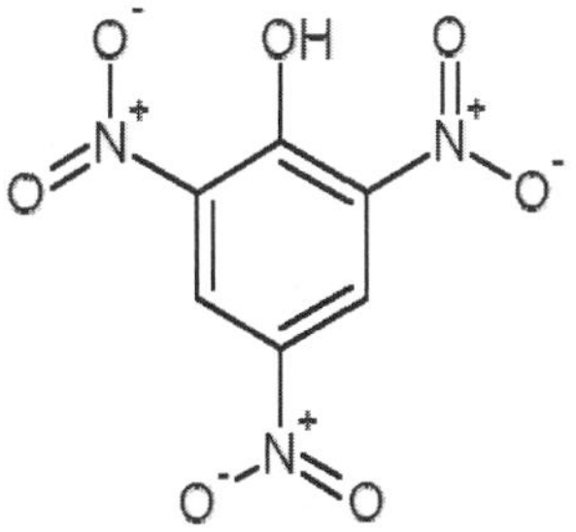

Exercise 20.3—Nomenclature of Alkyl Halides

(a) 2-bromobutane
(b) 3-chloro-2-methylpentane
(c) 1-choro-3-methylbutane

(d) 1,3-dichloro-3-methylbutane
(e) 1-bromo-4-chlorobutane
(f) 4-bromo-1-chloropentane

Exercise 20.5—Nomenclature of Amines

(a) primary amine

(b) secondary amine

(c) tertiary amine

(d) quaternary ammonium salt

Exercise 20.7—IR Spectroscopy of Alcohols

The conversion of cholesterol to cholest-5-ene-3-one involves the oxidation of the alcohol to a ketone. If the conversion were successful, we would see a disappearance of the O–H stretch in the 3200–3550 cm^{-1} range and the appearance of a new peak in the 1705–1725 cm^{-1} range corresponding to the C=O stretch for the ketone.

Exercise 20.9—^{13}C NMR Spectra

The ^{13}C NMR spectrum for *p*-ethoxyaniline would have the following peaks:

CH_3–CH_2–: ~10–30 ppm
CH_3–CH_2–: ~40–60 ppm
C aromatic: 6 peaks in the ~110–150 ppm range

Exercise 20.11—Nucleophilic Substitution Reactions

(a)

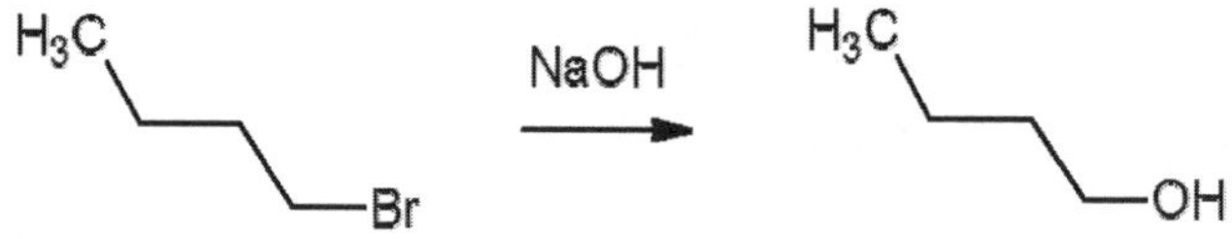

(b)

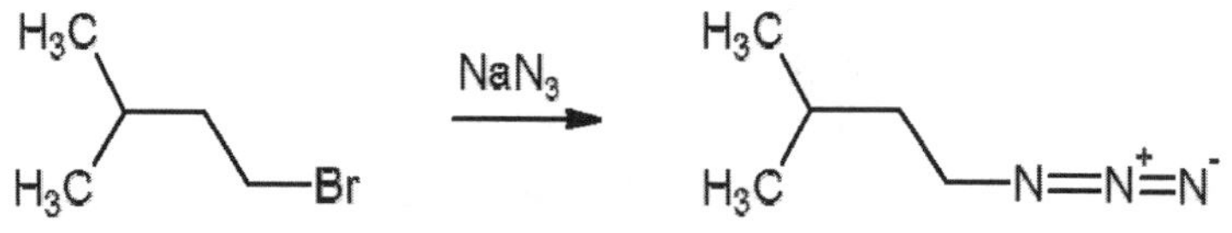

Exercise 20.13—S$_N$2 Reactions

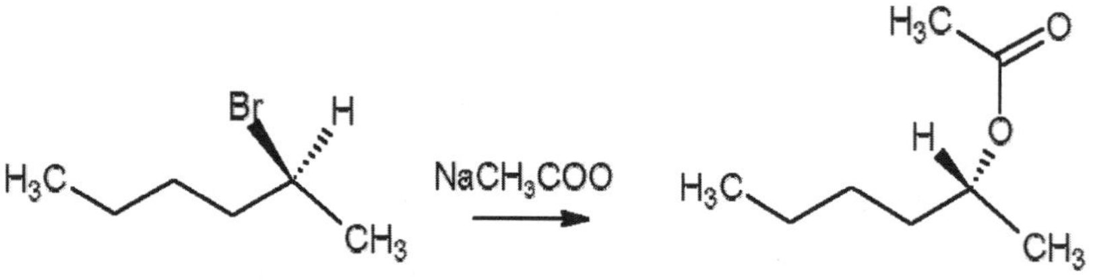

(S)-2-bromohexane

(R)-2-hexyl acetate

Exercise 20.15—Kinetics of S$_N$2 Reaction

(a) CN$^-$ (cyanide ion) reacts faster with $CH_3CH_2CH_2Br$ than $CH_3CH(Br)CH_3$. The back of the bromine-bearing carbon atom (where the cyanide reacts) is more accessible in a primary bromide than a secondary bromide. A smaller fraction of cyanide bromide collisions will produce a reaction in the case of the secondary bromide.

(b) Reaction of I⁻ with $(CH_3)_2CHCH_2Cl$ is faster. In fact, $H_2C=CHCl$ is unreactive with iodide. Vinylic halides do not undergo S_N2 substitution reactions.

Exercise 20.17—Kinetics of S_N1 Mechanism

(a) The rate determining step of S_N1 reactions is unimolecular. The rate of S_N1 substitution of *tert*-butyl alcohol and HBr is proportional only to the concentration of *tert*-butyl alcohol. Therefore, tripling the concentration of HBr has no effect on the rate of reaction.

(b) Halving the HBr concentration has no effect. However, doubling the *tert*-butyl alcohol concentration doubles the rate of reaction.

Exercise 20.19—Stereochemistry of S_N1 Reaction

The product is a 50/50 mixture (racemate) of the enantiomeric 2-phenylbutan-2-ols.

Exercise 20.21—Synthesis Using Grignard Reagents

Exercise 20.23—Williamson Synthesis of Ethers

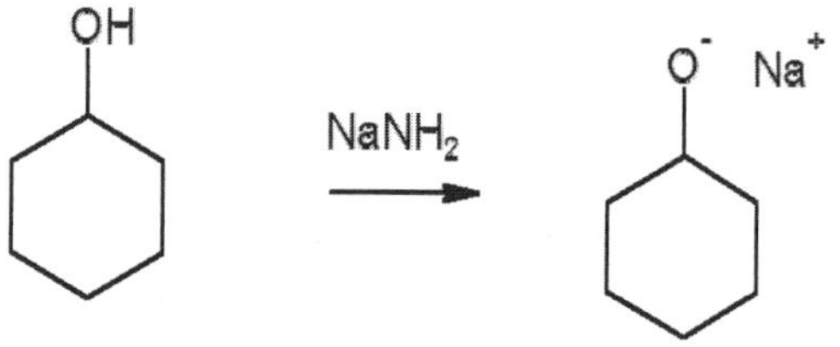

Could you make cyclohexyl ethyl ether by first forming sodium ethoxide (product of treatment of ethanol with sodium metal—hydrogen gas is the other product), then reacting it with cyclohexyl iodide? No. Since the alkyl iodide substrate in this case is much more sterically hindered, an E2 reaction to produce cyclohexene will compete with an S_N2 reaction, giving a much poorer yield of the substitution product.

Exercise 20.25—Dehydration of Alcohols

(a)

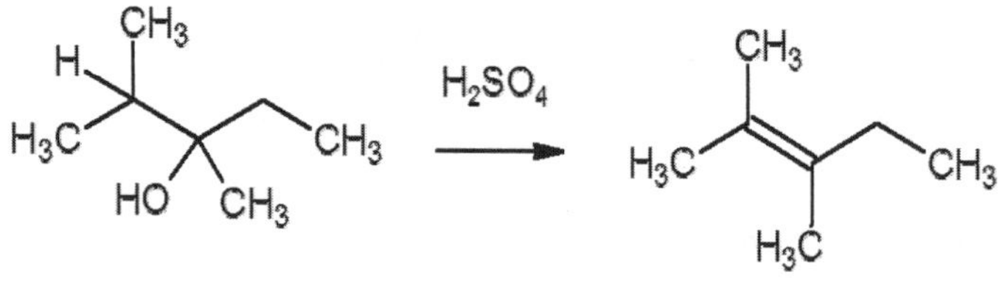

(b)

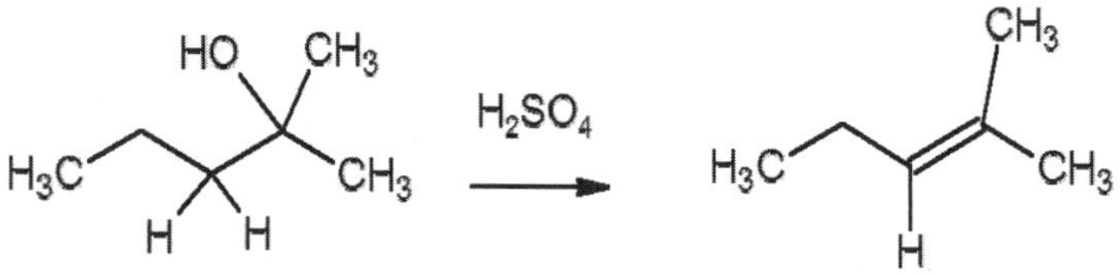

Note that the most substituted alkene is formed in each case. Concentrated sulfuric acid can achieve the reverse of the hydration of an alkene because the water eliminated on dehydration is quickly taken up by the solvent.

Exercise 20.27—Oxidation of Alcohols

(a)

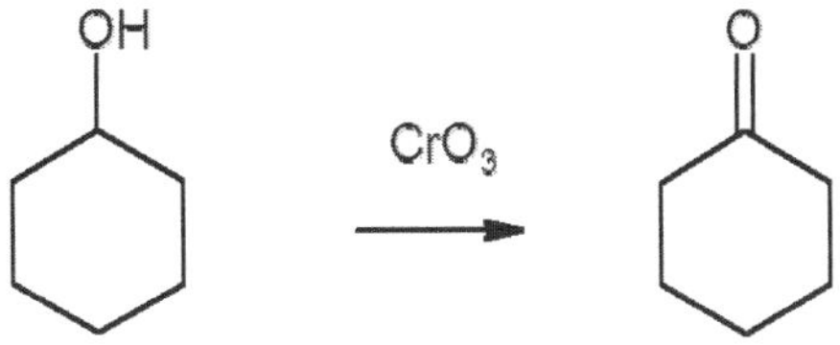

(b)

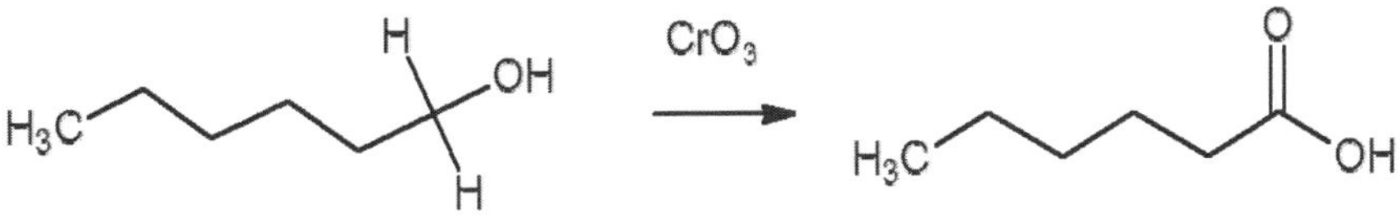

(c)

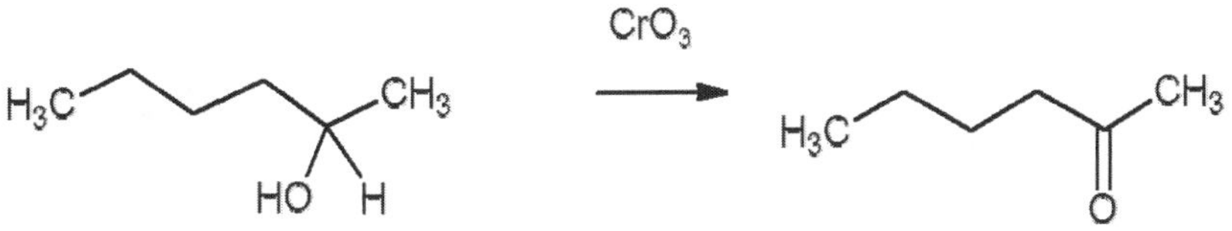

Exercise 20.29—Acid-Base Properties

With respect to basicity we have

(a) $CH_3CH_2NH_2 > CH_3CH_2CONH_2$

The nitrogen atom of the amide is much less basic than that of an amine because its electron pair can be delocalized by orbital overlap with the neighbouring π orbital of the C=O group.

(b) $NaOH > C_6H_5NH_2$

Hydroxide is a strong base, whereas amines are weak bases, like ammonia.

(c) $CH_3NHCH_3 > CH_3NHC_6H_5$

Phenyl amines have reduced basicity because the N lone pair is partially delocalized with the phenyl π system—i.e., it is less available to be donated to a proton.

(d) $(CH_3)_3N > CH_3OCH_3$

Just as ammonia is a stronger base than water, amines are stronger bases than ethers.

Exercise 20.31—Alkylation Reactions

(a)

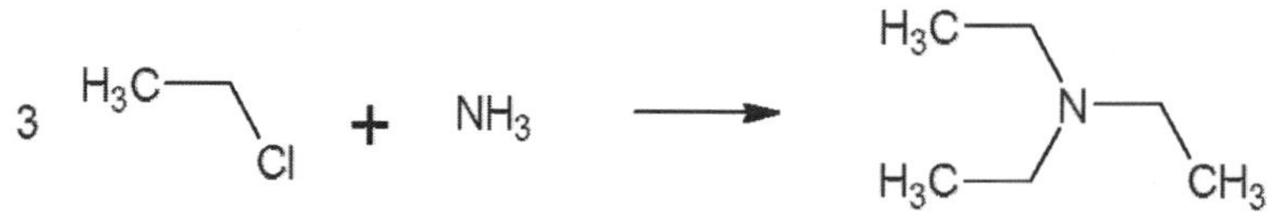

(b)

In part (a) some quaternary ammonium salt will also be formed as a side reaction.

REVIEW QUESTIONS

Section 20.1: Cyclodextrins

20.33

The hydroxyl groups on the outside of cyclodextrins make them soluble in water, allowing the drug encapsulated by the cyclodextrin to be delivered throughout the body.

20.35

The article describes the use of cyclodextrins to form aggregates and give rise to dispersed systems (i.e., not solutions) with more complex structures for dispersed drug delivery. Such dispersed structures include emulsions, micro and nano capsules and spheres, as well as liposomes and niosomes.

Sublingual literally means "under the tongue," referring to the administration of a drug through dissolution and absorption of a tablet held under the tongue. In the context of this case study, sublingual might be used as a metaphor meaning "under the ability-to-detect (i.e., taste) by the tongue."

20.37

Some types of wounds—e.g., exudating (oozing) wounds, venous leg ulcers, cancerous lesions, and wounds with necrotic (dead) tissue—produce unpleasant odours. Cyclodextrins have been used to treat this problem by encapsulating odour molecules. This is in addition to their use in increasing bioavailability of antibiotics to the wound.

Section 20.2: Overview of Alcohols, Amines, and Alkyl Halides

20.39

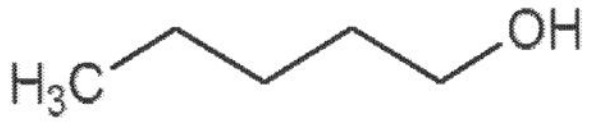

1-pentanol

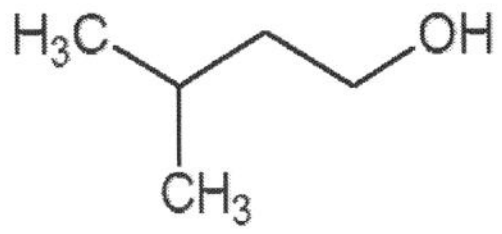

3-methylbutan-1-ol

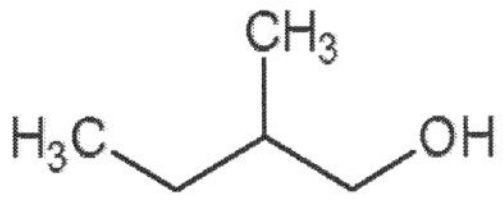

2-methylbutan-1-ol

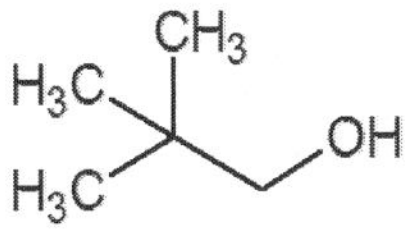

2,2-dimethylpropan-1-ol

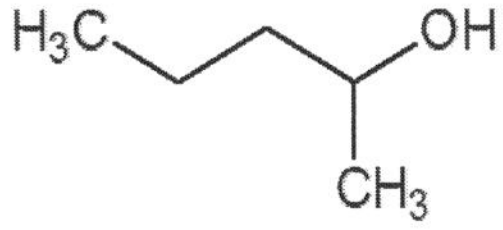

2-pentanol

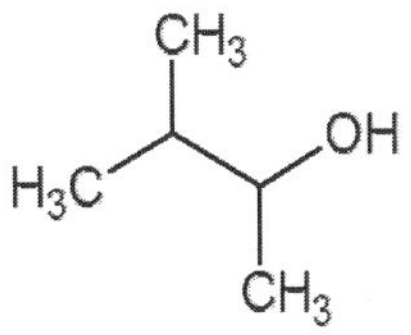

3-methylbutan-2-ol

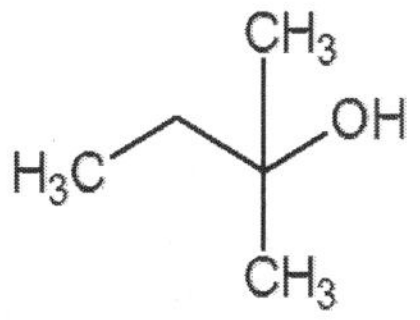

2-methylbutan-2-ol

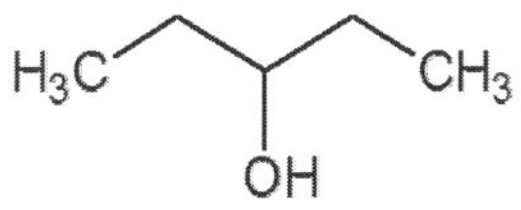

3-pentanol

20.41

(a) *N*-methylisopropylamine—it is a secondary amine
(b) *trans*-(2-methylcyclopentyl)amine—it is a primary amine
(c) *N*-isopropylaniline—it is a secondary amine

20.43

(a)

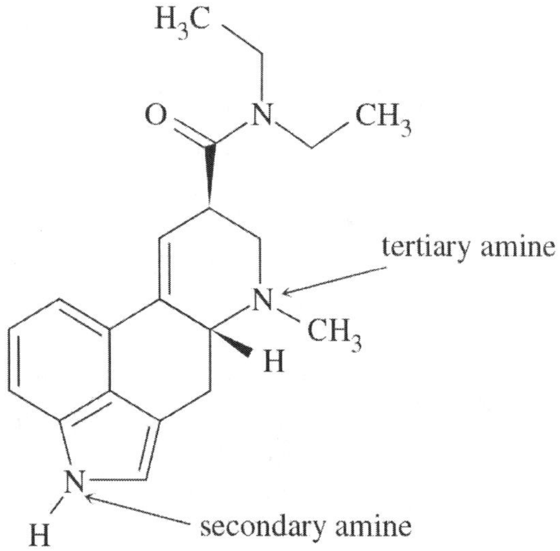

(b)

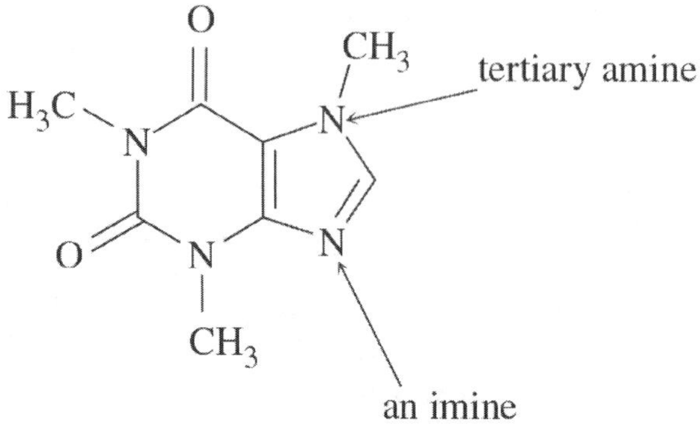

20.45

(a)

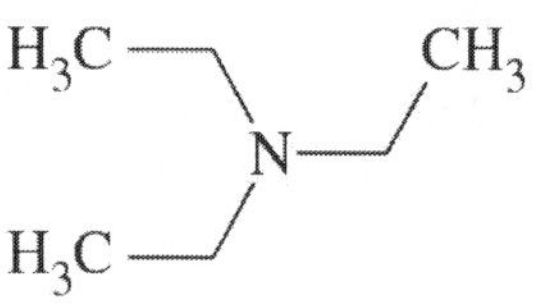

(b)

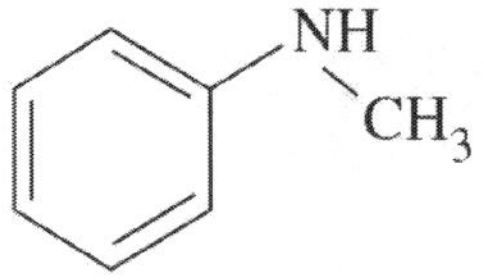

20.47

In the ^{1}H NMR spectrum, the benzyl H atoms of benzyl alcohol absorb downfield (around 4 ppm) compared with the methyl protons of cresol (around 2 ppm). Also, they integrate to twice the hydroxyl proton peak area, versus three times for the cresol methyl protons. Benzyl alcohol would show three aromatic proton peaks (relative integration of 2, 2 and 1) around 7–8 ppm (although they will be very close to each other in chemical shift), whereas cresol would show two peaks—both with relative integration of 2. In the ^{13}C NMR spectrum, benzyl alcohol would show a downfield (60–80 ppm) benzyl carbon peak (downfield due to attached O), whereas the methyl of cresol would be upfield—around 20 ppm. The aromatic carbons would show four peaks (around 120–140 ppm with intensities 1:2:2:1) in both cases, except that the cresol carbon attached to the oxygen atom should be a bit further downfield. The IR spectra of the two compounds show differences in the out of plane bending vibrations for C–H groups on the benzene ring, reflecting differences between a mono and disubstituted ring. The mass spectrum of both compounds should show a molecular ion peak at $m/z = 108$, and a peak at 107 corresponding to loss of an H radical. However, the benzyl alcohol radical cation can lose hydroxyl radical (molecular ion—17) to form the benzyl cation with an m/z value of 91. The benzyl cation is stabilized by the benzene ring delocalizing the positive charge, and it forms readily.

20.49

There are six distinct multiplets, consistent with the five distinct H atoms in the given structure. The peak integrals are consistent with the pattern, 1,1, 2, 2, 1 and 3 (from left to right). The two non-equivalent vinyl protons are assigned to the two downfield peaks—singlets at 4.85 and 4.75 ppm. Note that the vinyl protons are well-separated from other protons—no splitting is visible. The triplet at 3.7 ppm is assigned to the protons on the carbon atom attached to the OH group. These protons are deshielded by the electron-withdrawing oxygen atom, and split by the 2 neighbouring methylene protons. The neighbouring methylene protons appear as the triplet at 2.3 ppm. Note that the two triplets exhibit the same coupling constant because they split each other. The broader singlet at 2.15 ppm, which integrates to 1, is the hydroxyl H atom peak. Hydroxyl peaks are broader, and do not show coupling, because of rapid exchange with other OH groups in solution. The remaining singlet corresponds to the relatively isolated (split only by very small coupling constants) methyl protons.

20.51

(a) The ^{1}H NMR spectrum of this compound, C_3H_9NO, shows four multiplets—a triplet at 3.7 ppm, a triplet at 2.85 ppm, a broad peak at 2.7 ppm, and a quintet (or possibly a triplet of triplets with nearly equal coupling constants) at 1.7 ppm. The integrals are a little difficult to assign because of the very broad peak. However, the three multiplets appear to have the same integrals, 2, 2, and 2 (to be consistent with two triplets and a triplet of triplets). The remaining 3 H atoms appear to manifest as two overlapping broad peaks around 2.7 ppm. This is consistent with a hydroxyl H atom and two amino H atoms. These observations are consistent with the structure,

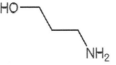

The triplet at 3.7 ppm corresponds to the methylene connected to O, the triplet at 2.85 ppm corresponds to the methylene bonded to N—note that N is less electronegative than O. The upfield peak corresponds to H atoms bonded to an alkyl carbon bonded to other alkyl carbons.

(b) Here, we see four multiplets—a triplet at 4.3 ppm, a singlet at 3.4 ppm, a doublet at 2.8 ppm, and a broad singlet 1.2 ppm integrating to 1, 6, 2, and 2, respectively. The triplet and doublet are consistent with methine and methylene H atoms, with the methine carbon bonded to (likely) two oxygen atoms. The singlet, integrating to 6, is also downfield—consistent with two equivalent methyl groups bonded to the two O atoms—on the other side of the methine group. The methylene H atom is also downfield—though not as far. This is consistent with being bonded to nitrogen—a primary amine, with the –NH$_2$ group giving the broad singlet integrating to 2.

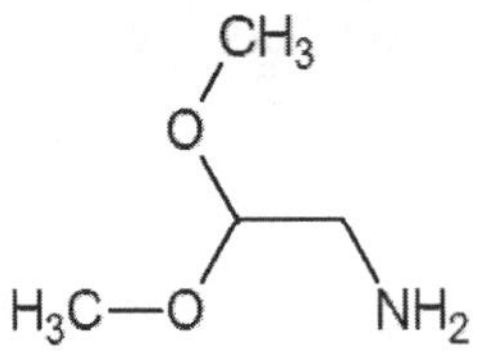

Section 20.3: Structure and Reactivity of Alkyl Halides

20.53

(a)

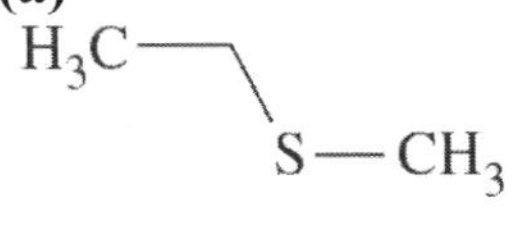

Nucleophile: CH₃S⁻   OH⁻

(b)

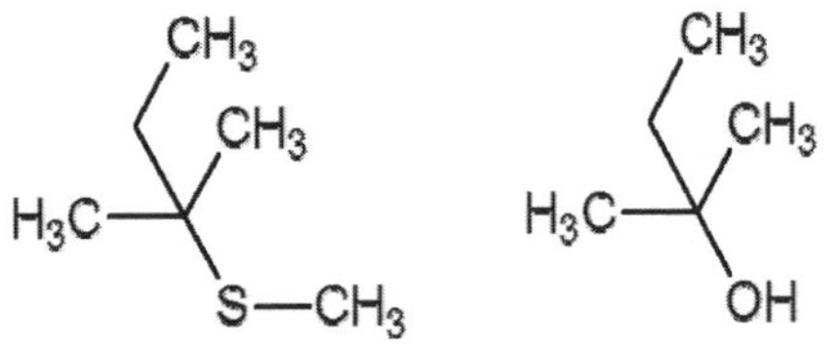

Nucleophile: CH₃S⁻   OH⁻
Under strongly basic conditions, OH⁻ will not give a substitution product. Instead, we get a mixture of the two elimination products, with the former—the more highly substituted alkene—predominating.

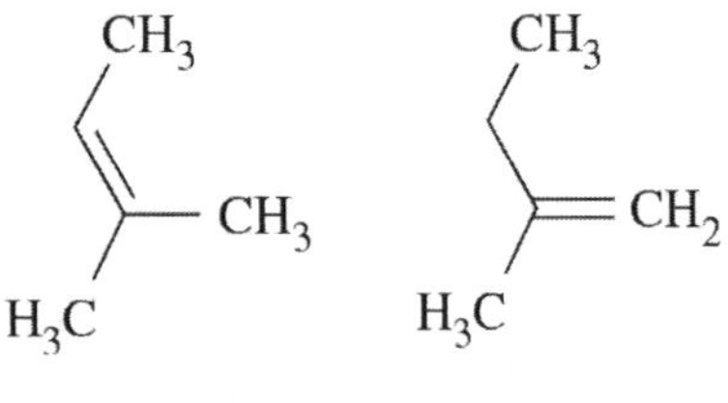

(c)

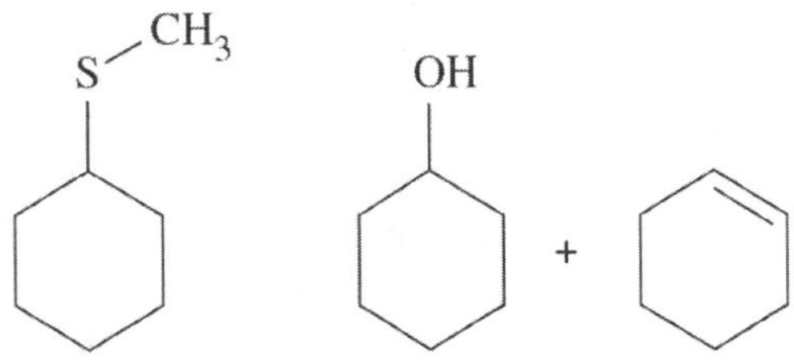

Nucleophile: CH₃S⁻   OH⁻
Note the elimination product with the strong base nucleophile, OH⁻.

20.55

S configuration at C–Cl carbon   R configuration at C–CN carbon

Note that S_N2 substitution is expected at the chlorine bearing carbon as opposed to the hydroxyl bearing carbon. This is because hydroxide ion is not a good leaving group. In the absence of acid to protonate the hydroxyl group, making water the leaving group, S_N2 substitution at the hydroxyl carbon will not go—in spite of the fact that the primary alcohol is less sterically hindered than the secondary chloride. Note the inversion of configuration characteristic of S_N2 reactions.

20.57

(a) Benzyl bromide, $C_6H_5CH_2Br$, reacts faster in an S_N2 reaction with OH^- than bromobenzene. Bromobenzene is an aryl bromide—it does not undergo this reaction.

(b) CH_3Cl reacts faster in an S_N2 reaction with OH^- than $(CH_3)_3CCl$, because it is primary whereas the latter is tertiary. *tert*-butyl chloride is significantly sterically hindered and does not follow an S_N2 mechanism.

(c) $H_2C=CHCH_2Br$ (allyl bromide) reacts faster in an S_N2 reaction with OH^- than $CH_3CH=CHBr$. The latter is a vinyl bromide that does not undergo S_N2 reactions.

20.59

S_N1 reaction rates are proportional only to the concentration of the alkyl halide (i.e., in the case of S_N1 reactions of alkyl halides)—not the nucleophile. The rate determining step is dissociation of the alkyl halide into halide and carbocation.

(a) If the concentration of $(CH_3)_3CBr$ is doubled and that of CH_3OH is halved, the rate increases two-fold.

(b) If the concentration of $(CH_3)_3CBr$ is halved and that of CH_3OH is doubled, the rate decreases two-fold—i.e., it is half as big.

(c) If the concentrations of both $(CH_3)_3CBr$ and CH_3OH are tripled, the rate increases three-fold.

(d) If the reaction temperature is lowered, the rate of reaction decreases—there is less thermal energy available allowing the alkyl halide to overcome the barrier to dissociation.

20.61

According to decreasing S_N2 reactivity, the reagents are ranked as follows:

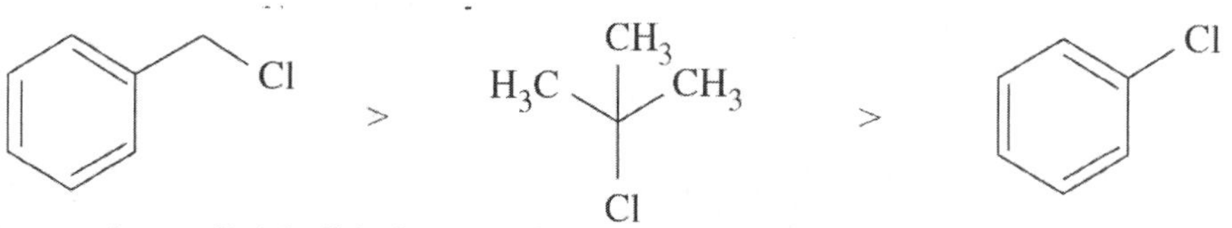

The primary (benzylic) halide has much greater S$_N$2 reactivity than a tertiary halide. Neither *tert*-butyl chloride or chlorobenzene are amenable to S$_N$2 reactions. Benzene rings are subject to electrophilic attack, not nucleophilic attack. According to decreasing S$_N$1 reactivity, the reagents are ranked as follows:

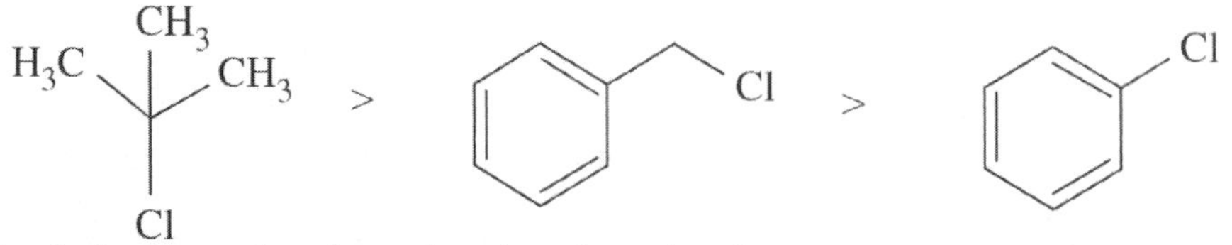

While both the benzyl and *tert*-butyl carbocation intermediates give stabilized carbocation intermediates, the *tert*-butyl carbocation has the greatest stabilization and reacts fastest. Phenyl chloride does not undergo S$_N$1 reactions.

Section 20.4: Structure and Reactivity of Alcohols

20.63

(a)

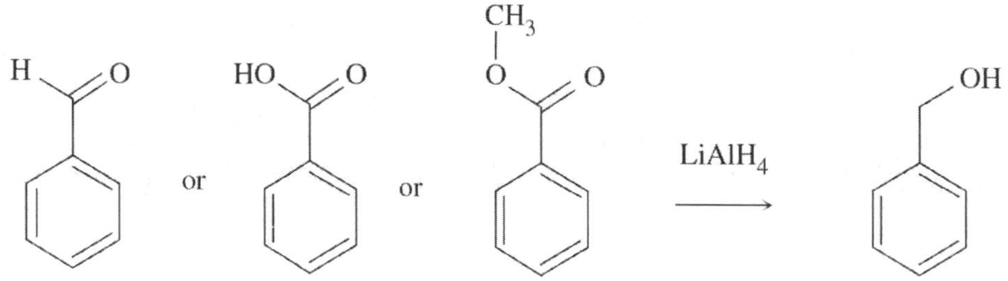

(b)

(c)

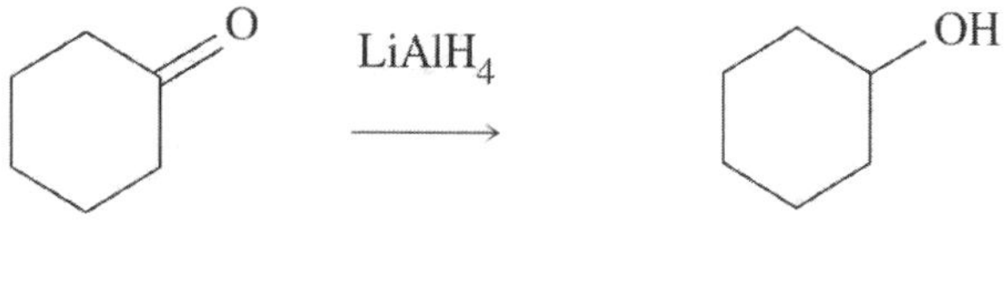

20.65

(a)

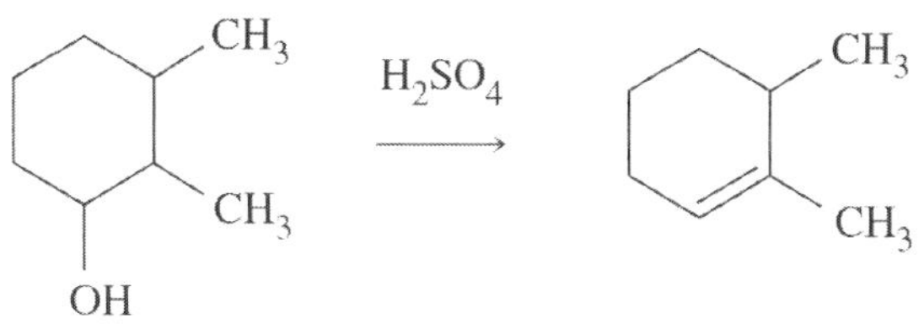

(b)

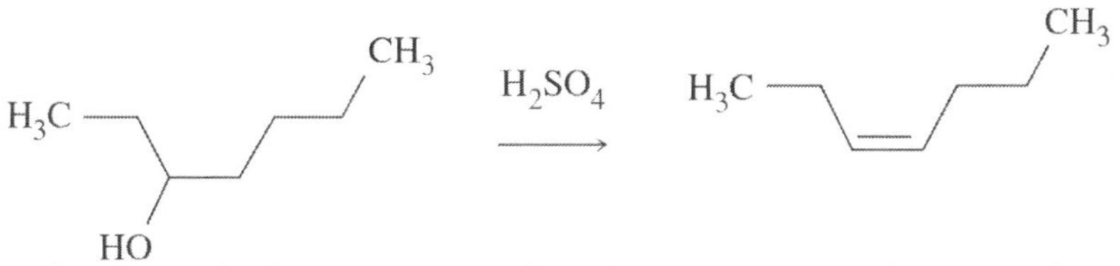

Note that in (a) had we used 1,2-dimethylcyclohexanol, the product would have been 1,2-dimethylcyclohexene—the more substituted of the two possible products.

20.67

(a)

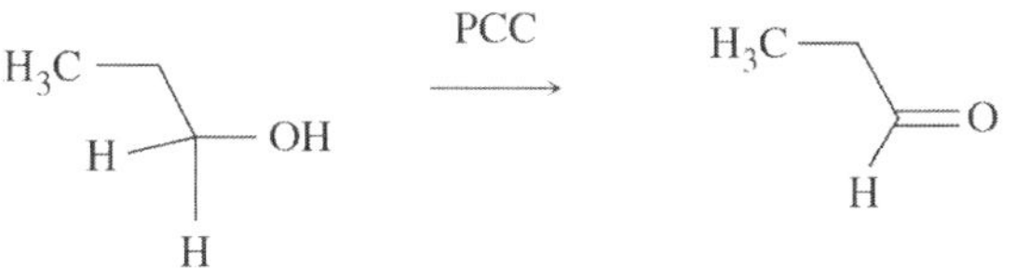

(c)

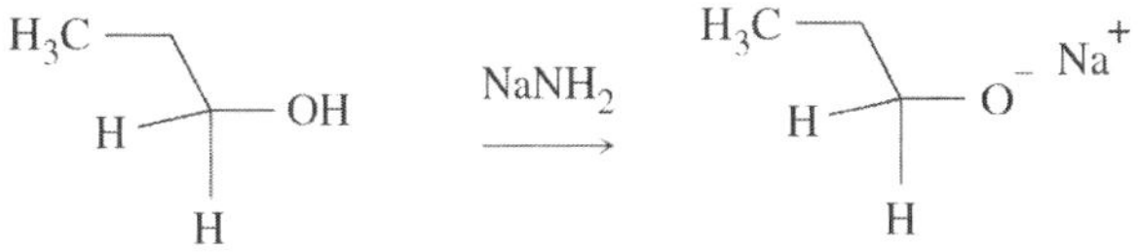

(b)

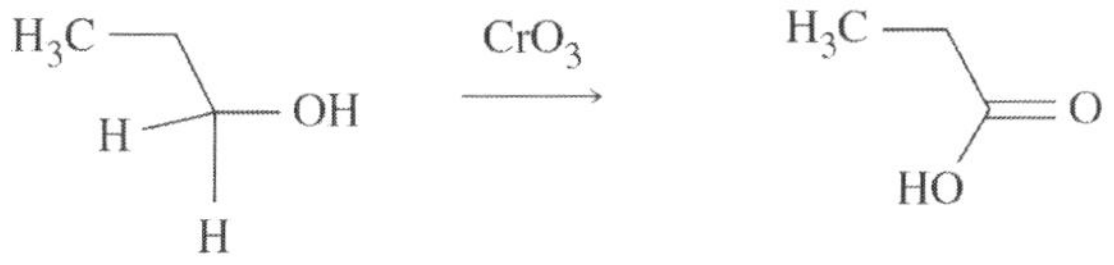

(d)

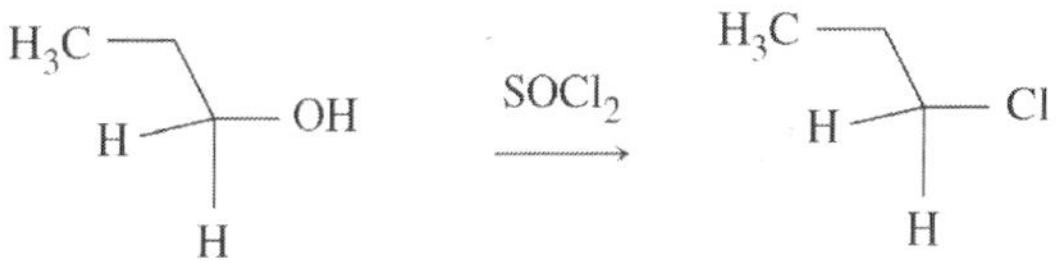

20.69

(a)

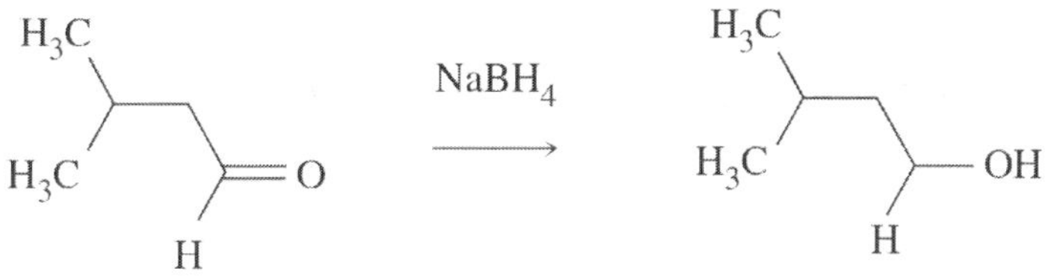

(b)

(c)

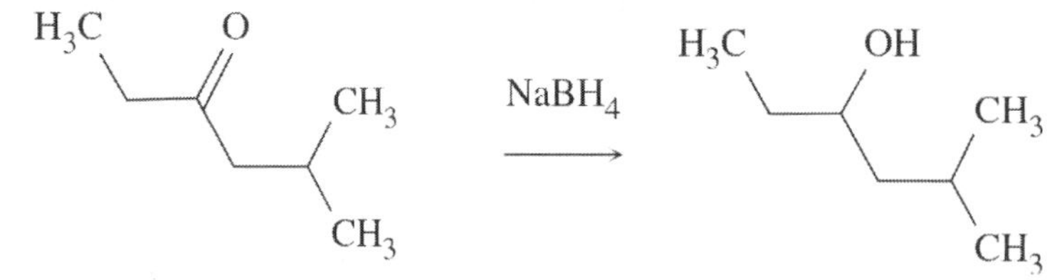

20.71

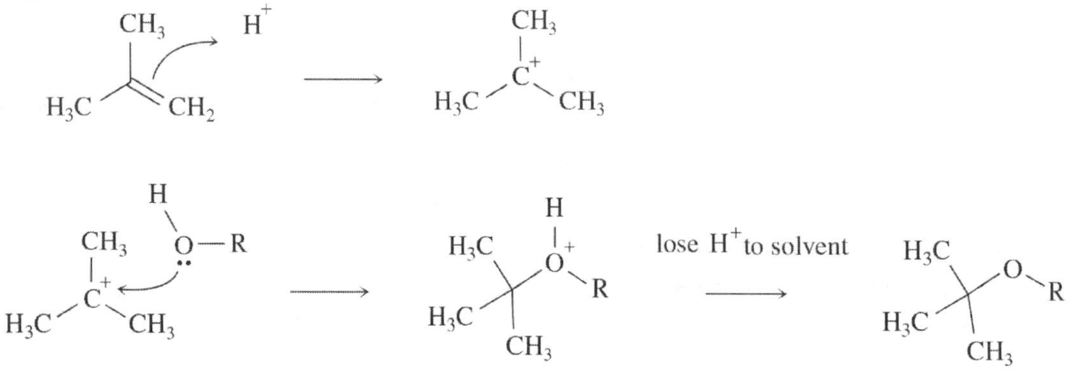

Section 20.5: Structure and Reactivity of Amines

20.73

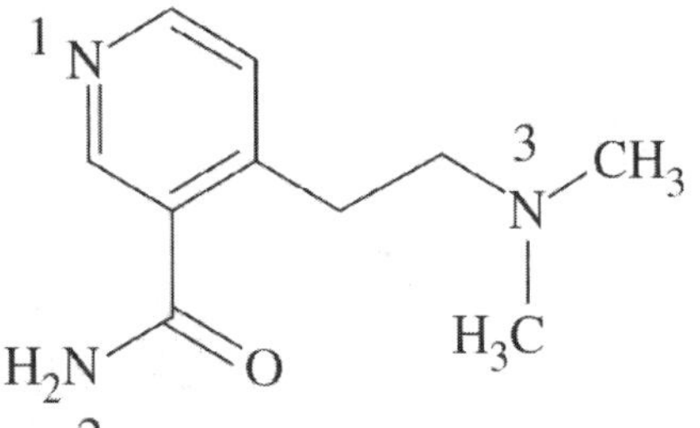

According to basicity, the three N's in this molecule are ranked as follows: 3 > 1 > 2. 3 is an alkyl amine, 1 an aromatic heterocyclic amine, and 2 an amide. Alkyl amines are more basic than aromatic heterocyclic amines such as pyridine, and amides are not very basic at all—because their lone pair of electrons are delocalized over the π system of the adjacent C=O group.

20.75

Trimethylamine has a lower boiling point than dimethylamine even though it has a higher molecular weight because it lacks N-H groups. It is therefore incapable of the intermolecular hydrogen bonding that occurs in dimethylamine.

SUMMARY AND CONCEPTUAL QUESTIONS

20.77

(a)

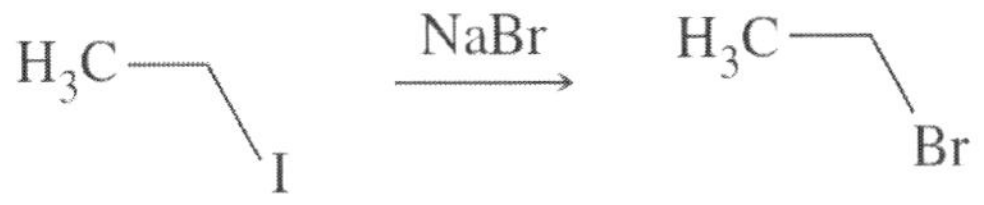

(b)

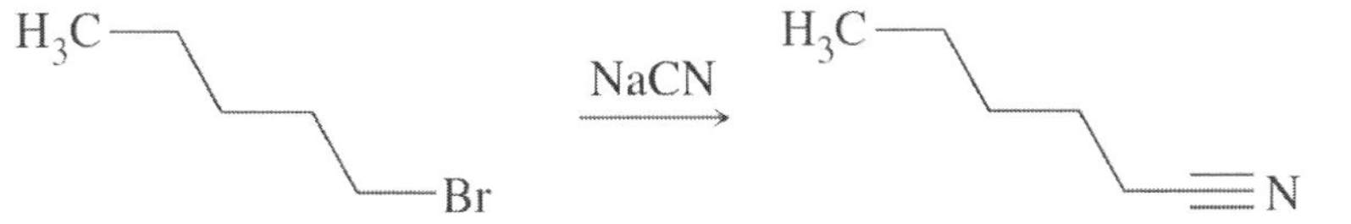

(c)

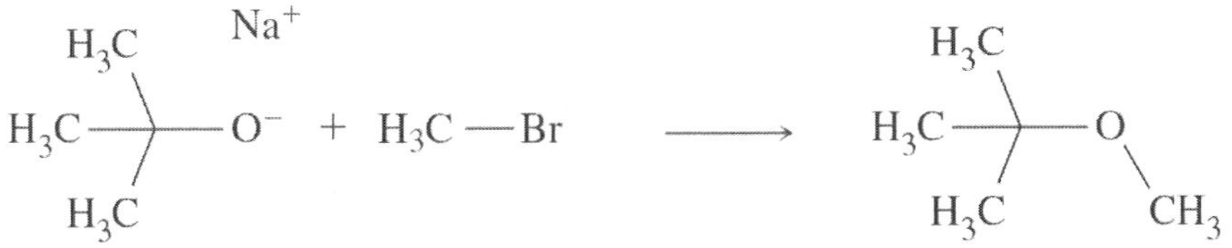

(d)

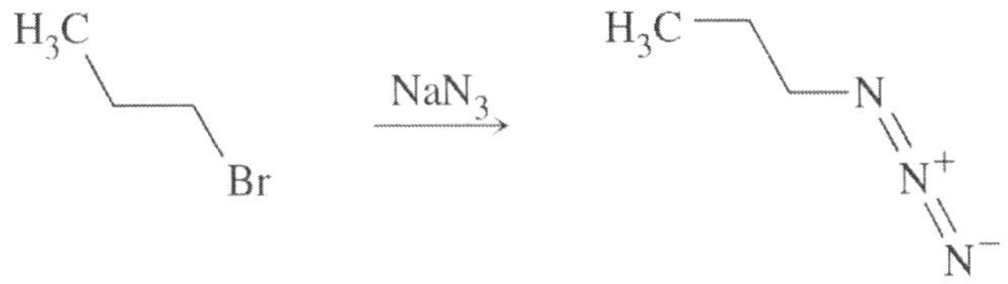

(e)

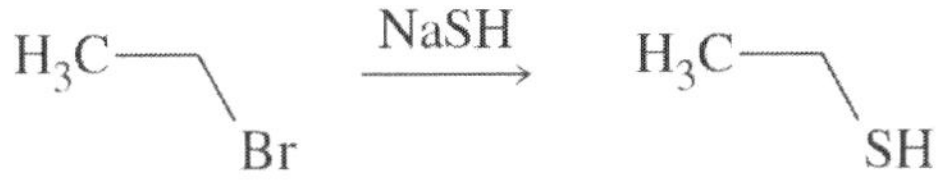

(f)

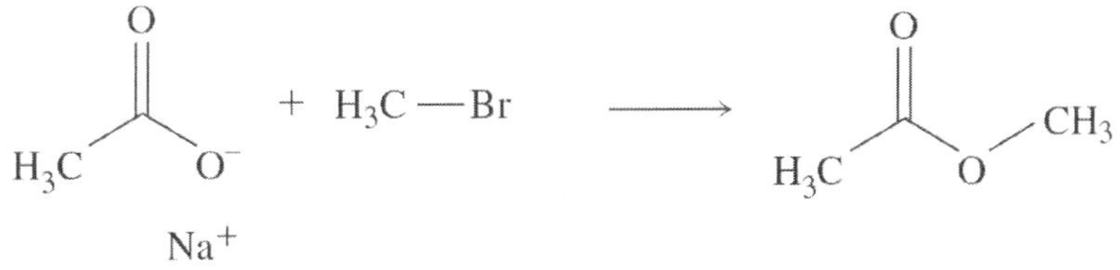

20.79

(a) 1-iodopropane   **(b)** butanenitrile   **(c)** 1-propanol
(d) Upon reaction with Mg in ether, we get propylmagnesium bromide. Subsequent reaction with water produces propane.
(e) methyl propyl ether

20.81

Arranged in order of decreasing reactivity in the S_N2 reaction, the isomers of C_4H_9Br are

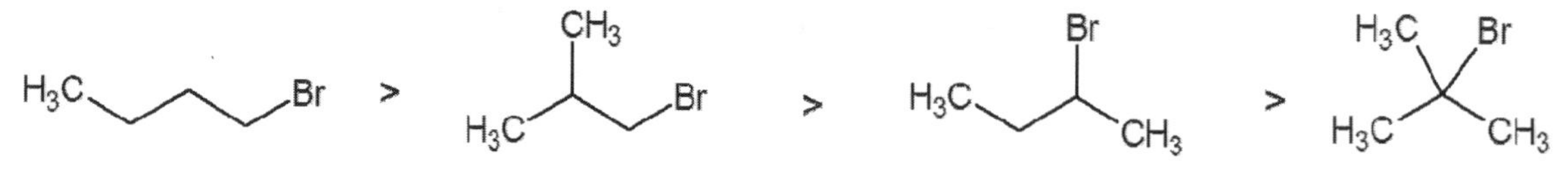

H₃C⌒⌒Br > (CH₃)(H₃C)CH–CH₂Br > H₃C–CH(Br)–CH₂CH₃ > (H₃C)(H₃C)(H₃C)C–Br

1-bromobutane

1-bromo-2-methylpropane

2-bromobutane

2-bromo-2-methylpropane

20.83

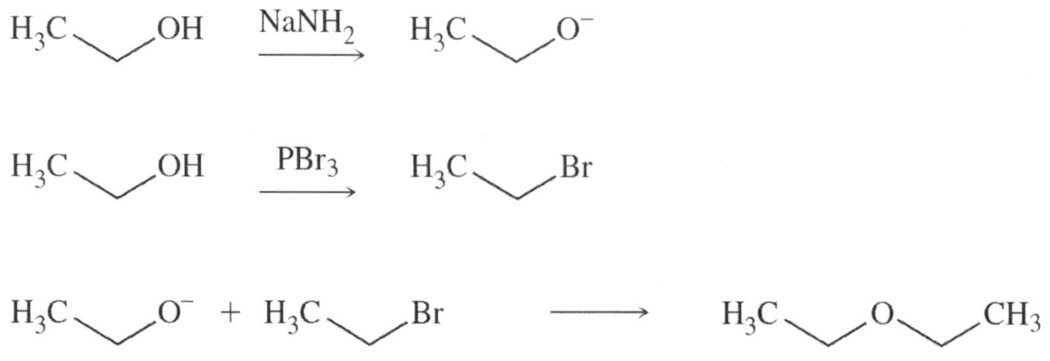

H₃C⌒OH —NaNH₂→ H₃C⌒O⁻

H₃C⌒OH —PBr₃→ H₃C⌒Br

H₃C⌒O⁻ + H₃C⌒Br ⟶ H₃C⌒O⌒CH₃

20.85

Treatment of (*R*)-2-bromohexane with NaBr yields *racemic* 2-bromohexane because the bromide substitutes for bromide according to the S$_N$2 mechanism, flipping the orientation of the 2-bromohexane to (*S*). However, the (*S*) in turn undergoes S$_N$2 substitution, producing the (*R*) bromide. After many substitutions, both (*R*) and (*S*) bromides are present in equal amounts.

20.87

(*S*)-Butan-2-ol racemizes to give (±)-butan-2-ol on standing in dilute sulfuric acid, because it slowly undergoes reversible S$_N$1 substitution with the water solvent. Since the carbocation intermediate can equally react with water from above or below, the substitution yields the racemic alcohol product. Eventually, the enantiomeric excess is lost.

20.89

Acetic acid, pentane-2,4-dione and phenol will all react, essentially to completion, with NaOH. These three species have lower pK_a than water.

20.91

Only acetic acid will react with NaHCO₃. Of the substances in Exercise 22.58, only acetic acid has a pK_a lower than that of carbonic acid.

20.93

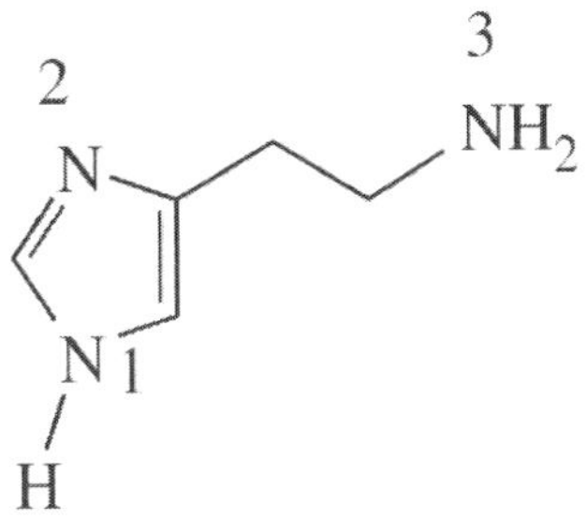

Order of basicity: $3 > 2 > 1$

With respect to basicity the N atoms are ordered as follows: $3 > 2 > 1$. Alkyl amines (N-3) are more basic than pyridine-type N atoms (N-2), which are more basic than pyrrole-type N atoms (N-1). The lone pair on N-1 is delocalized in the π system, making it unavailable for donation to a proton. This is in contrast to N-2, the pyridine-type N atom. Using the orbital hybrid model, we can consider the lone pair to be in an sp^2 orbital, not in the π system. However, the sp^2 lone pair is less available for donation than the sp^3 lone pair of N-3.

20.95

$CH_3CH_2NH_2$ is more basic than $CF_3CH_2NH_2$. The fluoro substituents are electron withdrawing—they reduce the availability of electrons at the nitrogen atom.

20.97

Triethylamine is more basic than aniline. The nitrogen lone pair in aniline is partially delocalized into the π system of the benzene ring, making it less available for donation to protons.

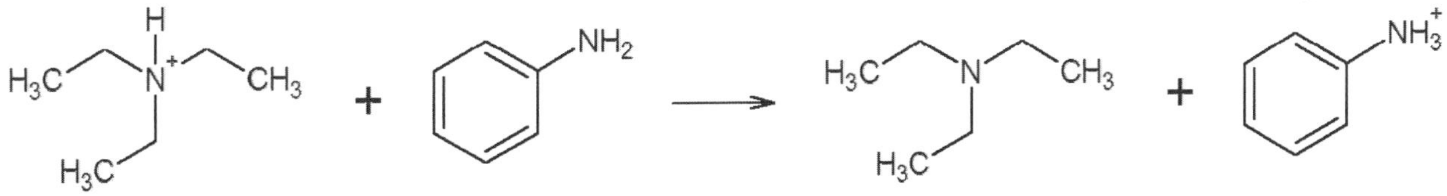

This reaction does not go in the direction indicated.

20.99

Stereocentres are marked with arrows below. There are four stereocentres, and as such there are $2^4 = 6$ stereoisomers.

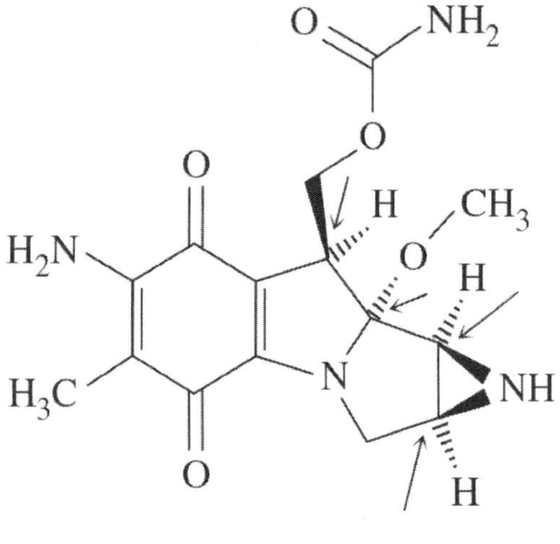

20.101

The ^{1}H NMR spectrum of compound, $C_{15}H_{24}O$, shows only four distinct types of protons in the ratio, 2:1:3:18. The lone inequivalent proton is identified as a hydroxyl proton because it disappears on adding D_2O. The downfield peak at 7 ppm with a relative integration of 2 is consistent with two equivalent phenyl protons—a phenyl ring is consistent with the four degrees of unsaturation in the formula. The very large (integrates to 18 protons) upfield peak—1.4 ppm—suggests 6 equivalent methyl groups. Since it is a singlet, these methyl protons must have no neighbouring protons different from them within three bonds. The peak at 2.2 ppm is also a singlet and has a relative integration of 3, suggesting a methyl group isolated from the rest of the molecule. Its chemical shift is consistent with benzylic protons. This must be a highly symmetrical molecule. The correct structure is BHT (butylated hydroxytoluene) or 2,6-di-*tert*-butyl-4-methylphenol (shown here on the left).

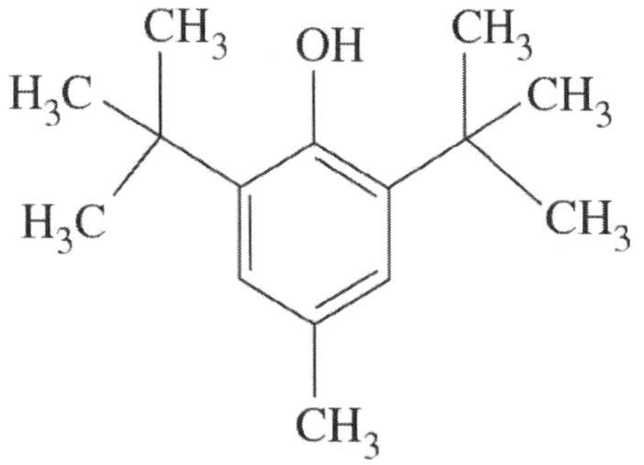

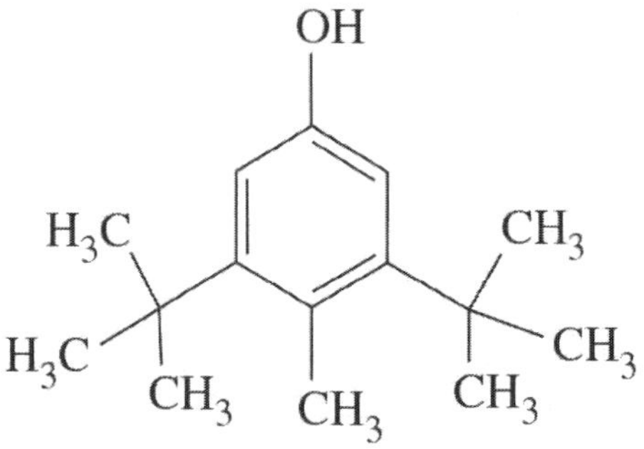

CHAPTER 21
Understanding Structure, Understanding Reactivity: Aldehydes and Ketones, Carboxylic Acid Derivatives

IN-CHAPTER EXERCISES

Exercise 21.1—Naming Aldehydes and Ketones

(a) 2-methylpentan-3-one
(b) 3-phenylpropanal
(c) octane-2,6-dione
(d) *trans*-2-methylcyclohexanecarbaldehyde
(e) pentanedial
(f) *cis*-2,5-dimethylcyclohexanone

Exercise 21.3—Naming Carboxylic Acids

(a)

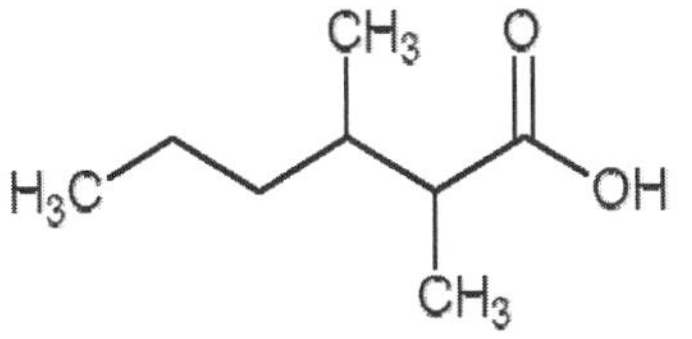

(b)

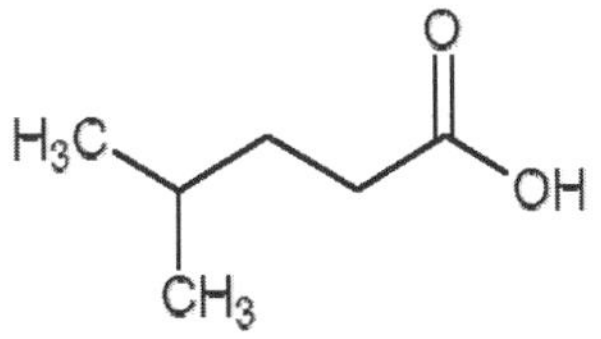

(c)

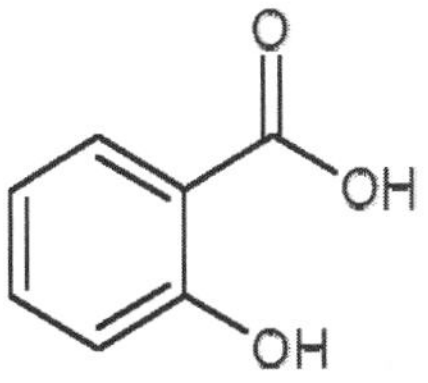

(d)

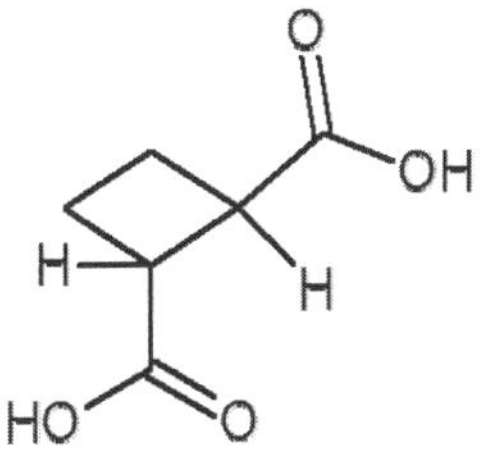

Exercise 21.5—IR Spectroscopy

(a) An IR absorption at 1735 cm^{-1} indicates a C=O stretch—most likely a saturated ester.
(b) An absorption at 1810 cm^{-1} is likely the C=O stretching band of an acyl halide.
(c) Absorptions at 2400-3300 cm^{-1} and 1710 cm^{-1} are associated with carboxylic acids. The O–H stretch has a very wide range.
(d) Absorption at 1715 cm^{-1} indicates a ketone C=O stretch (unless a carboxylic acid is suggested by an additional very broad O–H stretch—2400–3300 cm^{-1}).
(e) The two absorptions at 2716 cm^{-1} and 2817 cm^{-1} correspond to the C–H stretch of an aldehyde. The absorption at 1725 cm^{-1} is associated with the aldehyde carbonyl group.

Exercise 21.7—NMR Spectroscopy

Isomers cyclopentanecarboxylic acid and 4-hydroxycyclohexanone are possibly distinguished by their ^{13}C NMR spectra via the position of the carbonyl carbon absorption. The carboxylic acid carbonyl carbon will absorb around 180–185 ppm. The ketone carbonyl range is from 180–220 ppm. Thus, an absorption well downfield from 185 ppm indicates the ketone. Also, 4-hydroxycyclohexanone has a hydroxyl functionality and will have a C–O carbon absorption in the 50–90 ppm range. The other alkyl carbons will mostly absorb in the 10–30 ppm range—though the carbon next to the carbonyl (in both molecules) can be as far as 50 ppm downfield. In the ^{1}H NMR spectra, the molecules are easily distinguished by the single peak between 10 and 12 ppm for the proton of the carboxylic acid. A broad singlet is found for the alcohol O–H of 4-hydroxycyclohexanone, but it typically comes between 2 and 5 ppm.

Exercise 21.9—Nucleophilic Addition Reactions

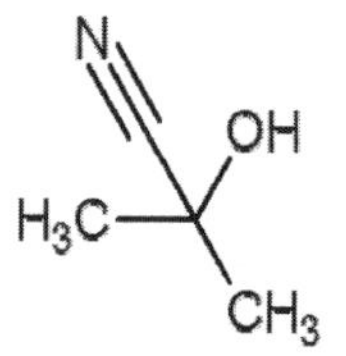

2-hydroxy-2methylpropanenitrile

Exercise 21.11—Addition of alcohols

1 mol of ethanol:

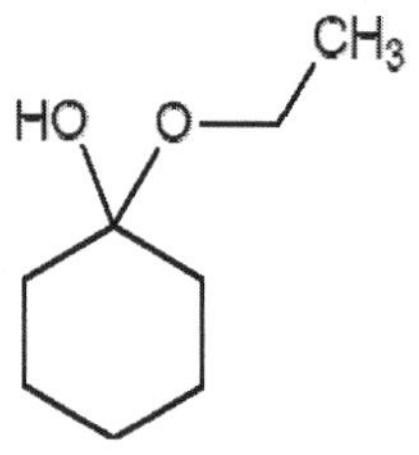

2 mol of ethanol:

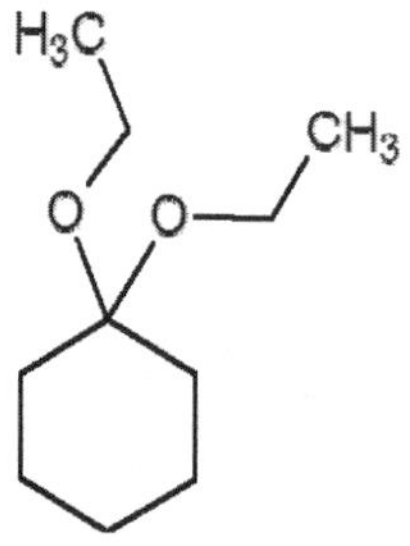

Exercise 21.13—Using Grignard Reagents

(a)

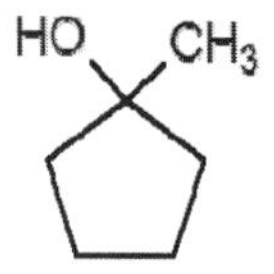

(b)

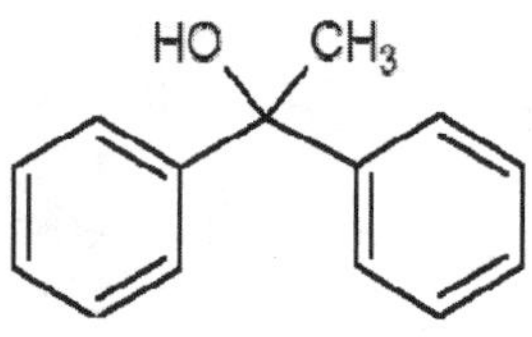

(c)

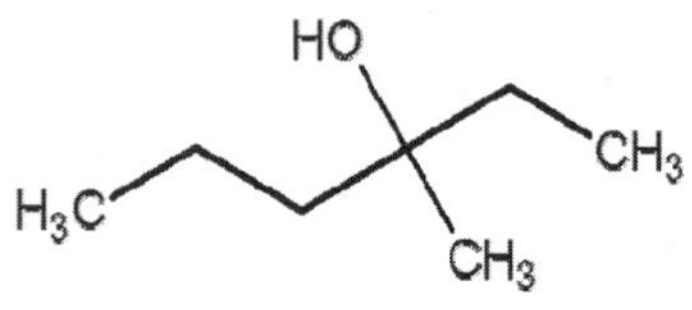

Exercise 21.15—Reaction with Amines

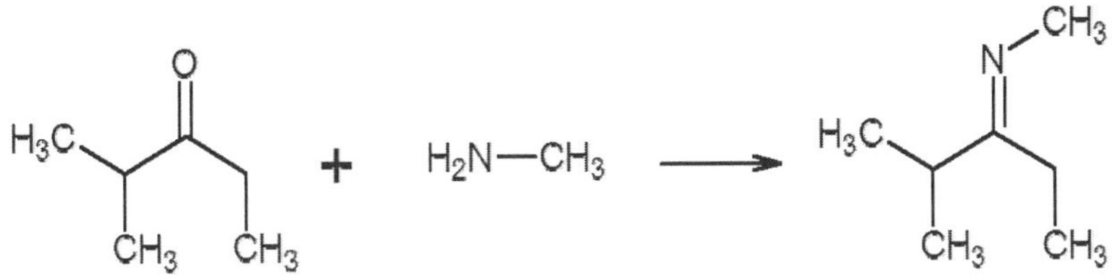

Exercise 21.17—Relative Acidity

According to increasing acidity, we have the following ranking:

methanol < phenol < *p*-nitrophenol < acetic acid < sulfuric acid

Example 21.19—Reactivity of Carboxylic Acid Derivatives

With respect to nucleophilic acyl substitution reactivity we have

(a) $CH_3COCl > CH_3CO_2CH_3$

(b) $CH_3CH_2CO_2CH_3 > (CH_3)_2CHCONH_2$

(c) $CH_3CO_2COCH_3 > CH_3CO_2CH_3$

(d) $CH_3CO_2CH_3 > CH_3CHO$

Note that acid chlorides and then anhydrides are the most reactive. The carbonyl carbon atom is highly positively polarized in these compounds, and they have good leaving groups—chloride and carboxylate. Esters are more reactive than amides—the amide leaving group is a stronger base than the alkoxide leaving group of an ester. Aldehydes are unreactive with respect to acyl substitution (but not addition) reactions—hydride is a very poor leaving group (it is a very strong base).

Exercise 21.21—Reactions of Acyl Chlorides

(a)

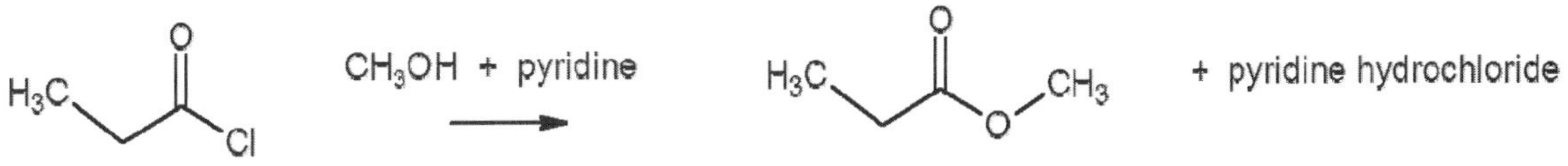

(b)

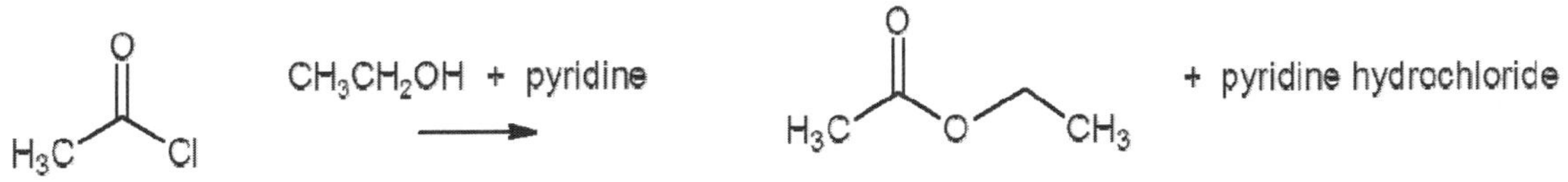

(c)

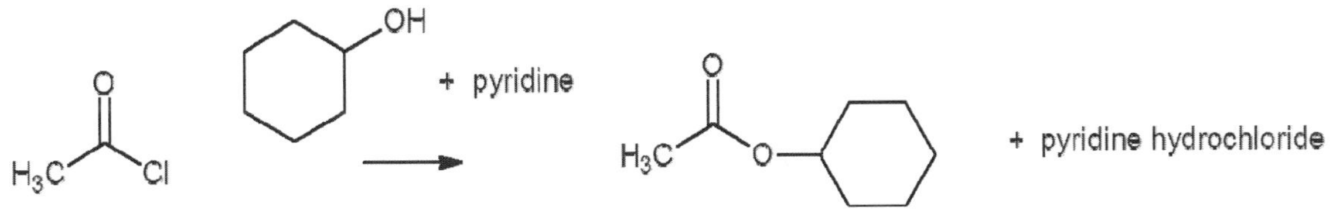

Exercise 21.23—Reactions of Acyl Chlorides

(a)

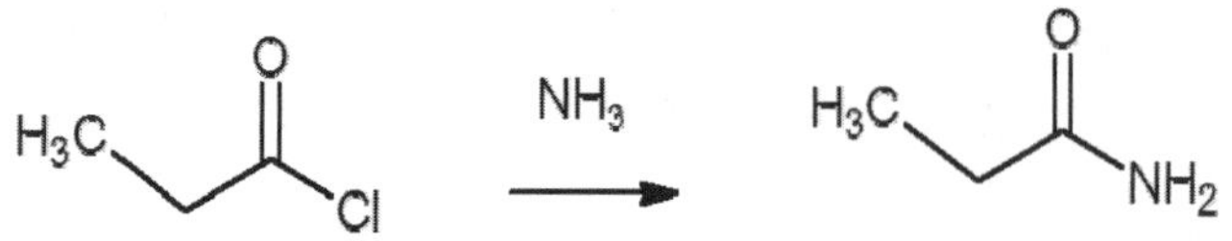

(b)

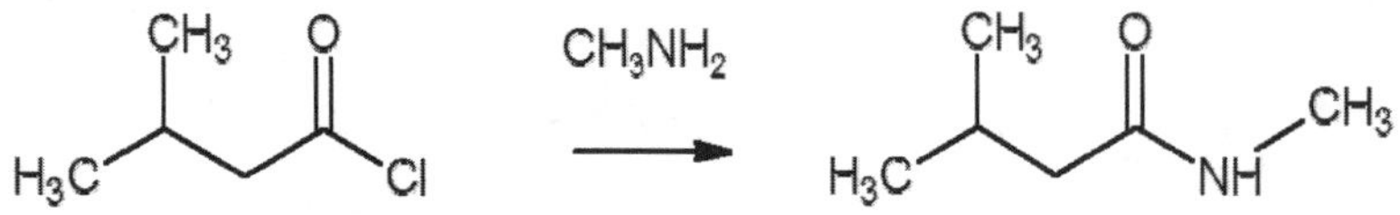

(c)

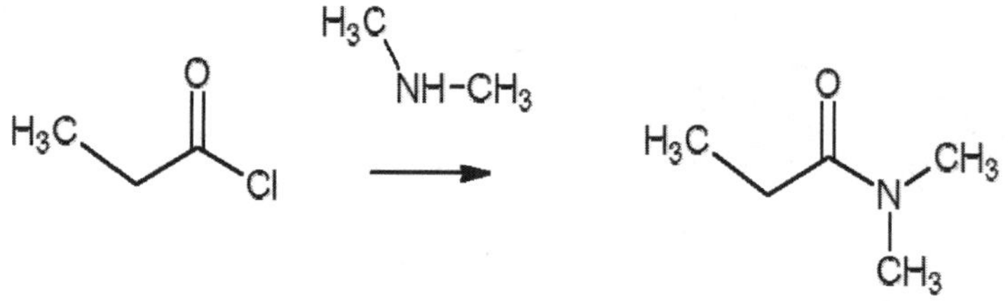

(d)

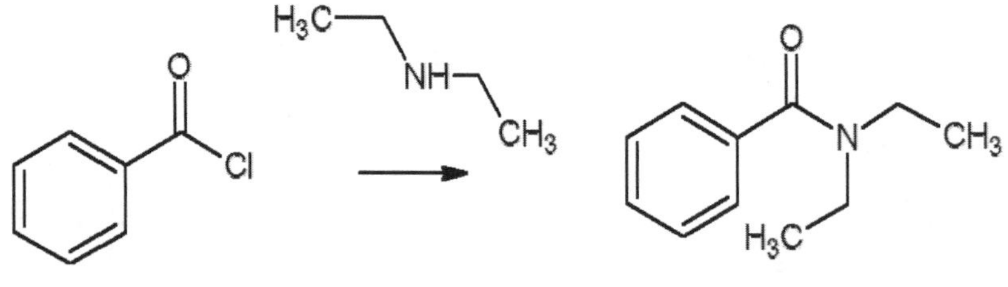

Exercise 21.25—Reactions of Acid Anhydrides

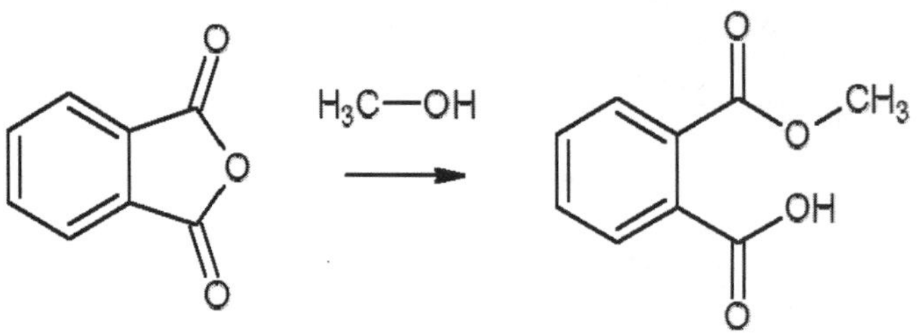

Exercise 21.27—Reactions of Esters

(a)

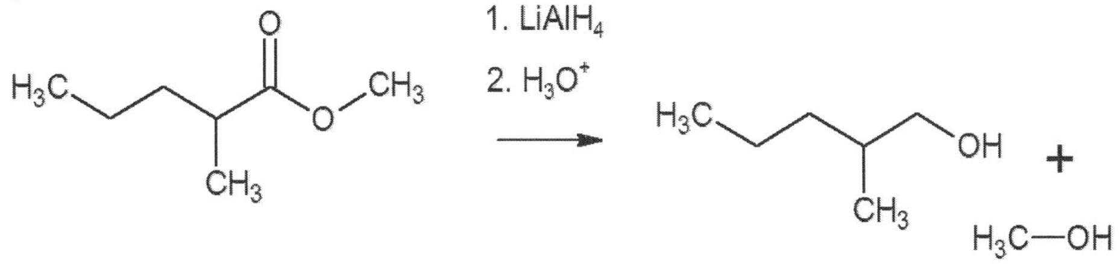

(b)

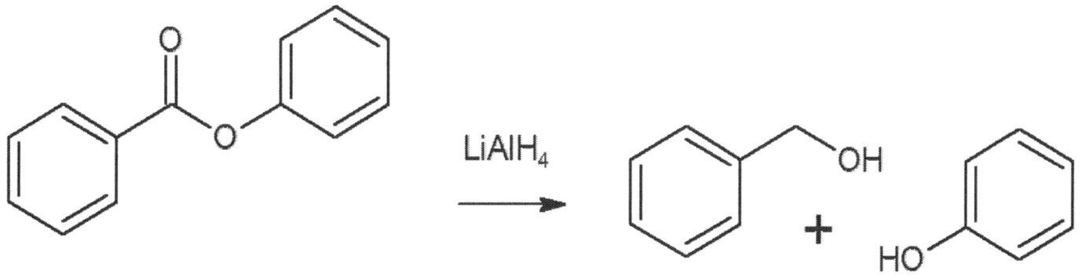

Exercise 21.29—Reactions of Amides

(a)

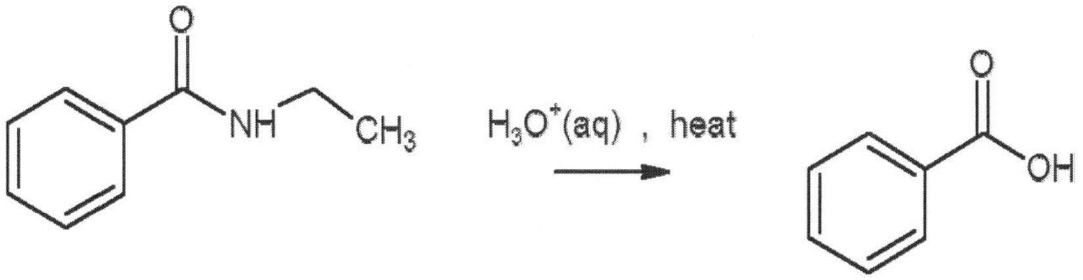

(b)

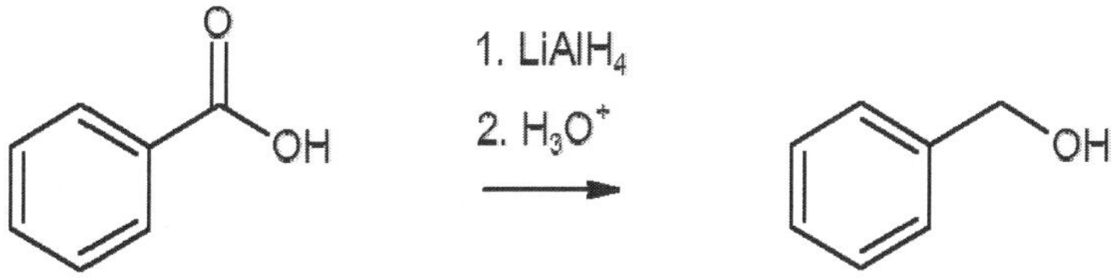

from (a)

(c)

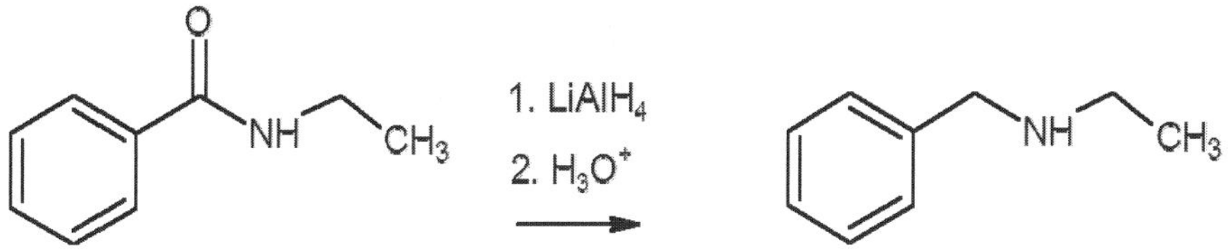

REVIEW QUESTIONS

Section 21.1: How Do Bacteria Tweet? Social Networking with Chemistry

21.31

(a) amide (specifically, acetamide): level 3
(b) ester (specifically, acetate): level 3
(c) aldehyde (level 2) & carboxylic acid (level 3)

21.33

Multiple answers possible. Some examples:

(a)

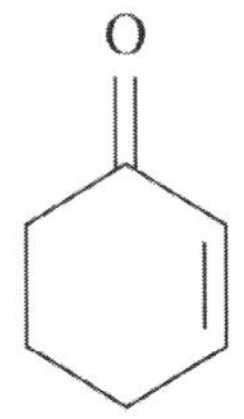

(b)

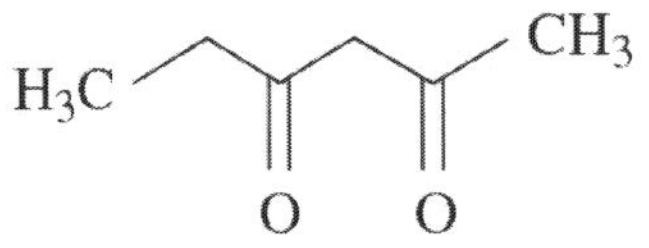

(c)

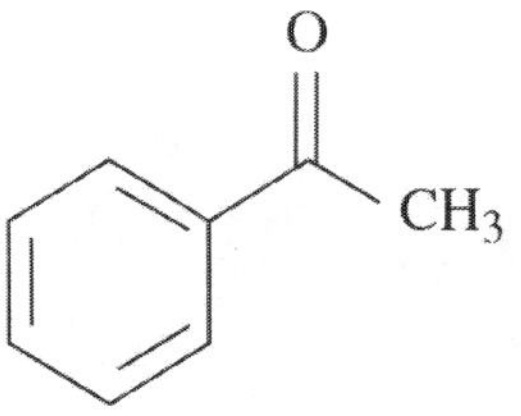

(d)

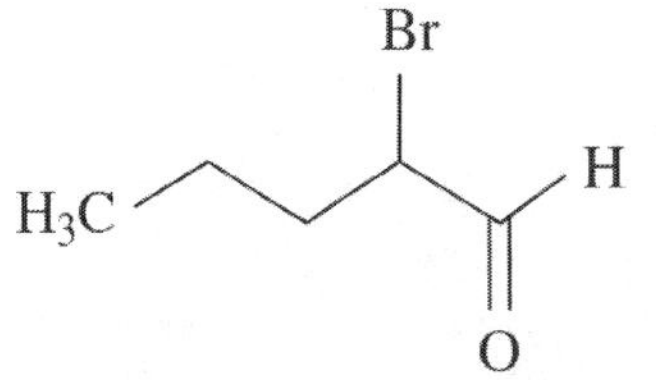

21.35

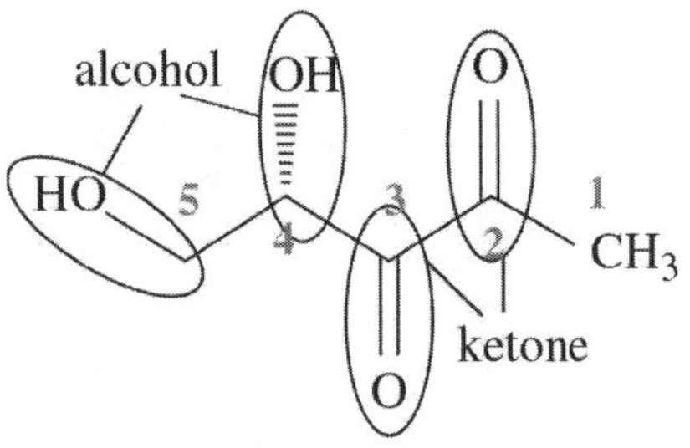

(4*S*)-DPD

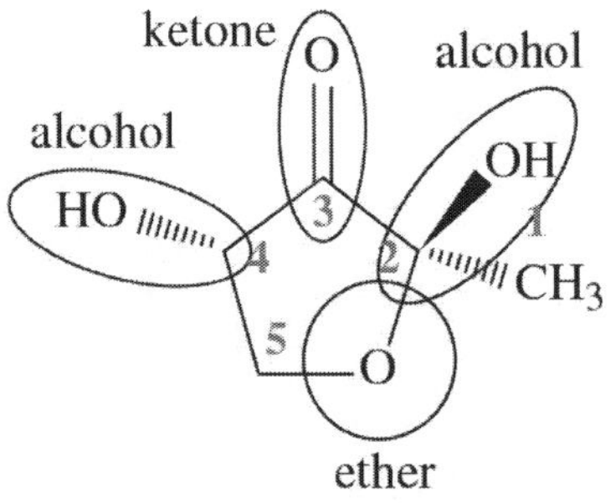

(2*S*,4*S*)-DHMF

The reaction is an addition reaction, with the terminal hydroxyl group adding across the carbonyl group at C-3.

21.37

(a) Yes, these molecules would be considered pheromones.

(b) Many examples possible. Some possibilities: Insects: gypsy moth pheromone ((+)-disparlure); honey bee alarm pheromone (isopentyl acetate). Humans: putative sex pheromomes: (e.g. androstenone, androstenol, androstadione).

(c) Gypsy moth pheromone: emitted by female to attract males; honey bee alarm pheromone: alerts other bees to danger; androstenone, androstenol, androstadione: functions unclear, but appear to be involved in attraction.

(d) (+)-disparlure (epoxide, no carbonyl present)

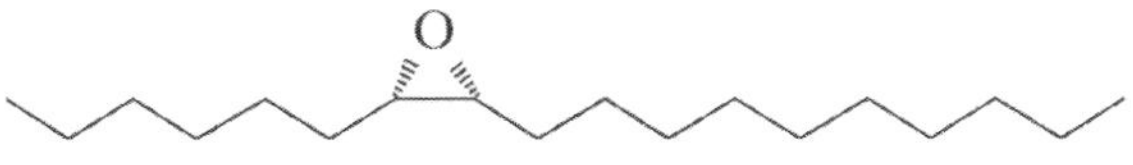

isopentyl acetate (ester, contains carbonyl group)

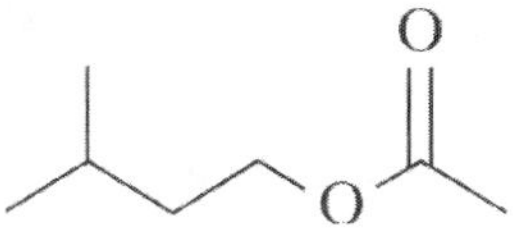

Human pheromones:

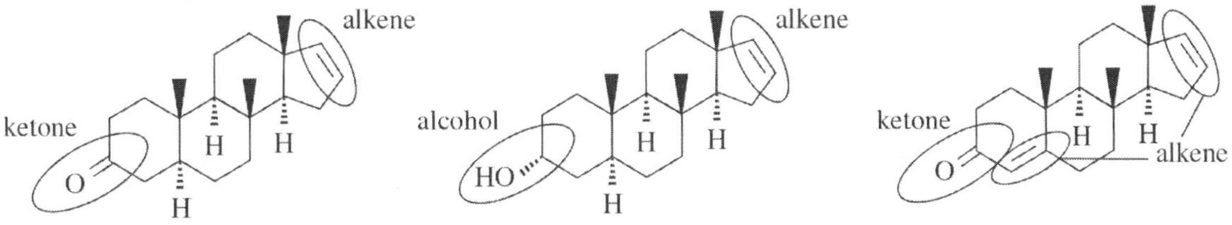

Section 21.2: Overview of Carbonyl Compounds

21.39

(a)

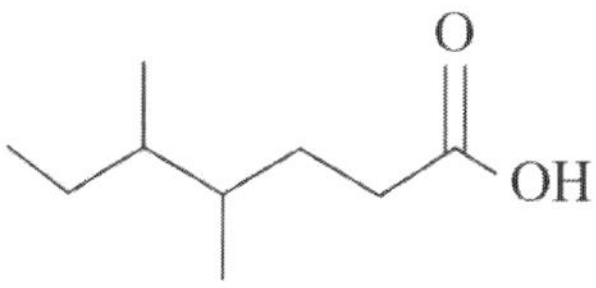

(b)

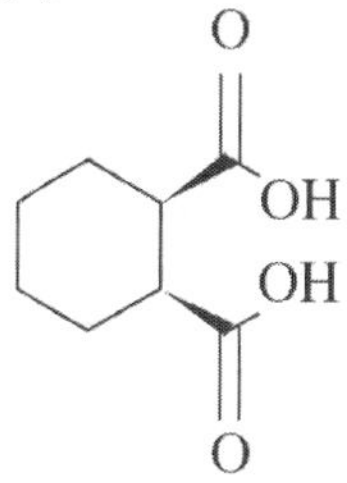

(c)

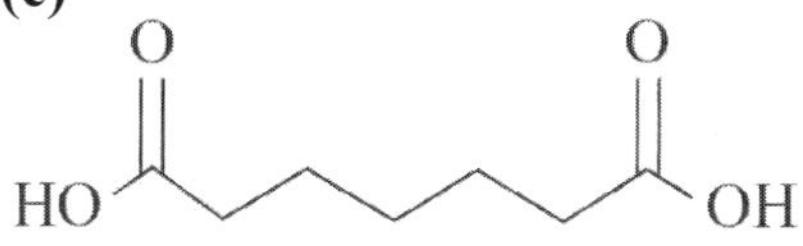

(d)

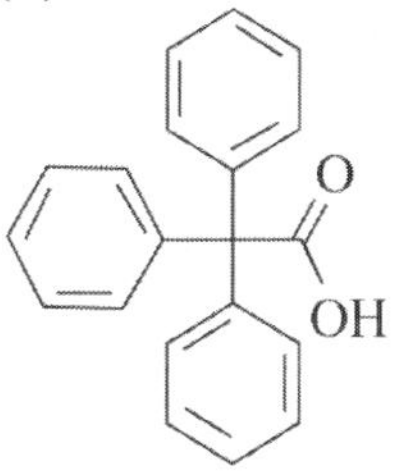

(e)

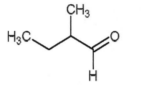

(f)

(g)

21.41

pentanal

3-methylbutanal

2-methylbutanal

2,2-dimethylpropanal

pentan-2-one

3-methylbutan-2-one

pentan-3-one

Of the above seven compounds, only 2-methylbutanal is chiral.

21.43

(a) *p*-methylbenzamide
(b) 4-ethylhex-2-enenitrile
(c) dimethyl butanedioate
(d) isopropyl 3-phenylpropanoate

(e) phenyl benzoate
(f) *N*-methyl 3-bromobutanamide
(g) 3,5-dibromobenzoyl chloride
(h) 1-cyanocyclopentene

21.45

(a)

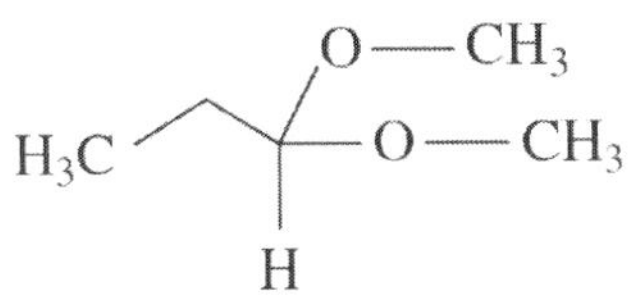

H₃C—CH(—O—CH₃)(—O—CH₃) structure

an acetal

(c)

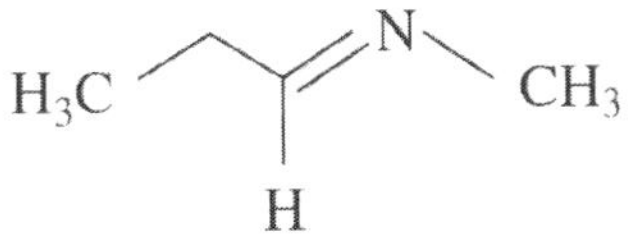

H₃C—CH₂—C(=N—CH₃)—H structure

an imine

(b)

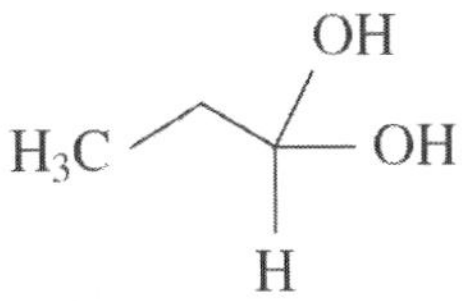

H₃C—CH₂—C(OH)(OH)—H structure

a gem diol

(d)

H₃C—CH₂—C(—O—CH₃)(—OH)—H structure

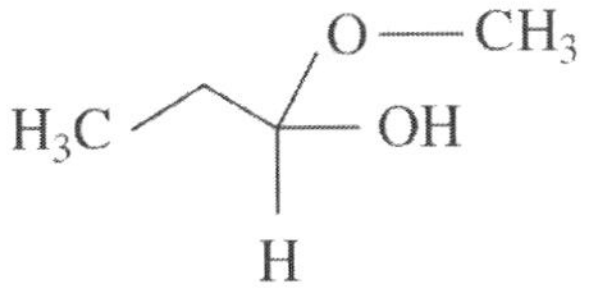

a hemiacetal

21.47

In carboxylic acid derivatives, the carbonyl carbon is bonded to two electronegative atoms (as opposed to one in aldehydes and ketones), making it more electrophilic and therefore more prone to attack by nucleophiles than is the case for aldehydes and ketones. Also, the singly-bonded heteroatom group (C–**OR** in an ester or C–**NR₂** in an amide) can depart as a leaving group under the appropriate conditions (e.g., acid or base catalysis), while the substituents on the carbonyl group of aldehydes and ketones generally will not act as leaving groups.

21.49

Here, the IR absorption for the C=O stretching vibration is a little lower, typical of a ketone. The ¹H NMR spectrum shows no aldehyde H atom peak—suggesting a ketone. The two downfield alkyl H peaks (2.4 and 2.1 ppm) are characteristic of a methine (integrates to 1) H and three methyl H's on either side of the carbonyl group. The doublet at 1.2 ppm integrating to six looks like two equivalent methyl groups split by the downfield methine H (it appears as a septet—i.e., split by six equivalent H's).

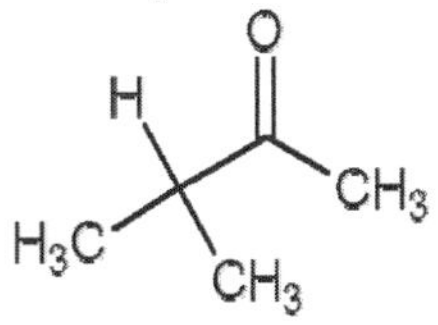

structure of compound B

B

21.51

The IR spectrum of A suggests a carboxylic acid. The –COOH accounts for two of the three O atoms. The ¹H NMR spectrum shows four peaks with relative integrals of 1, 2, 2, and 3 at 11.1, 4.1, 3.7, and 1.2, respectively. The downfield peak is clearly the acidic proton. The latter two peaks, a quartet and a triplet with relative integrations of 2 and 3, respectively, are indicative of an ethyl group. The downfield chemical shift of the two peaks at 4.1 and 3.7 suggests methylene hydrogen atoms adjacent to an oxygen atom. The structure is

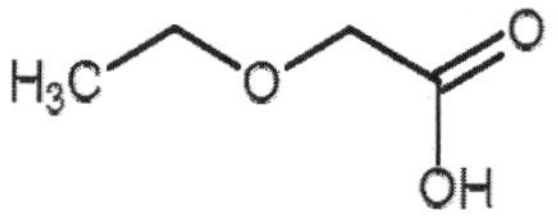

Section 21.3: Structure and Reactivity of Aldehydes and Ketones

21.53

(a)

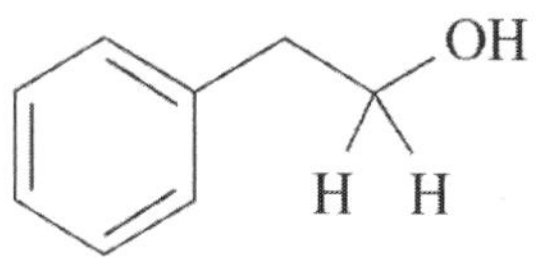

(c)

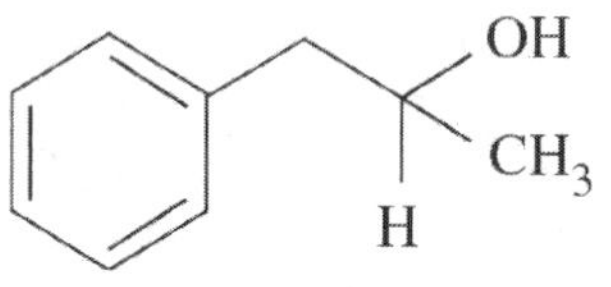

(b)

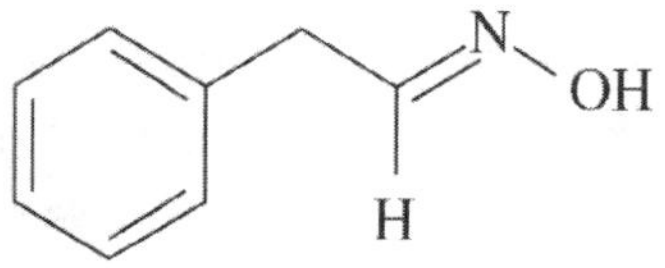

(d)

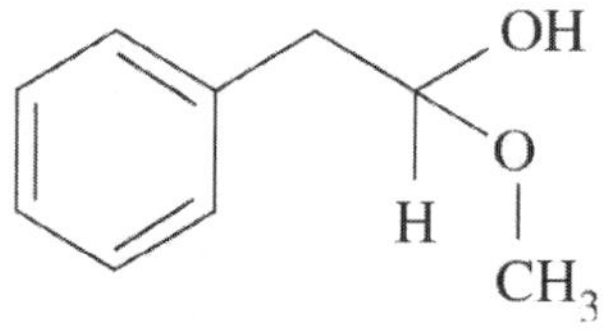

21.55

(a)

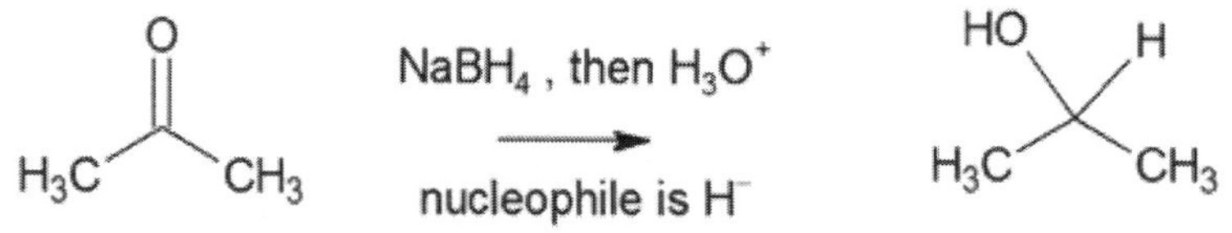

(b)

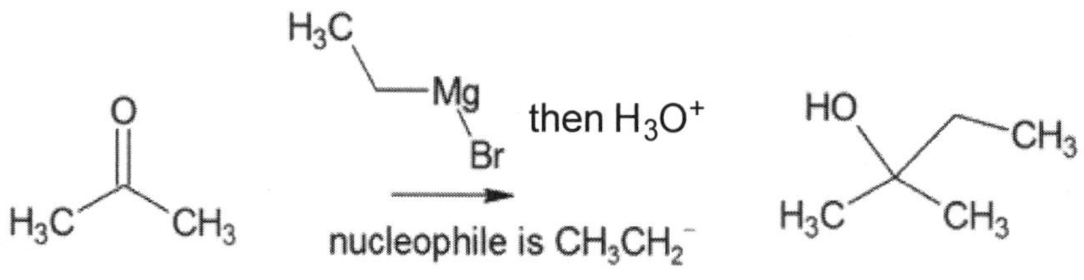

(c)

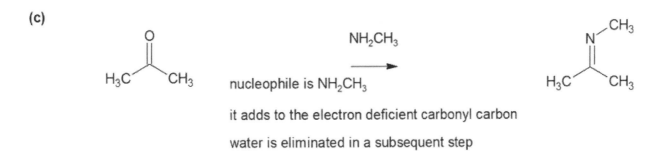

nucleophile is NH₂CH₃

it adds to the electron deficient carbonyl carbon

water is eliminated in a subsequent step

(d)

nucleophile is CH₃S⁻

attacking with an S
lone pair

21.57

The reaction of phenylmagnesium bromide with butan-2-one produces a racemate of (*R*) and (*S*)
2-phenylbutan-2-ol.

21.59

(a)

H₃O⁺ catalyst

a hemiacetal

H₃O⁺ catalyst

(b)

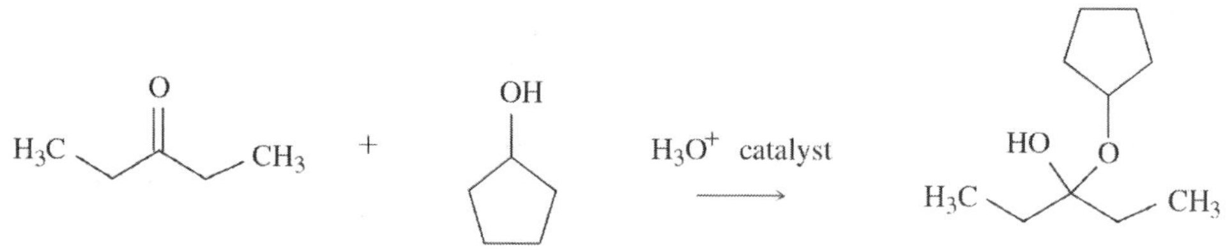

a hemiacetal

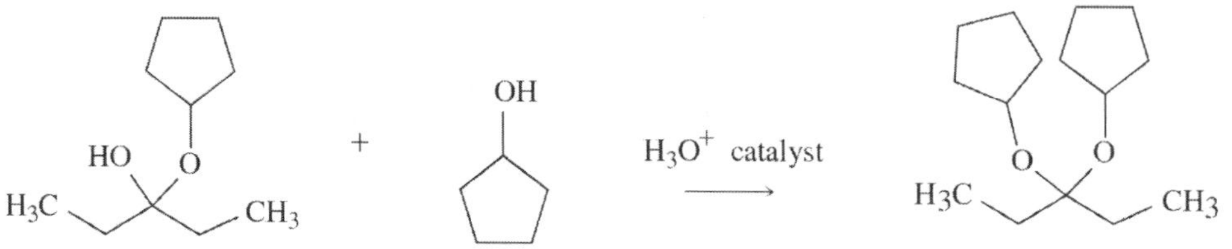

an acetal

21.61

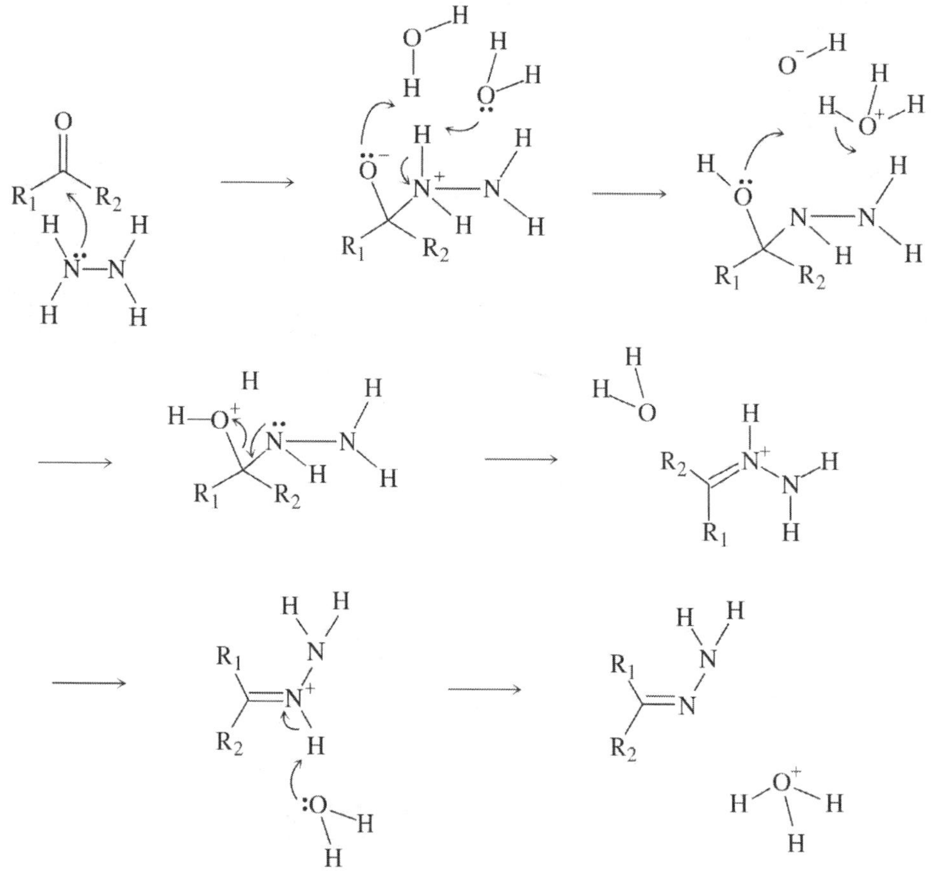

add another ketone with
the same steps

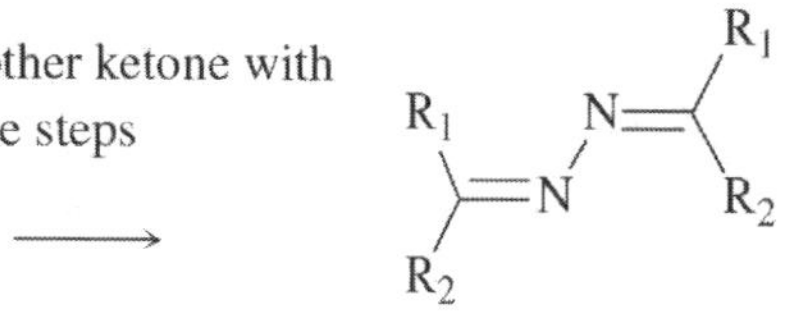

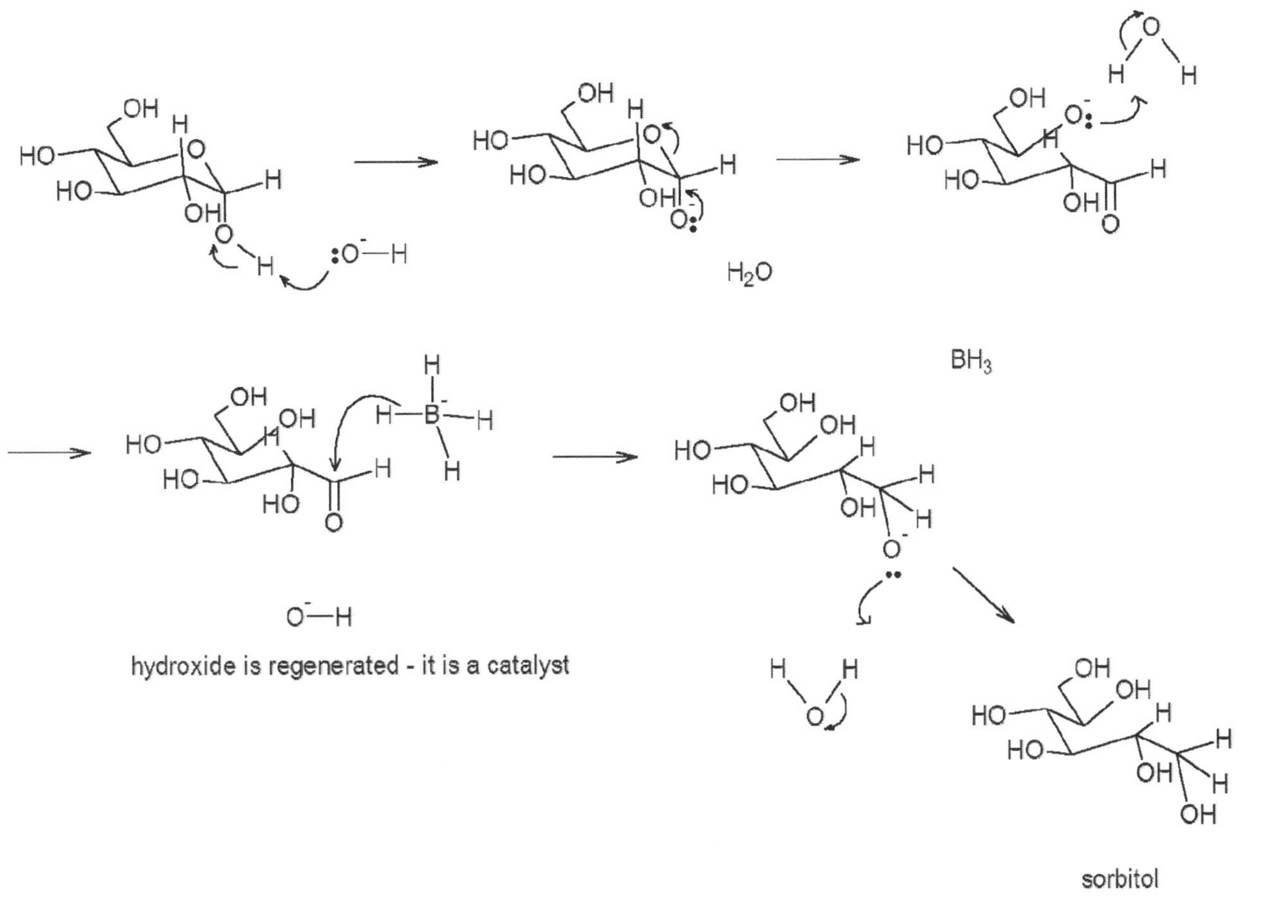
H2O
BH3
hydroxide is regenerated - it is a catalyst
sorbitol

21.63

(a)

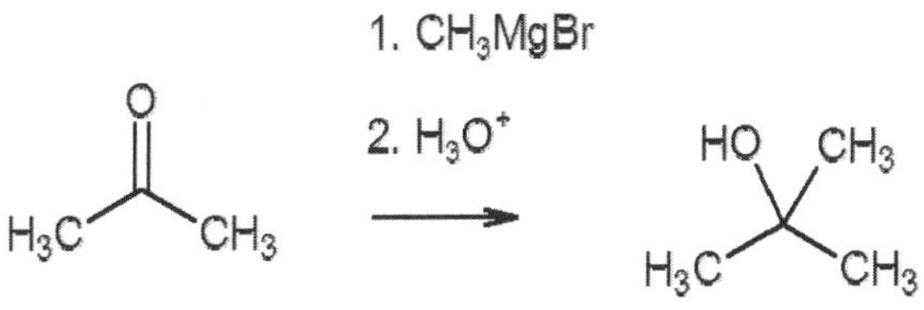
1. CH3MgBr
2. H3O+

(b)

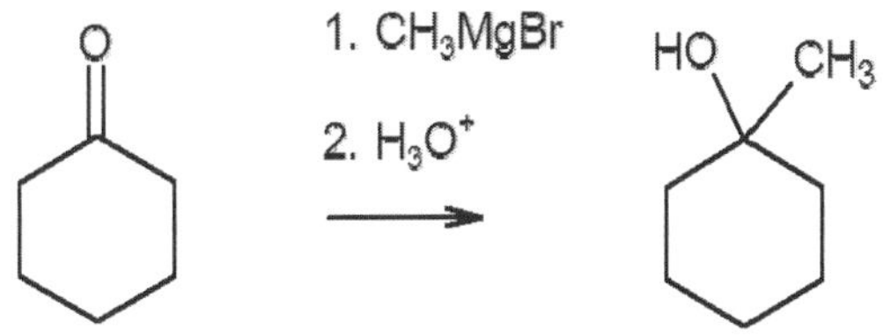
1. CH3MgBr
2. H3O+

(c)

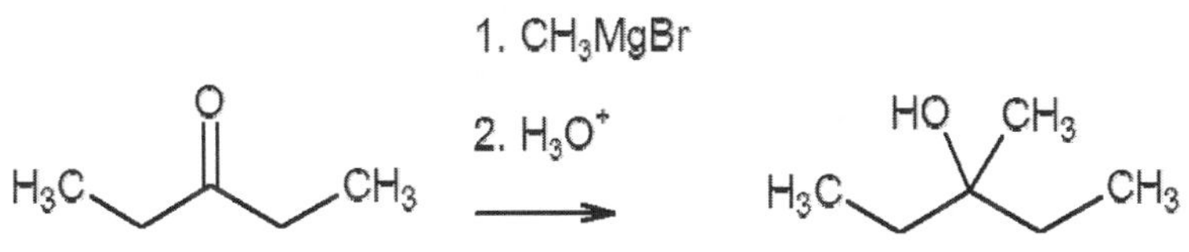
1. CH3MgBr
2. H3O+

21.65

(a)

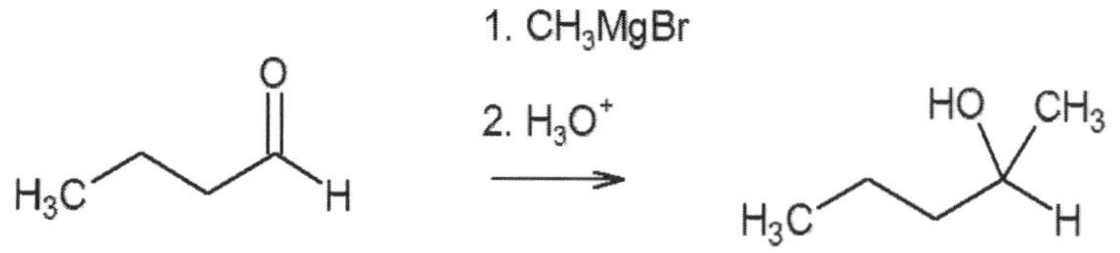
1. CH3MgBr
2. H3O+
H3C
O
H
HO
CH3
H

or

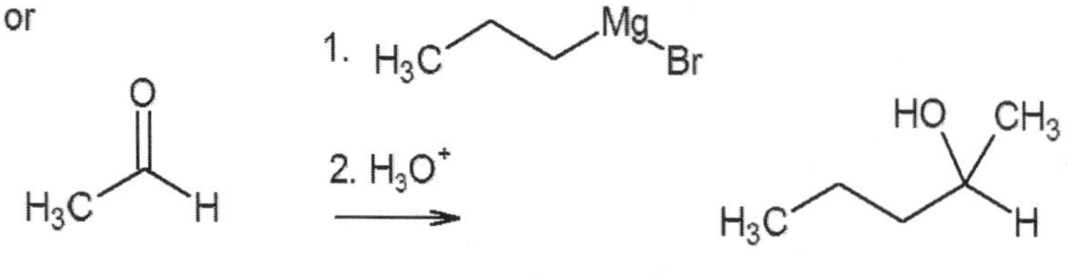
1. H3C
Mg
Br
2. H3O+
O
H3C
H
HO
CH3
H

(b)

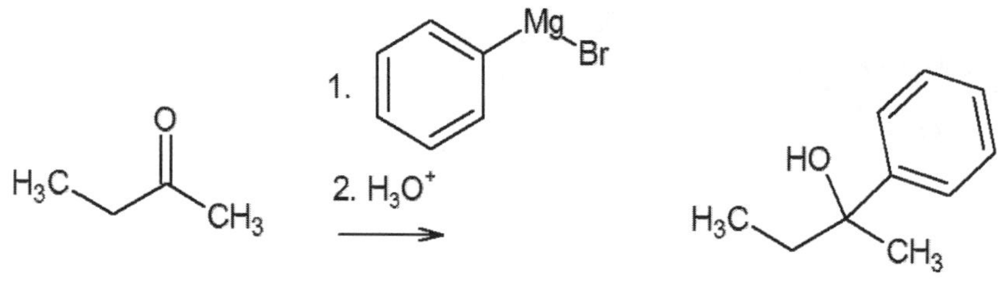
1.
Mg
Br
2. H3O+
O
H3C
CH3
HO

(c)

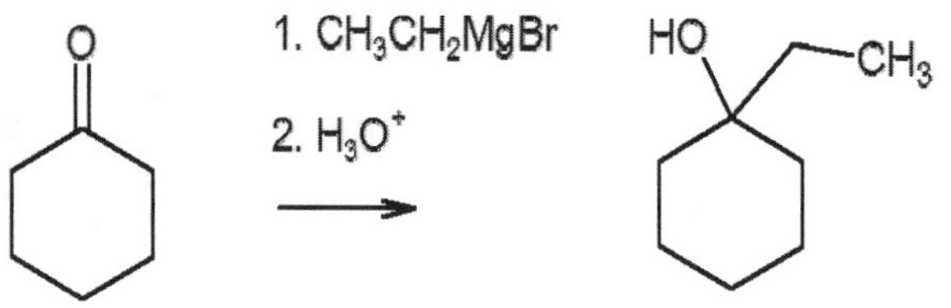
1. CH3CH2MgBr
2. H3O+
O
HO
CH3

(d)

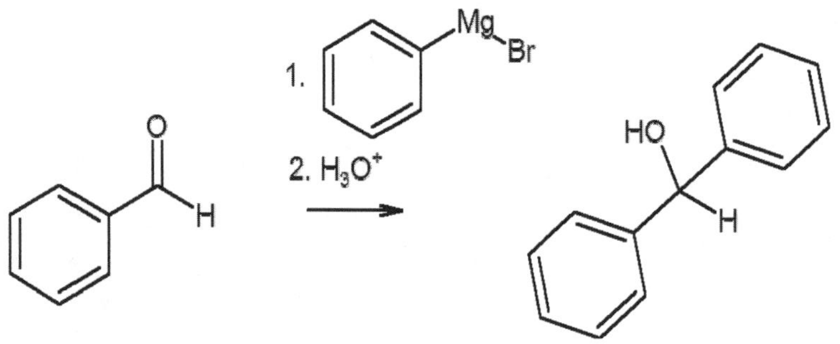
1.
Mg
Br
2. H3O+
O
H
HO

21.67

(a)

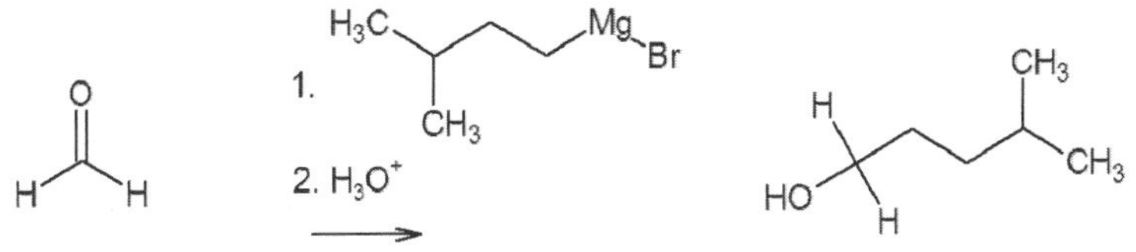

(b)

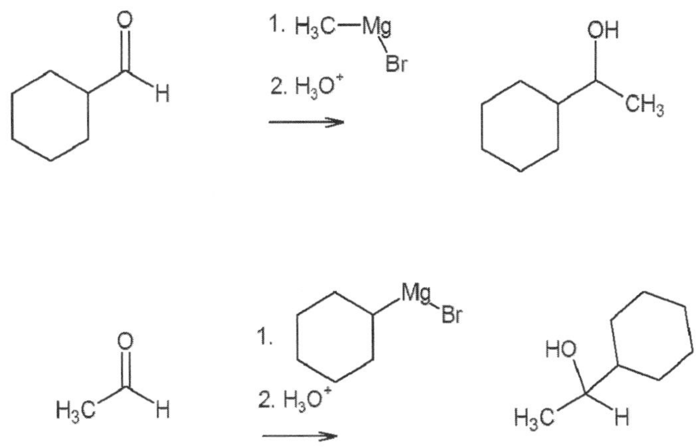

(c)

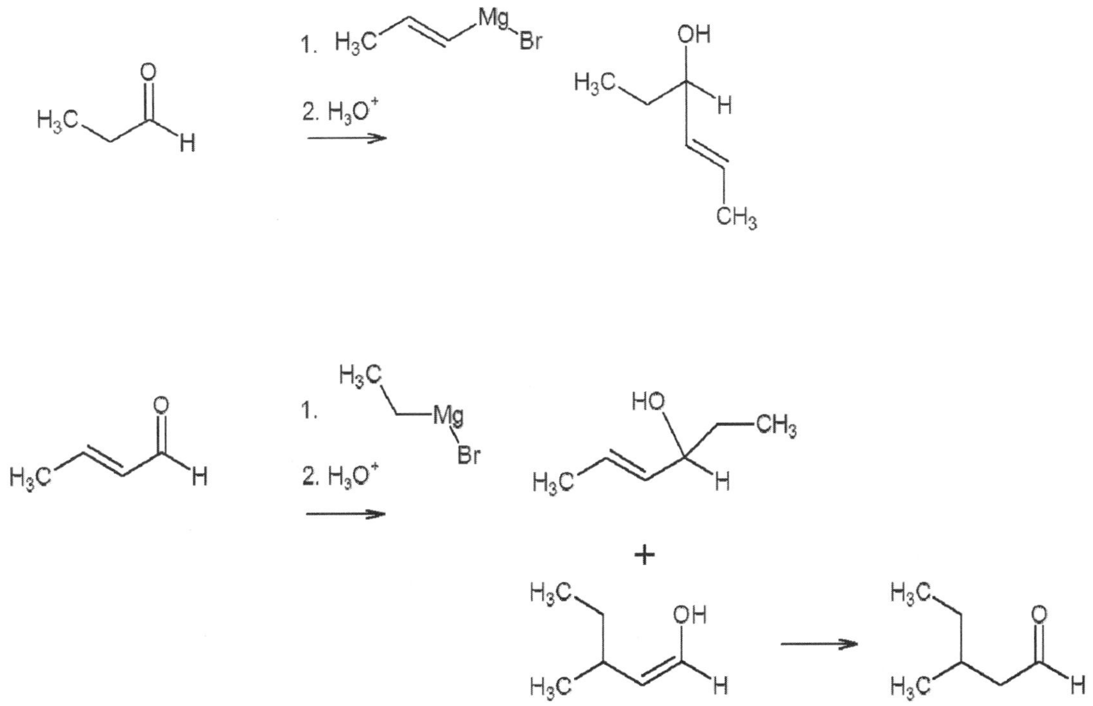

this choice leads to an additional aldehyde product
which could then react with the Grignard reagent to
form yet additional products

Section 21.4: Structure and Reactivity of Carboxylic Acid Derivatives

21.69

Tartaric acid, with the lower pK_a value, is the stronger acid.

21.71

The electron withdrawing chlorine substituent stabilizes the conjugate base anion by sharing some of the negative charge. The effect is greatest when the chlorine is next to the carboxyl group—i.e., as in 2-chlorobutanoic acid (the lowest pK_a value in this series)—and it drops off as the chlorine substituent gets further away. Nevertheless, 4-chlorobutanoic acid is still a (slightly) stronger acid than butanoic acid.

21.73

Based on the electrostatic potential maps of methyl thioacetate and methyl acetate, we can expect methyl acetate to be more reactive with respect to nucleophilic acyl substitution. The sulfur atom in the thioacetate compound is not as electron-withdrawing. Consequently, the carbonyl carbon does not have as great a partial positive charge, and will be less attracted to nucleophiles. A second factor to consider, however, is that the thiomethyl group is a weaker base and better leaving group than the methoxyl group, and methyl thioacetate is somewhat more reactive than methyl acetate.

21.75

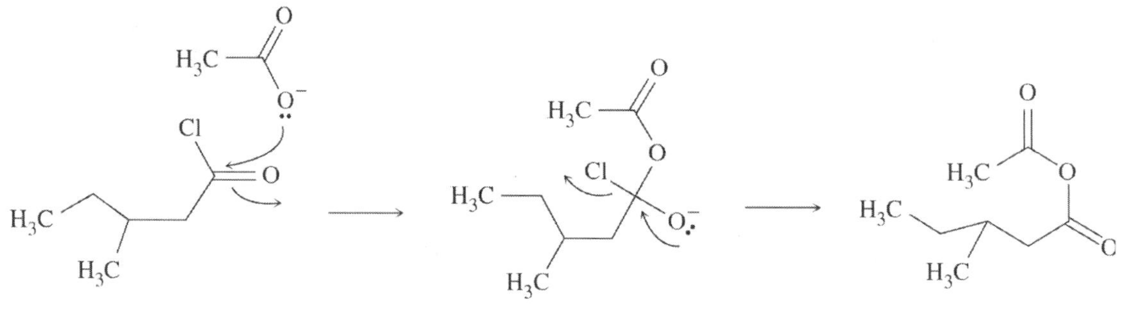

21.77

(a)

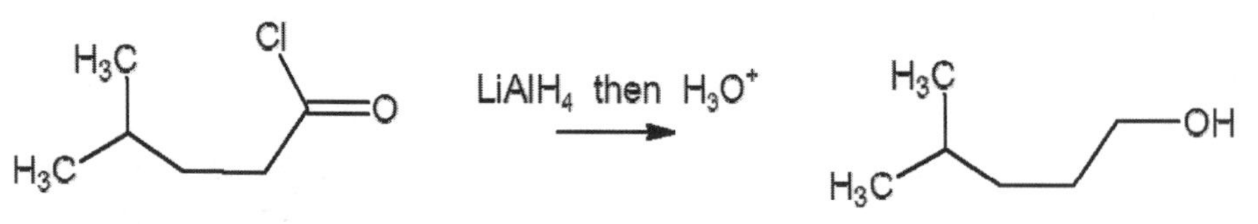

(b)

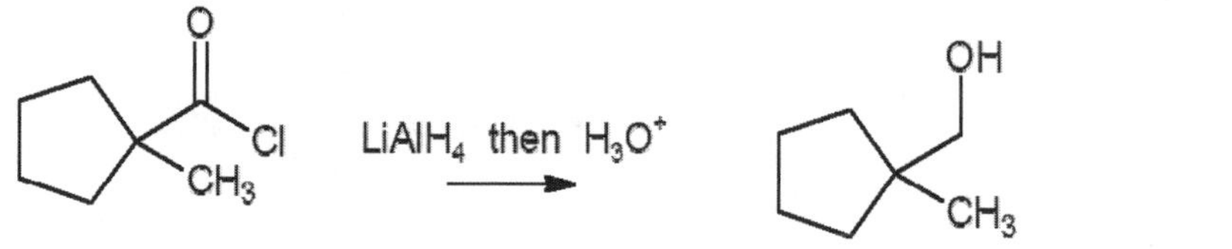

21.79

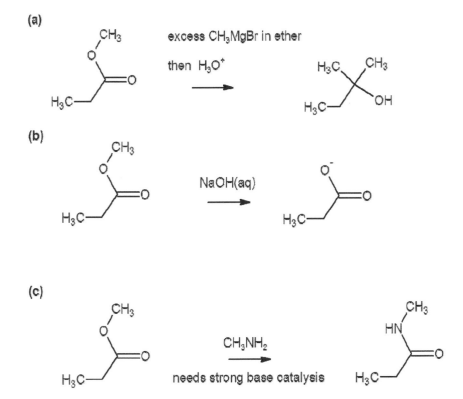

21.81

Phenyl Grignard reagent substitutes twice at the carbonate carbon, with methoxide ion as the leaving group, to give benzophenone. Then, the phenyl Grignard reagent adds to the carbonyl carbon of benzophenone to give triphenylmethanol.

21.83

(a)

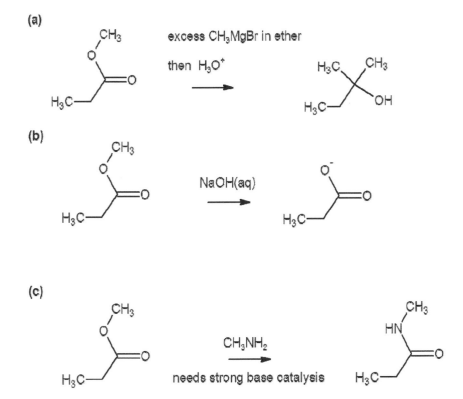

(b)

(c)

(d)

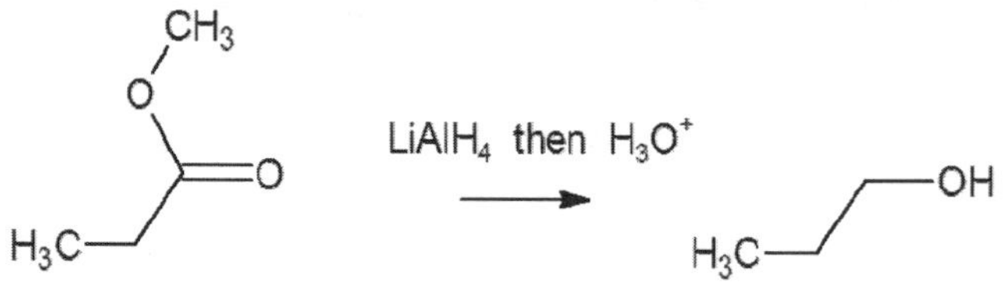
CH3
O
O
H3C
LiAlH4 then H3O+
OH
H3C

(e)

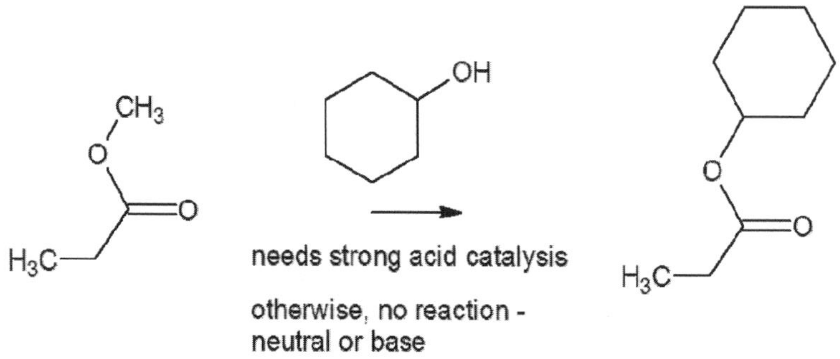
CH3
O
O
H3C
OH
needs strong acid catalysis
otherwise, no reaction -
neutral or base
O
O
H3C

(f)

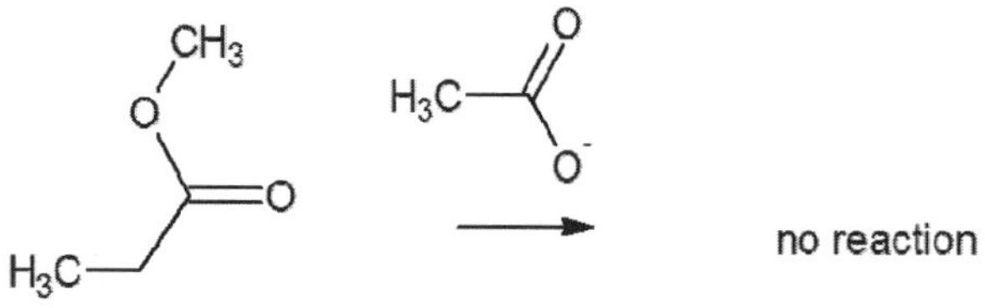
CH3
O
O
H3C
O
H3C
O-
no reaction

21.85

(a)

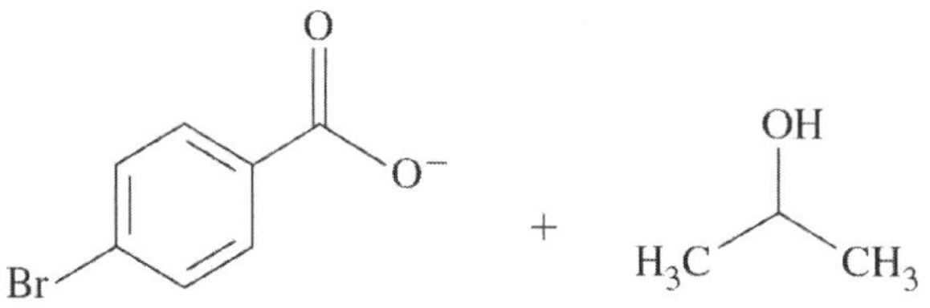
O
O−
Br
+
OH
H3C
CH3

(b)

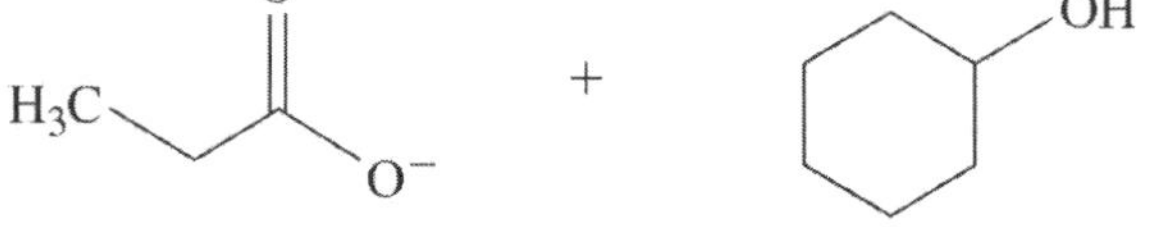
O
H3C
O−
+
OH

Section 21.5: Bacterial Cross-Talk Revisited

21.87

These two stereoisomers are not mirror images of each other. Therefore, they are diastereomers.

SUMMARY AND CONCEPTUAL QUESTIONS

21.89

(a)

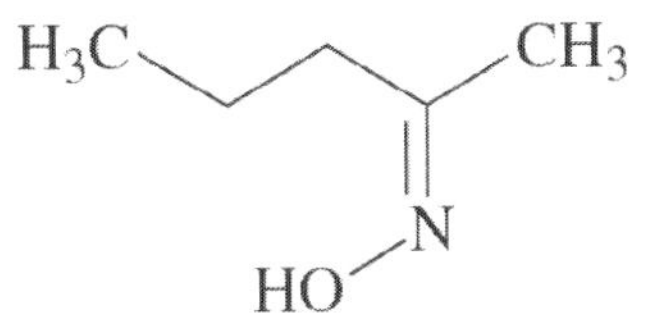

(b)

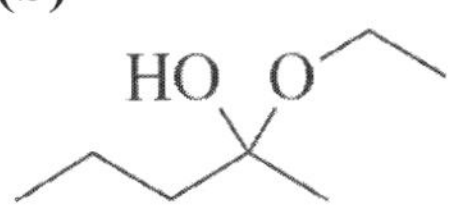

21.91

Thioacetal formation is analogous to acetal formation, with thiols adding to a ketone rather than alcohols adding to a ketone.

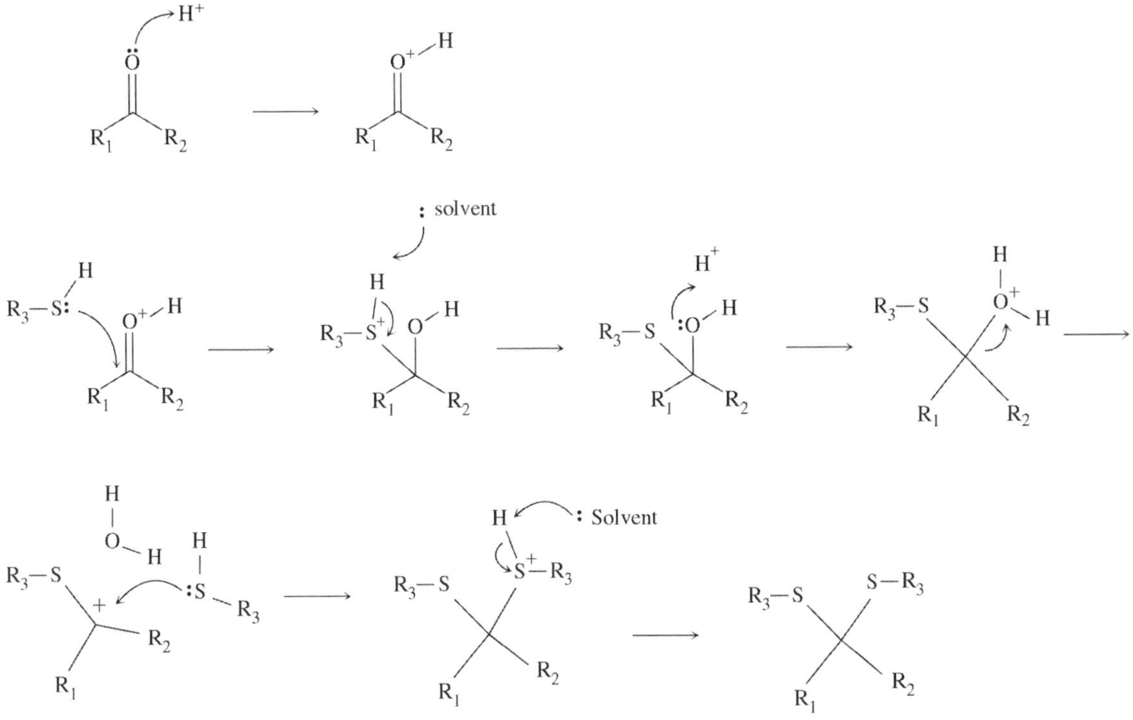

21.93

Acetic acid boils at 118°C, whereas its ethyl ester boils at 77°C—even though it is a larger molecule—because acetic acid molecules hydrogen bond to each other, with the carboxyl O–H group on one molecule H-bonding to the C=O oxygen atom of another. Ethyl acetate has no H-bonding.

21.95

(a)

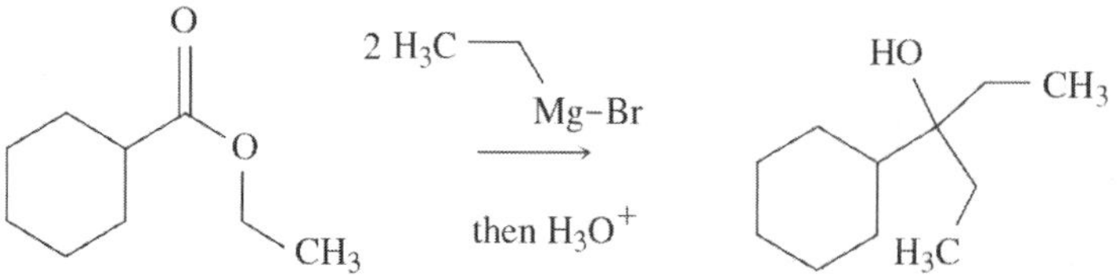

(b)

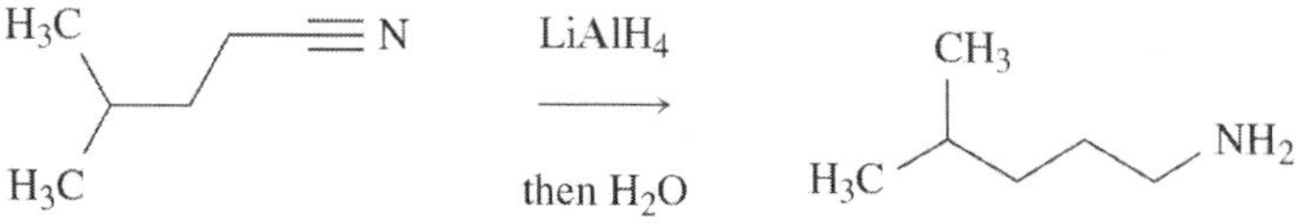

(c)

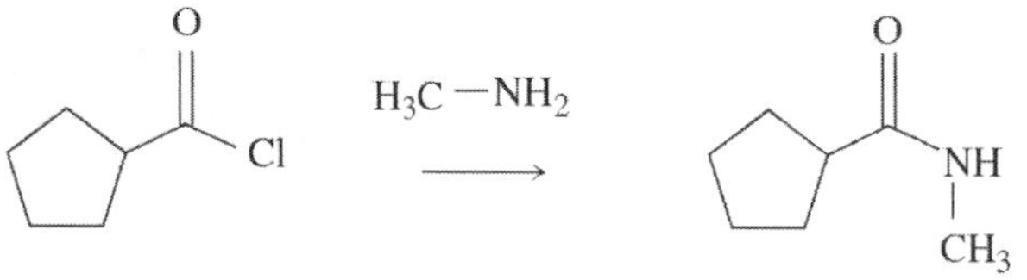

(d)

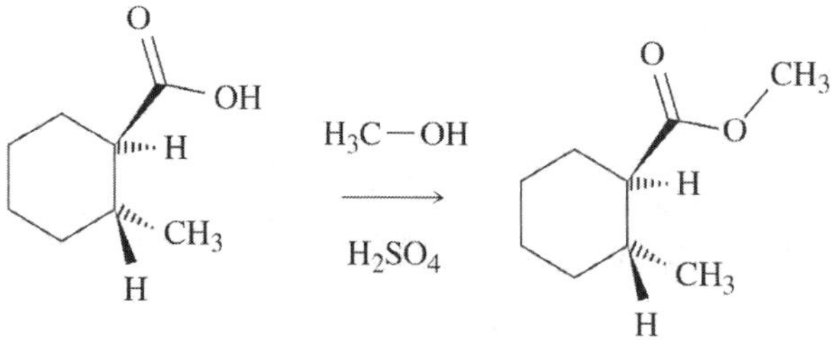

(e)

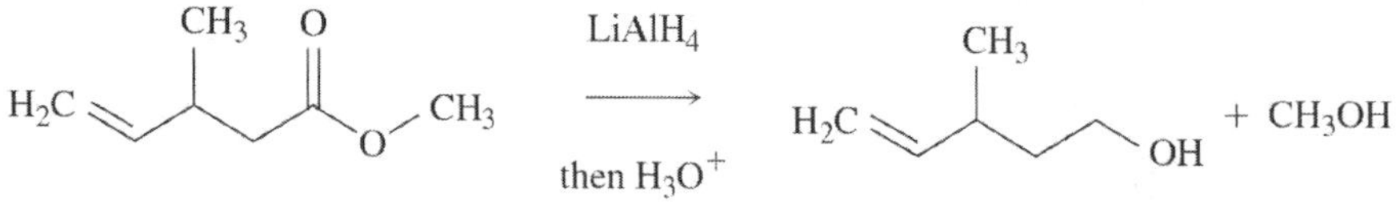

(f)

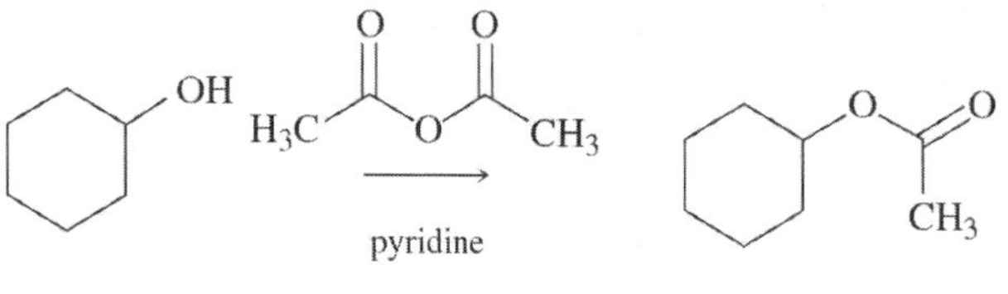

(g)

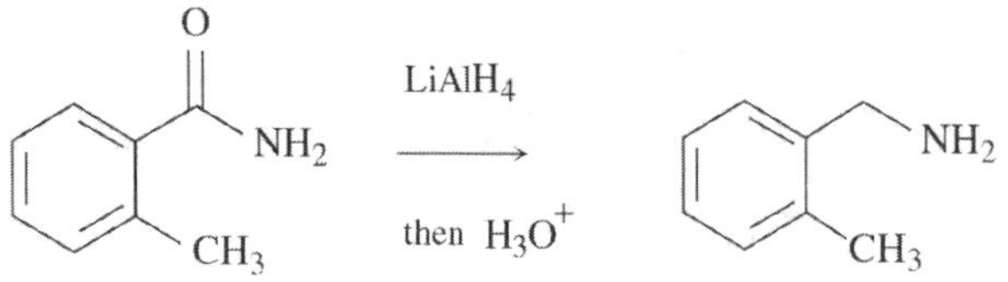

(h)

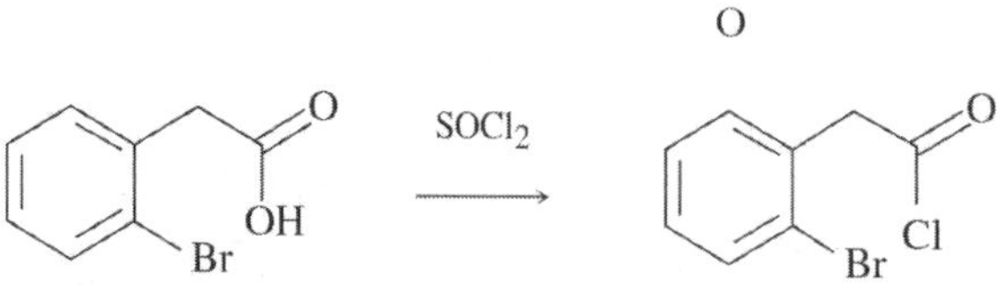

21.97

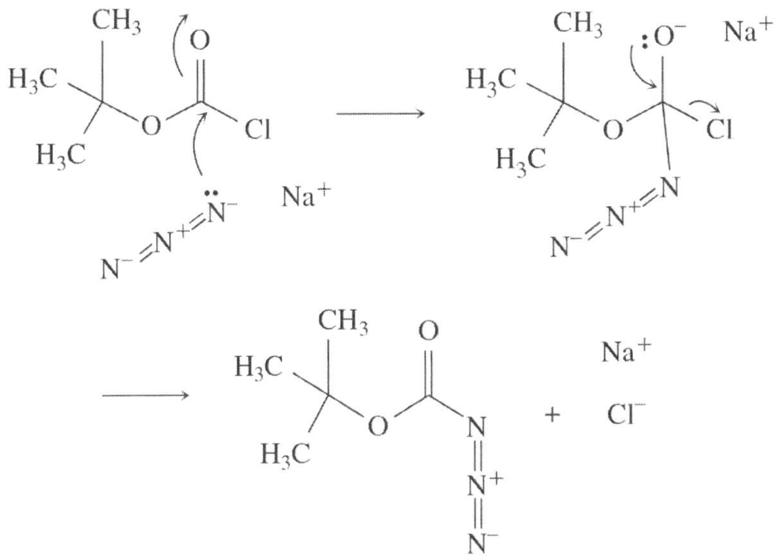

21.99

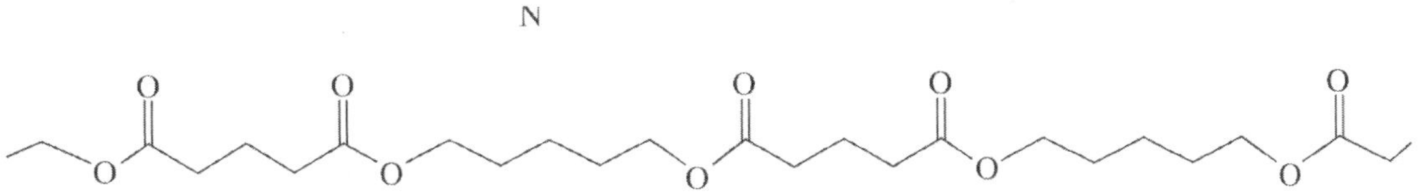

21.101

Spongistatin has 6 alcohol, 2 ether, 1 ketone, 3 ester, 1 hemiacetal, and 2 acetal groups. There are no carboxylic acids or aldehydes.

21.103

Step 1: This is a nucleophilic substitution reaction of an amine on an acyl chloride to make an amide.

Step 2: This is an S$_N$2 reaction in which an amine nucleophile displaces a chloride ion leaving group on a primary substrate.

CHAPTER 22
Main Group Elements and Their Compounds

IN-CHAPTER EXERCISES

Exercise 22.1—Influence of Charge Densities of Cations on Properties

Be^{2+} is more polarizing than Ba^{2+} because it is smaller. It can get much closer to other species, enhancing the effect of its +2 charge.

Exercise 22.3—Hydrogen and Its Compounds

$$2\,H_2(g) + O_2(g) \rightarrow 2\,H_2O(\ell)$$
$$H_2(g) + Cl_2(g) \rightarrow 2\,HCl(g)$$
$$3\,H_2(g) + N_2(g) \rightarrow 2\,NH_3(g)$$

Exercise 22.5—The Alkali Metals, Group 1

Suppose you chose sodium. The reaction with chlorine is
$$2\,Na(s) + Cl_2(g) \rightarrow 2\,NaCl(s)$$
Since all alkali metals have the same valence of 1, the other alkali metals react with chlorine with the same stoichiometry. As sodium is very electropositive and chlorine is very electronegative, the above reaction is exothermic, producing positive and negative ions bound electrostatically—i.e., an ionic product. The product is a colourless solid at room temperature—it is made from ions with closed valence shells. Colour is generally associated with open shell systems, which have smaller spacings between electronic energy levels. Ionic materials are generally high-melting solids, except in the case of specially constructed materials with very large polyatomic ions (a large spacing between ions reduces the Coulomb attraction). Sodium chloride is soluble in water, as the ions are readily solvated by polar water molecules.

Exercise 22.7—The Alkaline Earth Elements, Group 2

Suppose you chose barium. The reaction with oxygen is
$$2\,Ba(s) + O_2(g) \rightarrow 2\,BaO(s)$$
Since all alkaline earth metals have the same valence of 2, the other alkaline earth metals react with oxygen with the same stoichiometry. As barium is very electropositive and oxygen is very electronegative, the above reaction is exothermic producing positive and negative ions bound electrostatically—i.e., an ionic product. The product is a colourless solid at room temperature—it is made from ions with closed valence shells. Ionic materials are generally high-melting solids. In this case, because both ions are doubly charged, we expect a quite high melting point. Barium oxide is somewhat soluble in water, as the ions are readily solvated by polar water molecules. However, the lattice energy of the solid is very high because of the doubly charged ions. Had we chosen magnesium for the element, because Mg^{2+} is so much smaller than Ba^{2+}, the lattice energy is so high that MgO is quite insoluble in water—solubility increases as we descend the

group since the cations get bigger. Be^{2+} is so small and polarizing that its oxide has significant covalent character, and cannot be considered ionic.

Exercise 22.9—The Group 13 Elements

(a) In HCl solution, we have
$$Ga(OH)_3(s) + 3 H_3O^+(aq) \rightarrow Ga^{3+}(aq) + 6 H_2O(\ell)$$
In NaOH solution,
$$Ga(OH)_3(s) + OH^-(aq) \rightarrow [Ga(OH)_4]^-(aq)$$

(b) $Ga^{3+}(aq)$, with $K_a = 1.2 \times 10^{-3}$, is a stronger acid than $Al^{3+}(aq)$, with $K_a = 1.3 \times 10^{-5}$.

(c) Complete reaction of each mole of $Ga(OH)_3(s)$ with $H_3O^+(aq)$ ions (see part (a) above) requires 3 mol of $H_3O^+(aq)$ ions.
Amount of $Ga(OH)_3(s) = 1.25$ g / (120.75 g/mol) = 0.0104 mol
required amount of $H_3O^+(aq)$ ions = 3×0.0104 mol = 0.0312 mol
volume of HCl solution required = 0.0312 mol / 0.0112 mol L^{-1} = 2.79 L

Exercise 22.11—The Group 14 Elements

Both SiO_2 and CO_2 are covalently bonded, as group 14 elements—with a valence of 4—would be far too polarizing as +4 cations to form ionic compounds. However, whereas the period 2 element carbon forms stable double bonds, period 3 element silicon does not. Consequently, to achieve a valence of 4, silicon forms a network of single bonds with oxygen—i.e., we get a covalently bonded network solid. An entire crystal of SiO_2 is, in a sense, a single molecule. To melt (or break) a crystal of SiO_2 requires the breaking of strong covalent bonds. This requires high temperature (or a strong force). In contrast, carbon achieves its valence of 4 by forming two doubles with oxygen atoms. The small, non-polar, CO_2 molecule is bound to other such molecules by weak dispersion forces that are easily broken. Carbon dioxide is gas at ordinary temperatures.

Exercise 22.13—The Group 15 Elements

(a) The Lewis structure of N_2O shown in Table 22.8 of the text (also shown below) is not the only structure possible. Two structures, that are not unreasonable, are shown here. However, the first structure (the one shown in Table 22.8 of the text) is the most important because the negative formal charge is on the more electronegative atom—oxygen as opposed to nitrogen.

(b) $NH_4NO_3(s)$ decomposes explosively to give $N_2O(g)$ and $H_2O(g)$. The reaction is exothermic.

Exercise 22.15—The Group 16 Elements

$$\overset{\ominus}{:}\!\overset{..}{\underset{..}{S}} - \overset{..}{\underset{..}{S}}\!\overset{\ominus}{:}$$

disulfide ion

Exercise 22.17—Noble Gases

(a) Using Slater's rules, we get

$Z^*(Ar) = +18 - [(7 \times 0.35) + (8 \times 0.85) + (2 \times 1.0)] = +18 - 11.25 = +6.75$

The valence electrons of Ar are held very tightly. In contrast, the valence electron of K experiences a much smaller effective nuclear charge.

$Z^*(K) = +19 - [(0 \times 0.35) + (8 \times 0.85) + (10 \times 1.0)] = +19 - 16.8 = +2.2$

Although the valence electrons of neighbouring Cl are also held tightly, the effective nuclear charge in this case is not quite as large.

$Z^*(Cl) = +17 - [(6 \times 0.35) + (8 \times 0.85) + (2 \times 1.0)] = +17 - 10.9 = +6.1$

(b) Using Slater's rules, we get

$Z^*(Ar^-) = +18 - [(0 \times 0.35) + (8 \times 0.85) + (10 \times 1.0)] = +18 - 16.8 = +1.2$

The valence electron of Ar^- is held quite weakly. In contrast, the valence electrons of Cl^- experience a much larger effective nuclear charge.

$Z^*(Cl^-) = +17 - [(7 \times 0.35) + (8 \times 0.85) + (2 \times 1.0)] = +17 - 11.25 = +5.75$

Even the exotic K^- ion's valence electrons are held more tightly.

$Z^*(K^-) = +19 - [(1 \times 0.35) + (8 \times 0.85) + (10 \times 1.0)] = +19 - 17.15 = +1.85$

Exercise 22.19—Noble Gases

The ionization energies of xenon and krypton are much less than that of neon. So xenon and krypton atoms can enter into reactions by complete removal of electrons to form cations (with highly electronegative elements) or by partial removal to share in covalent bonds. Neon atoms attract their valence electrons too strongly for either of these to happen with common reagents—and yet their electron affinity is too low for them to "grab" electrons from other species to form anions.

REVIEW QUESTIONS

Section 22.1: Sulfur Chemistry and Life on the Edge

22.21

Extremophiles have an increasingly important role in environmental clean-up (e.g., oil spills, pesticides) and in the production of biofuels. Enzymes from extremophiles are also used for many purposes—e.g., laundry detergents, food pasteurization, and starch liquefaction (breakdown of starch into fermentable sugars).

There has also been speculation that the earliest life forms on Earth may have lived in the extreme environments of hydrothermal vents. Some types of extremophiles are also being investigated for potential use in terraforming (i.e., the transformation of other planets to develop conditions that support life).

Section 22.2: The Main Group Elements

22.23

N_2 and N_2O are neither acidic nor basic—they can neither donate nor accept H^+.
NO_3^- is basic—it can serve as a proton acceptor: $NO_3^- + HA \rightarrow HNO_3 + A^-$
NO_2^- is also basic—$NO_2^- + HA \rightarrow HNO_2 + A^-$
NH_4^+ is acidic—it can donate H^+: $NH_4^+ + B^- \rightarrow NH_3 + HB$

22.25

Many examples possible. Some possibilities, with oxidation state of chlorine in brackets:
HCl (–1) acidic ($HCl + B^- \rightarrow HB + Cl^-$)
$HClO$ (+1) acidic ($HClO + B^- \rightarrow HB + ClO^-$)
$HClO_2$ (+3) acidic ($HClO_2 + B^- \rightarrow HB + ClO_2^-$)
$HClO_3$ (+5) acidic ($HClO_3 + B^- \rightarrow HB + ClO_3^-$)
$HClO_4$ (+7) acidic ($HClO_4 + B^- \rightarrow HB + ClO_4^-$)
$NaClO$ (+1) basic ($NaClO \rightarrow Na^+ + ClO^-$; $ClO^- + HA \rightarrow HClO + A^-$)

Section 22.3: Charge Density of Cations: An Explanatory Concept

22.27

(a) If $BeCl_2$, BCl_3, $AlCl_3$, $TiCl_4$, and $FeCl_3$ were covalent, the cations would be Be^{2+}, B^{3+}, Al^{3+}, Ti^{4+}, and Fe^{3+}. These species are very small and have a high charge. They are consequently very polarizing, and there are no anions that could remain intact next to such cations. In contrast, K^+ ions and Ag^+ ions, which are singly charged and not so small, are not nearly as polarizing. Chloride anions retain their independent anionic form next to these cations—their compounds are ionic.

(b) Although Al^{3+} ions are highly polarizing, giving covalent character to bonds with halides, very small fluoride ions are rather non-polarizable, so AlF_3 is essentially an ionic compound—the bonding has very little covalent character. Its contrast, chloride and especially bromide are much more polarizable. The compounds, $AlCl_3$ and $AlBr_3$ have significantly more covalent character. These compounds more readily melt to form liquids of mobile $AlCl_3$, and $AlBr_3$ molecules.

(c) Singly charged cations, K^+, Ag^+, and NH_4^+, are not very polarizing. They do not attract water molecules strongly enough to hold them as coordinating ligands in the ionic crystal lattices of KCl, $AgNO_3$ and NH_4NO_3. In contrast, the polarizing doubly and triply charged cations, Mg^{2+}, Fe^{3+}, and Cr^{3+}, retain coordinated water ligands when their salts are precipitated from aqueous solution—i.e., they crystallize as hydrated compounds.

(d) The smaller Li^+ ion is more polarizing than the larger Rb^+ ion. Consequently, Li^+ holds water ligands about it rather tightly, and its mobility in water is that of $[Li(H_2O)_4]^+$ rather than that of free lithium ions. A rubidium ion is insufficiently polarizing to drag water molecules along with it. Its mobility is that of the free cation.

22.29

(a) The molar enthalpy change of hydration of Be^{2+} (2455 kJ mol^{-1}) is greater than that of Ba^{2+} (1275 kJ mol^{-1}) because the former is much smaller. The charges of these ions are the same, but the Coulomb energy of attraction to the negative end of the hydrating water molecules is larger for the smaller beryllium ion.

(b) The molar enthalpy change of hydration of Fe^{3+} (4340 kJ mol^{-1}) is greater than that of Fe^{2+} (1890 kJ mol^{-1}) because the former has a greater charge—it is smaller too since there is one less electron to engage in shielding (electron-electron repulsions).

22.31

The Pb^{4+} cation of lead(IV) chloride is much more polarizing than the Pb^{2+} cation of lead(II) chloride. Pb^{4+} polarizes the chlorides to such an extent that the bonds in lead(IV) chloride have significant covalent character and lead(IV) chloride can exist as individual neutral molecules in a non-polar solvent.

Section 22.4: Hydrogen

22.33

$CH_4(g) + H_2O(g) \rightarrow CO(g) + 3 H_2(g)$
$\Delta H \approx 4\,D(C-H) + 2\,D(O-H) - [D(C\equiv O) + 3\,D(H-H)]$
$= 4 \times 413 + 2 \times 463 - [1046 + 3 \times 436]$ kJ mol^{-1} = 224 kJ mol^{-1}

The reaction is endothermic because, while the number of product bonds equals the number of reactant bonds (think of the CO bond as 3 bonds), the CO triple bond is less than 3 times either of the reactant single bond energies—i.e., weaker bonds are formed from stronger bonds).

22.35

$$C(s) + H_2O(g) \rightarrow CO(g) + H_2(g)$$
$$\Delta_r H° = \Delta_f H°[H_2(g)] + \Delta_f H°[CO(g)] - (\Delta_f H°[C(s)] + \Delta_f H°[H_2O(g)])$$
$$= \{0 + (-110.525) - (0 + (-241.83))\} \text{ kJ mol}^{-1} = 131.30 \text{ kJ mol}^{-1}$$
$$\Delta_r S° = S°[H_2(g)] + S°[CO(g)] - (S°[C(s)] + S°[H_2O(g)])$$
$$= \{130.7 + 197.674 - (5.6 + 188.84)\} \text{ J K mol}^{-1} = 133.9 \text{ J K mol}^{-1}$$
$$\Delta_r G° = \Delta_r H° - T \Delta_r S° = 131.30 \text{ kJ} - (298.15 \text{ K}) \times (133.9 \text{ J K mol}^{-1}) \times (1/1000 \text{ kJ/J}) =$$
$$91.4 \text{ kJ mol}^{-1}$$

Section 22.5: The Alkali Metals, Group 1

22.37

$$Na(s) + \tfrac{1}{2}F_2(g) \rightarrow NaF(s)$$
$$Na(s) + \tfrac{1}{2}Cl_2(g) \rightarrow NaCl(s)$$
$$Na(s) + \tfrac{1}{2}Br_2(\ell) \rightarrow NaBr(s)$$
$$Na(s) + \tfrac{1}{2}I_2(s) \rightarrow NaI(s)$$

All alkali metal halides are ionic solids. They are crystalline, brittle, dissolve in water and are colourless (M^+ and X^- are closed shell species).

Section 22.6: The Alkaline Earth Elements, Group 2

22.39

$$Mg(s) + \tfrac{1}{2}O_2(g) \rightarrow MgO(s)$$
$$3 Mg(s) + N_2(g) \rightarrow Mg_3N_2(s)$$

22.41

Limestone, $CaCO_3$, is used in agriculture. It is added to fields to neutralize acidic soil, to provide a source of Ca^{2+}, an essential nutrient for plants (and animals). Since there is generally a magnesium impurity, it also provides a source of Mg^{2+}, another essential nutrient. Limestone is also used to make lime, CaO, which is used to make bricks and mortar.

22.43

Start with a balanced chemical reaction,
$$CaO(s) + SO_2(g) \rightarrow CaSO_3(s)$$
Amount of $CaO(s)$, $n = 1.2 \times 10^6 \text{ g} / 56.08 \text{ g mol}^{-1} = 2.1 \times 10^4 \text{ mol}$
Mass of $SO_2(g)$ that can be removed $= 2.1 \times 10^4 \text{ mol} \times 64.06 \text{ g mol}^{-1} = 1.3 \times 10^6 \text{ g} = 1.3 \times 10^3 \text{ kg}$

22.45

Because the equilibrium constant for the above reaction is so large, $\sim 10^7$, it can be used to obtain magnesium hydroxide solid (wherein magnesium ions are concentrated in an ionic solid) from sea water with very small concentrations. Magnesium is effectively concentrated by many orders of magnitude in the process.

22.47

Beryllium is a light-weight, high-strength metal that is stable at high temperatures and has a low thermal expansivity. Consequently, beryllium and beryllium alloys are used in airplane (especially high speed planes), missile, and spacecraft construction. It is also used in high-energy physics experiments to make windows and filters. Because of its low density, it is relatively transparent to X-rays, for example, yet has the high strength and stiffness of much heavier metals. Beryllium dust generated in manufacturing processes is very toxic. People breathing this dust over a period of time can develop berylliosis—an inflammatory disease of the lungs wherein sufferers experience coughing, shortness of breath, fever, and weight loss.

Section 22.7 Boron, Aluminum, and the Group 13 Elements

22.49

$2\ BH_4^- + 2\ I_2 \rightarrow B_2H_6 + 2\ HI + 2\ I^-$

Iodine gets reduced and the borohydride ion, BH_4^-, gets oxidized (specifically, one of the H atoms goes from oxidation state -1 to $+1$).

22.51

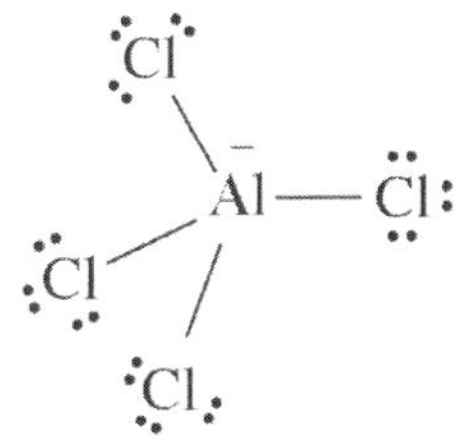

$AlCl_4^-$ is tetrahedral—it has the AX_4 electron domain configuration.

22.53

(a) $BCl_3(g) + 3\ H_2O(\ell) \rightarrow B(OH)_3(s) + 3\ H^+(aq) + 3\ Cl^-(aq)$
   or $BCl_3(g) + 6\ H_2O(\ell) \rightarrow B(OH)_3(s) + 3\ H_3O^+(aq) + 3\ Cl^-(aq)$

(b) $\Delta_r H^\circ =$
   $\Delta_f H^\circ[B(OH)_3(s)] + 3 \times \Delta_f H^\circ[HCl(aq)] - (\Delta_f H^\circ[BCl_3(g)] + 3 \times \Delta_f H^\circ[H_2O(\ell)])$
   $= \{-1094 + 3(-167.159) - ((-403) + 3(-285.83))\}\ kJ\ mol^{-1} = -335\ kJ\ mol^{-1}$

Section 22.8: Silicon and the Group 14 Elements

22.55

Elemental silicon can be prepared from sand by heating purified silica (i.e., sand) and coke.

$$SiO_2(s) + 2\,C(s) \rightarrow Si(\ell) + 2\,CO(g)$$

Molten silicon is drawn from the bottom of the furnace. Its purity is not sufficient for electronics applications. Consequently, the elemental silicon is reacted with chlorine to make silicon tetrachloride.

$$Si(s) + 2\,Cl_2(g) \rightarrow SiCl_4(\ell)$$

The low boiling silicon tetrachloride (boiling point = 57.6°C) is purified by distillation, then reacted with extremely pure magnesium (or zinc).

$$SiCl_4(g) + 2\,Mg(s) \rightarrow 2\,MgCl_2 + Si(s)$$

The magnesium chloride product is washed away with water—it is water soluble, while silicon is not. Silicon produced in this way is further purified by zone refining wherein a narrow cylinder of silicon is heated and melted in one segment along the cylinder. The heating slowly traverses the length of the cylinder. Impurities concentrate in the melted segment, leaving higher purity silicon behind. This is because of melting-point depression—impure silicon has a lower melting point than pure silicon.

22.57

(a) $Si + 2\,CH_3Cl \rightarrow (CH_3)_2SiCl_2$

(b) Amount of Si = 2.65 g / 28.09 g mol^{-1} = 0.0943 mol
Amount of chloromethane needed = 2×0.0943 mol = 0.1886 mol
Pressure of chloromethane needed,
$p = nRT/V = 0.1886$ mol $\times 8.314$ L kPa K^{-1} mol$^{-1} \times 298.15$ K / 5.60 L = 83.5 kPa

(c) Amount of $(CH_3)_2SiCl_2$ produced, n = 0.0943 mol
Mass of $(CH_3)_2SiCl_2$ produced, m = 0.0943 mol $\times$ 129.06 g mol^{-1} = 12.2 g

Section 22.9: Group 15 Elements and Their Compounds

22.59

$$Ca(OH)_2(s) + H_3PO_4(aq) \rightarrow CaHPO_4(s) + 2\,H_2O(\ell)$$

22.61

$$2\,NO_2(g) \longrightarrow N_2O_4(g)$$
$$\Delta_r H° = \Delta_f H°[N_2O_4(g)] - 2 \times \Delta_f H°[NO_2(g)]$$
$$= (9.08 - 2 \times 33.1)\ \text{kJ mol}^{-1} = -57.1\ \text{kJ mol}^{-1}$$
$$\Delta_r G° = \Delta_f G°[N_2O_4(g)] - 2 \times \Delta_f G°[NO_2(g)]$$
$$= (97.73 - 2 \times 51.23)\ \text{kJ mol}^{-1} = -4.73\ \text{kJ mol}^{-1}$$

This reaction is exothermic and at 25°C and with both substances at 1 bar pressure, the reaction is spontaneous in the direction written.

22.63

Half reactions (it is convenient to use the shorthand, $H^+(aq)$):

$N_2H_5^+(aq) \rightarrow N_2(g) + 5\,H^+(aq) + 4\,e^-$

$IO_3^-(aq) + 6\,H^+(aq) + 5\,e^- \rightarrow \frac{1}{2}I_2(s) + 3\,H_2O(\ell)$

Balance overall reaction by balancing the electrons (i.e., $5 \times$ 1st reaction $+ 4 \times$ 2nd reaction):

$5\,N_2H_5^+(aq) + 4\,IO_3^-(aq) \rightarrow 5\,N_2(g) + H^+(aq) + 2\,I_2(s) + 12\,H_2O(\ell)$

$E° = E°[IO_3^-(aq)\,|\,I_2(s)] - E°[N_2H_5^+(aq)\,|\,N_2(g)]$
$= 1.195 - (-0.23)\ \text{V} = 1.43\ \text{V}$

22.65

(a) Lewis structures for N_2O_3

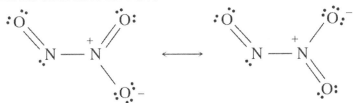

Here, we see that two almost equivalent resonance structures can be drawn, giving the two N–O bonds on the right bond orders of about 1½. Resonance structures showing the N–O bond on the left as anything but a double bond are not at all dominant. This latter bond is thus more of a double bond. This is consistent with its observed shorter length.

(b) $\Delta_rS° = (\Delta_rH° - \Delta_rG°) / T =$
$= (40.5 - (-1.59)\ \text{kJ mol}^{-1}) \times 10^3\ \text{J kJ}^{-1} / 298\ \text{K} = 141\ \text{J K}^{-1}\ \text{mol}^{-1}$

$K = \exp(-\Delta_rG° / RT) =$
$= \exp(-(-1.59 \times 10^3\ \text{J mol}^{-1}) / (8.314\ \text{J K}^{-1}\ \text{mol}^{-1} \times 298\ \text{K})) = 2.00$

(c)

$\Delta_rH° = \Delta_fH°[NO(g)] + \Delta_fH°[NO_2(g)] - \Delta_fH°[N_2O_3(g)] = 40.5\ \text{kJ mol}^{-1}$

Therefore,

$\Delta_fH°[N_2O_3(g)] = \Delta_fH°[NO(g)] + \Delta_fH°[NO_2(g)] - 40.5\ \text{kJ mol}^{-1}$
$= (90.29 + 33.1 - 40.5)\ \text{kJ mol}^{-1} = 82.9\ \text{kJ mol}^{-1}$

Section 22.10: The Group 16 Elements and Their Compounds

22.67

$2\,ZnS(s) + 3\,O_2(g) \longrightarrow 2\,ZnO(s) + 2\,SO_2(g)$

$\Delta_rH° = 2 \times \Delta_fH°[ZnO(s)] + 2 \times \Delta_fH°[SO_2(g)] - (2 \times \Delta_fH°[ZnS(s)] + 3 \times \Delta_fH°[O_2(g)])$
$= \{2 \times (-348.28) + 2 \times (-296.84) - (2 \times (-205.98) + 3 \times 0)\}\ \text{kJ mol}^{-1}$
$= -878.28\ \text{kJ mol}^{-1}$

$\Delta_rG° = 2\,\Delta_fG°[ZnO(s)] + 2\,\Delta_fG°[SO_2(g)] - (2\,\Delta_fG°[ZnS(s)] + 3\,\Delta_fG°[O_2(g)])$
$= \{2 \times (-318.30) + 2 \times (-300.13) - (2 \times (-201.29) + 3 \times 0)\}\ \text{kJ mol}^{-1}$
$= -834.28\ \text{kJ mol}^{-1}$

This reaction is product-favoured—$\Delta_r G° < 0$—and it is exothermic which means that it is less product-favoured at higher temperature.

22.69

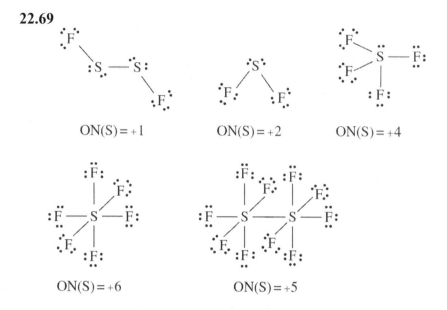

| $ON(S) = +1$ | $ON(S) = +2$ | $ON(S) = +4$ |

| $ON(S) = +6$ | $ON(S) = +5$ |

Section 22.11: The Halogens, Group 17

22.71

Any species with reduction potential greater than that of chlorine (i.e., > 1.36 V) can be used. For example,

$$F_2(g) + 2 Cl^-(aq) \rightarrow 2 F^-(aq) + Cl_2(g)$$
$$H_2O_2(aq) + 2 H^+(aq) + 2 Cl^-(aq) \rightarrow 2 H_2O(\ell) + Cl_2(g)$$
$$PbO_2(s) + SO_4^{2-}(aq) + 4 H^+(aq) + 2 Cl^-(aq) \rightarrow PbSO_4(s) + 2 H_2O(\ell) + Cl_2(g)$$
$$2 MnO_4^-(aq) + 16 H^+(aq) + 10 Cl^-(aq) \rightarrow 2 Mn^{2+}(aq) + 8 H_2O(\ell) + 5 Cl_2(g)$$

22.73

The reduction potential of $MnO_4^-(aq)$ is 1.51 V under standard conditions. Since this is greater than 1.44 V, it is NOT possible to oxidize $Mn^{2+}(aq)$ with $BrO_3^-(aq)$.

22.75

22.77

Amount of CaF₂ required = 1.0×10^6 L × 2.0×10^{-5} mol L^{-1} = 20 mol
Mass of CaF₂ required = 20 mol × 78.08 g mol^{-1} = 1560 g = 1.6 kg (to two significant figures)

SUMMARY AND CONCEPTUAL QUESTIONS

22.79

$P_4O_{10}(s)$ and $SO_3(g)$ are acidic oxides—potassium and sulfur are non-metals.
$$P_4(s) \ + \ 5\,O_2(g) \ \rightarrow \ P_4O_{10}(s)$$
$$S_8(s) \ + \ 12\,O_2(g) \ \rightarrow \ 8\,SO_3(g)$$
The acidic oxides react with water to form acidic solutions:
$$P_4O_{10}(s) \ + \ 6\,H_2O(\ell) \ \rightarrow \ 4\,H_3PO_4(aq) \text{ phosphoric acid, which ionizes in aqueous solution}$$
to form $H_3O^+(aq)$.
$$SO_3(g) \ + \ H_2O(\ell) \ \rightarrow \ H_2SO_4\,(aq) \text{ sulfuric acid, which ionizes in aqueous solution to form}$$
$H_3O^+(aq)$.

22.81

S^{2-}, Cl^-, K^+, and Ca^{2+} ions are monatomic ions with the same electron configuration as argon.

22.83

In order of increasing basicity, we have
$CO_2 \ < \ SiO_2 \ < \ SnO_2$

22.85

(a) $2\,Na(s) \ + \ Br_2(\ell) \ \rightarrow \ 2\,NaBr(s)$
(b) $2\,Mg(s) \ + \ O_2(g) \ \rightarrow \ 2\,MgO(s)$
(c) $2\,Al(s) \ + \ 3\,F_2(g) \ \rightarrow \ 2\,AlF_3(s)$
(d) $C(s) \ + \ O_2(g) \ \rightarrow \ CO_2(g)$

22.87

(a) Electrolysis of aqueous sodium chloride:
$2\,Na^+(aq) \ + \ 2\,Cl^-\,(aq) \ + \ 2\,H_2O(\ell) \ \rightarrow \ 2\,Na^+(aq) \ + \ 2\,OH^-\,(aq) \ + \ Cl_2(g) \ + \ H_2(g)$
(b) Electrolysis of aqueous sodium chloride is not the only source of sodium hydroxide, or of hydrogen. It is, however, the principal source of chlorine.

22.89

(a) Li(s) and Be(s) are metals
B(s) is a metalloid

C(s), N_2(g), O_2(g), F_2(g) and Ne(g) are non-metallic substances.

(b) + (c) Li(s), Be(s) and B(s) are silver-white shiny, with metallic appearance. C is a black solid (graphite) or a lustrous transparent very hard crystalline material (diamond). N_2(g), O_2(g), F_2(g) and Ne(g) are all gases, colourless except for a yellow tinge in the case of fluorine.

22.91

(a) $2 KClO_3(s) + heat \rightarrow 2 KCl(s) + 3 O_2(g)$

(b) $2 H_2S(g) + 3 O_2(g) \rightarrow 2 H_2O(\ell) + 2 SO_2(g)$

(c) $2 Na(s) + O_2(g) \rightarrow Na_2O_2(s)$

(d) $P_4(s) + 3 OH^-(aq) + 3 H_2O(\ell) \rightarrow PH_3(g) + 3 H_2PO_2^-(aq)$

(e) $NH_4NO_3(s) + heat \rightarrow N_2O(g) + 2 H_2O(g)$

(f) $2 In(s) + 3 Br_2(\ell) \rightarrow 2 InBr_3(s)$

(g) $SnCl_4(\ell) + 4 H_2O(\ell) \rightarrow Sn(OH)_4(aq) + 4 H^+(aq) + 4Cl^-(aq)$

22.93

Dry, inert powders are used to extinguish sodium fires. A class D fire extinguisher is required. The worst thing you can do is put water on the fire, as the sodium will react violently with water.

22.95

The amount of gas produced by heating 1.00 g A is

$$n(\text{gas}) = \frac{pV}{RT} = \frac{(27.86 \text{ kPa})(0.450 \text{ L})}{(8.314 \text{ L kPa K}^{-1} \text{ mol}^{-1})(298 \text{ K})} = 0.00506 \text{ mol}$$

Bubbling this gas through $Ca(OH)_2$(aq) gives a white solid, C. This suggests the gas is carbon dioxide. Carbon dioxide reacts in aqueous calcium hydroxide to form insoluble calcium carbonate, $CaCO_3$(s). CO_2(g) is also a typical product in the thermal decomposition of carbonates, suggesting A is a carbonate. The other decomposition product, B, would then be an oxide. This is consistent with it forming a basic aqueous solution. If the gas were CO_2, then the mass of 0.00506 mol would be

0.00506 mol $\times$ 44.01 g mol^{-1} = 0.223 g

This leaves $1.00 - 0.223 = 0.777$ g as the mass of the oxide. If we take the decomposition reaction to be (M is replaced by M_2 in case of a univalent metal, etc.)

$MCO_3(s) + heat \rightarrow MO(s) + CO_2(g)$,

then the molar mass of MO(s) is determined to be

0.777 g / 0.00506 mol = 154 g mol^{-1}

Subtracting the molar mass of O leaves the molar mass of M,

154 g mol^{-1} $-$ 16 g mol^{-1} = 138 g mol^{-1}

This is pretty close to the molar mass of barium. This identification is verified by the green coloured flame, characteristic of barium. Also, the sulfate precipitate is characteristic of barium.

A = $BaCO_3$(s)   B = BaO(s)   C = $CaCO_3$(s)   D = $BaCl_2$(s)   E = $BaSO_4$(s)

CHAPTER 23
Transition Elements and Their Compounds

IN-CHAPTER EXERCISES

Exercise 23.1—Ligands

(a) methylamine, CH_3NH_2, (b) methyl nitrile, CH_3CN, (c) azide ions, N_3^- and (e) bromide ions, Br^- are monodentate ligands.

(d) en (ethylenediamine, $H_2NCH_2CH_2NH_2$) and (f) phen (phenanthroline, $C_{12}H_8N_2$) are bidentate (special case of polydentate) ligands—i.e., they can form two coordination bonds with metals.

Exercise 23.3—Formulas of Coordination Complexes

(a) A complex ion composed of one Co^{3+} ion, three ammonia molecules, and three Cl^- ions, has the formula
$[Co(NH_3)_3Cl_3]$
The ammonia ligands are neutral and there are three chlorides each with a charge of -1. So the net charge coming from the ligands is -3. Since the cobalt ion has a charge of $+3$, the net charge on the complex is zero—it is not an ion.

(b) (i) The net charge on the complex in $K_3[Co(NO_2)_6]$ is -3. We deduce this because each of the three potassium ions has a charge of $+1$. Since each of the nitrite, NO_2^-, ions has a -1 charge, and there are 6 of them in the complex, the net contribution to the complex charge from the ligands is -6. A net charge of -3 for the complex results because the cobalt ion has a charge of $+3$. The coordination number of cobalt is 6: it is bonded to 6 donor atoms.

(ii) The $[Mn(NH_3)_4Cl_2]$ complex is neutral. The contribution to this charge of zero from the ligands is -2, arising solely from the two -1 charged chloride ions (the ammonia ligands are neutral). Thus, the charge on the manganese ion must be $+2$. The coordination number of manganese is 6: it is bonded to 6 donor atoms.

Exercise 23.5—The Chelate Effect

(i) $\Delta_rG° = \Delta_rH° - T\Delta_rS° = -12.1$ kJ mol^{-1} $- 298$ K $\times 185$ J K^{-1} mol^{-1} / 1000 J kJ^{-1} = -67.2 kJ mol^{-1}
The entropy term accounts for the bigger part of this negative change in free energy. The large positive entropy of this process results because 6 ligands are freed with only 3 bidentate ligands taking their place. The final arrangement of atoms is more constrained.

(ii) $K = \exp(-\Delta_rG°/RT) = \exp(67.2 \times 10^3$ J mol^{-1} / (8.314 J K^{-1} mol^{-1} $\times 298$ K)) = 6.02×10^{11}
This reaction is very product-favoured: at equilibrium, the concentrations of product species will heavily dominate the concentrations of reactants species.

Exercise 23.7—Constitutional Isomerism in Transition Metal Complexes

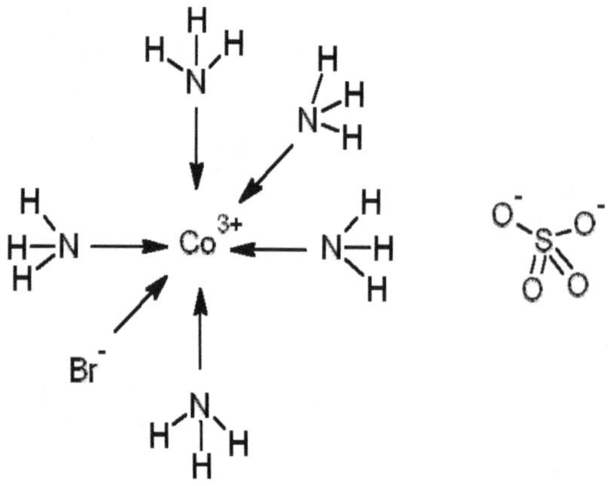

dark violet isomer (A)

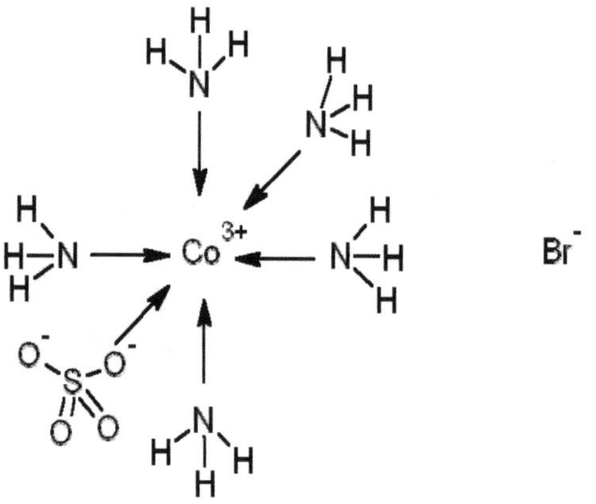

violet-red isomer (B)

(A) = $[Co(NH_3)_5 Br]SO_4$
(B) = $[Co(NH_3)_5 SO_4]Br$
Upon dissolution of (A):
$$[Co(NH_3)_5 Br]SO_4(s) \rightarrow [Co(NH_3)_5 Br]^{2+}(aq) + SO_4^{2-}(aq)$$
Upon dissolution of (B):
$$[Co(NH_3)_5 SO_4]Br(s) \rightarrow [Co(NH_3)_5 SO_4]^+(aq) + Br^-(aq)$$
Adding $Ba^{2+}(aq)$ to the (A) solution gives a barium sulfate precipitate.
$$Ba^{2+}(aq) + SO_4^{2-}(aq) \rightarrow BaSO_4(s)$$
Adding $Ba^{2+}(aq)$ to the (B) solution gives no reaction because the sulfate ions are tied up in the complex and are not available to form the precipitate.

Exercise 23.9—Isomerism in Coordination Complexes

(a) $[Co(NH_3)_2Cl_4]$ has *cis* and *trans* isomers with the ammonia ligands next to each other or on opposite sides of the cobalt ion. Neither of these has enantiomers (neither has chiral molecules).

(b) $[Pt(en)Cl_2]$ does not have *cis* and *trans* isomers: the ethylenediamine ligand must occupy adjacent positions on the metal as, therefore, must the Cl^- ion ligands. There is only one way to do this. The complex has a plane of symmetry, so there are no enantiomers.

(c) $[Co(NH_3)_5Cl]^{2+}$ has no isomers since there is only one chloride ligand—putting it in any of the 6 positions produces the same complex ion. The complex ion has a plane of symmetry, so there are no enantiomers.

(d) There are two enantiomers of $[Ru(phen)_3]^{3+}$ ions. Each phen occupies adjacent positions on the metal ion, limiting the number of possible isomers. If we draw (or build) the structure of this complex ion, it can be seen that it is not superimposable upon its mirror image. It is chiral, and there are enantiomers.

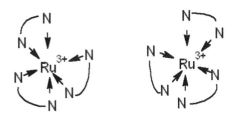

(e) $[MnCl_4]^{2-}$ has no isomers—all the ligands are the same.

(f) $[Co(NH_3)_5NO_2]^{2+}$ There are two linkage isomers with this formula—one with the NO_2^- ion ligand bound to the Co^{3+} ion through a lone pair on the N atom, and one with the NO_2^- ion ligand bound through a lone pair on an O atom.

Exercise 23.11—Colours of Coordination Complexes

According to the size of the crystal field splitting, these species are ordered in terms of energy of photons absorbed as follows:
$[Ti(OH_2)_6]^{3+} < [Ti(NH_3)_6]^{3+} < [Ti(CN)_6]^{3-}$
Here, the strength of the splitting is determined by the ligand, as these species are otherwise the same.

Exercise 23.13—Magnetism of Transition Metal Complexes

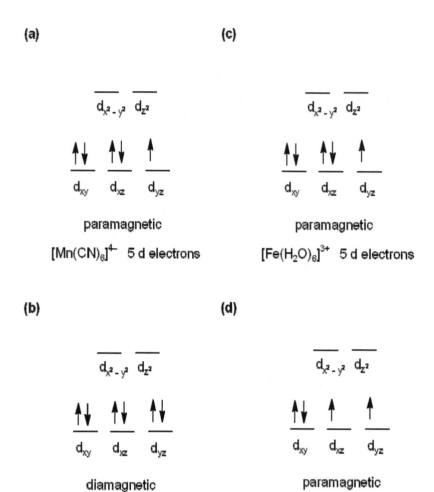

(a)

$d_{x^2-y^2}$ d_{z^2}

d_{xy} d_{xz} d_{yz}

paramagnetic

$[Mn(CN)_6]^{4-}$ 5 d electrons

(c)

$d_{x^2-y^2}$ d_{z^2}

d_{xy} d_{xz} d_{yz}

paramagnetic

$[Fe(H_2O)_6]^{3+}$ 5 d electrons

(b)

$d_{x^2-y^2}$ d_{z^2}

d_{xy} d_{xz} d_{yz}

diamagnetic

$[Co(NH_3)_6]^{3+}$ 6 d electrons

(d)

$d_{x^2-y^2}$ d_{z^2}

d_{xy} d_{xz} d_{yz}

paramagnetic

$[Cr(en)_3]^{2+}$ 4 d electrons

REVIEW QUESTIONS

Section 23.1: The Serendipitous Discovery of Cisplatin, an Anti-Cancer Drug

23.15

The *cis* isomer of $PtCl(OH_2)(NH_3)_2^+$ tends to bind to two adjacent guanosine nucleotides on the same strand because the two groups to be replaced ($-OH_2^+$ and $-Cl$) by the guanosine nitrogens are in close proximity to each other (on the same side). In contrast, the reacting groups of the *trans* isomer point in opposite directions. Due to this geometry, the *trans* isomer is not able to react with adjacent guanosine nucleotides and must instead bind to nucleotides on different strands or more distant nucleotides on the same strand.

Section 23.2: The *d*-Block Elements and Compounds

23.17

(a) $[Ar]\, 3d^6$ Fe^{2+} and Co^{3+}

(b) $[Ar]\, 3d^{10}$ Cu^+ and Zn^{2+}

(c) $[Ar]\, 3d^5$ Mn^{2+} and Fe^{3+}

(d) $[Ar]\, 3d^8$ Co^+ and Ni^{2+}

Section 23.3: Metallurgy

23.19

(a) Here, chromium goes from oxidation state +3 to 0, while aluminum goes from 0 to +3. The number of electrons gained by each chromium atom equals the number of electrons lost by each aluminum atom. Thus, it is just a matter of balancing the chromium and aluminum atoms.

$$Cr_2O_3(s) + 2\,Al(s) \longrightarrow Al_2O_3(s) + 2\,Cr(s)$$

(b) Each mol of $TiCl_4(\ell)$ gains 4 mol electrons, while 2 mol electrons are removed from each mol of magnesium atoms. Therefore, there is a 1:2 reaction stoichiometry.

$$TiCl_4(\ell) + 2\,Mg(s) \longrightarrow Ti(s) + 2\,MgCl_2(s)$$

(c) Ag goes from the +1 to the 0 state, while Zn goes from 0 to +2. Therefore, there is a 2:1 reaction stoichiometry.

$$2\left[Ag(CN)_2\right]^-(aq) + Zn(s) \longrightarrow 2\,Ag(s) + \left[Zn(CN)_4\right]^{2-}(aq)$$

(d) Mn goes from an average oxidation state of 8/3 to 0. Al goes from 0 to +3
Therefore, there is a $3 : 8/3 = 9 : 8$ ratio of amounts of manganese atoms (in Mn_3O_4) to aluminum atoms in Al(s) that react.

$$3\,Mn_3O_4(s) + 8\,Al(s) \longrightarrow 9\,Mn(s) + 4\,Al_2O_3(s)$$

Section 23.4: Coordination Compounds

23.21

$[Cr(en)_2(NH_3)_2]^{3+}$ Since both en and NH_3 are neutral, the complex has the charge of the chromium ion, +3.

23.23

First, dissolve a known amount of $CrCl_3$ in a known volume of water. Actually, we would require a volume of water less than desired final volume of solution, then top up the volume with water to account for the change in volume resulting from dissolution. This process is repeated several times, except that sodium chloride is also added to the solution. The resulting solutions all have the same (known) total concentration of Cr^{3+}, and different (known) total concentrations of chloride. Excess $AgNO_3(aq)$ is then added to each of these solutions. There will be a precipitate of silver chloride. After filtering, drying, and weighing the precipitate, we can determine how much chloride was available for precipitation in each solution. Since the total amount of chloride is known for each solution, we can determine the amount of chloride that was not available for precipitation in each case. This is the amount of chloride that was part of the complex. Since the total concentration of chromium is known in each case, we now know the ratio of chromium to complex-bound chloride for each solution. These values are then plotted against free chloride concentration—determined by the amount of precipitate. This plot should show whole number plateaus in ranges of chloride concentrations determined by the successive formation constants for the various species. We should be able to identify the ranges that correspond to $[Cr(H_2O)_6]^{3+}$, $[Cr(H_2O)_5Cl]^{2+}$, and $[Cr(H_2O)_4Cl_2]^+$. These correspond to 0, 1, and 2 bound chloride per chromium. According to the information given, we need to add diethylether to obtain $[Cr(H_2O)_3Cl_3]$. Thus, an additional experiment wherein diethylether is added to a solution from the experiment described—before the precipitation reaction—in order to obtain the neutral complex species. The neutral compound will partition into the diethylether layer. It can be separated off, and the neutral complex can be obtained by evaporating the ether. Also, once the solutions with specific charged species are identified, we can evaporate the water and crystallize the distinct ionic solid compounds. There may be other ways. Discuss as a group.

Section 23.5: Complexation Equilibria, Stability of Complexes

23.25

(a)

$$[Cd(OH_2)_6]^{2+} + en \rightleftharpoons [Cd(OH_2)_4(en)]^{2+} + 2H_2O(\ell) \qquad K_{f1} = 2.5\times10^5$$

$$[Cd(OH_2)_4(en)]^{2+} + en \rightleftharpoons [Cd(OH_2)_2(en)_2]^{2+} + 2H_2O(\ell) \qquad K_{f2} = 3.2\times10^4$$

$$\underline{[Cd(OH_2)_2(en)_2]^{2+} + en \rightleftharpoons [Cd(en)_3]^{2+} + 2H_2O(\ell) \qquad K_{f3} = 63}$$

$$[Cd(OH_2)_6]^{2+} + 3\,en \rightleftharpoons [Cd(en)_3]^{2+} + 6H_2O(\ell) \qquad K_f = K_{f1}\times K_{f2}\times K_{f3} = 5.0\times10^{11}$$

(b)

$$\frac{[Cd(OH_2)_4(en)]^{2+}}{[Cd(OH_2)_6]^{2+}[en]} = K_{f1} = 2.5\times10^5$$

$$\frac{[Cd(OH_2)_4(en)]^{2+}}{[Cd(OH_2)_6]^{2+}} = K_{f1}\times[en] = 2.5\times10^5\times1\times10^{-3} = 250$$

(c) $[Cd(OH_2)_4(en)]^{2+} = [Cd(OH_2)_2(en)_2]^{2+}$ when $[en] = 1/K_{f2} = 3.1\times10^{-5}$ M

23.27

Formation constants for the formation of complexes with ammonia.

Complex	Co^{2+}	Ni^{2+}	Cu^{2+}	Zn^{2+}
K_f also known as β	7.7×10^4	1×10^9	1×10^{13}	2.9×10^9

Here we see the order Co^{2+} < Ni^{2+} < Cu^{2+} > Zn^{2+} for the stability of the ammine complexes. This is consistent with the Irving-Williams series.

Section 23.7: Isomerism in Coordination Complexes

23.29

The square-planar complex, $[Pt(NH_3)(CN)Cl_2]^-$, has two stereoisomers—*cis* and *trans*.

23.31

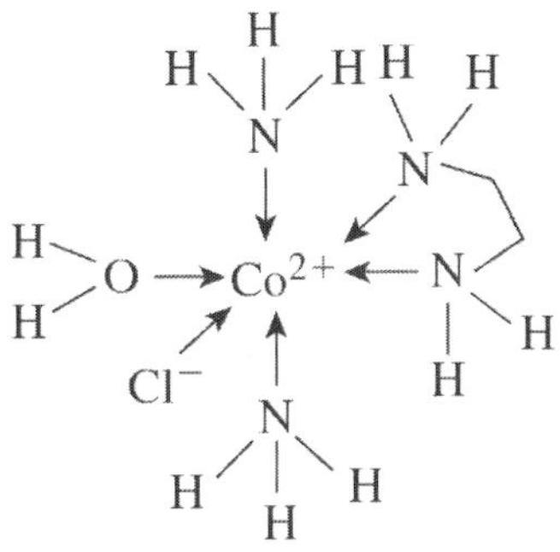

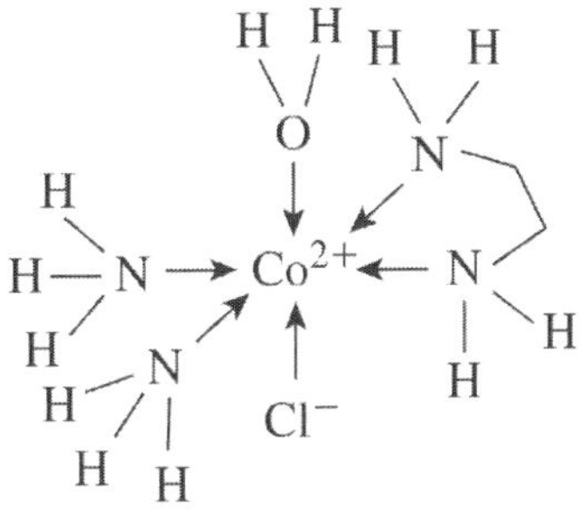

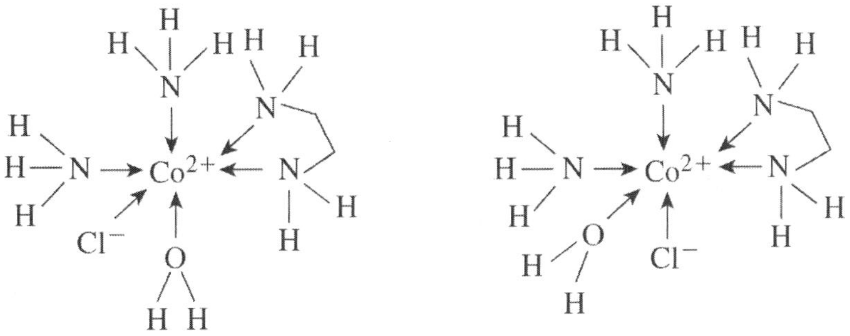

enantiomers

23.33
(a)

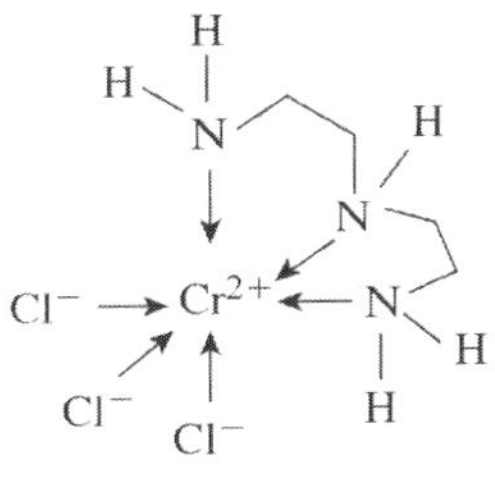

fac 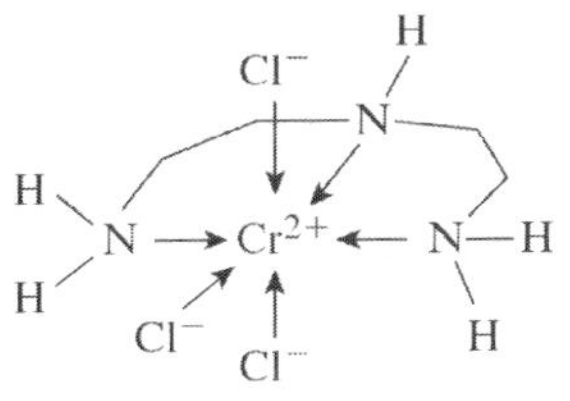   *mer*

(b)

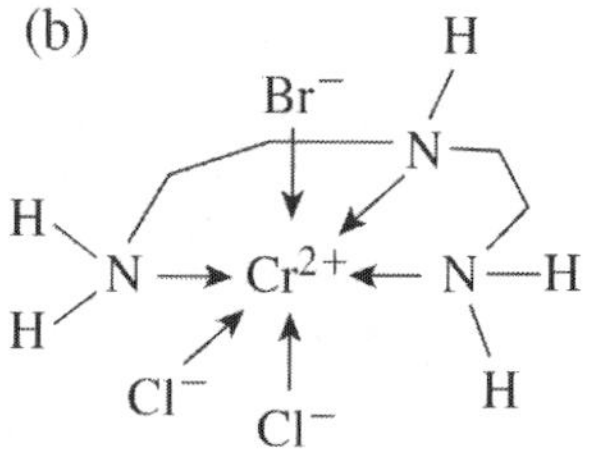

cis-mer 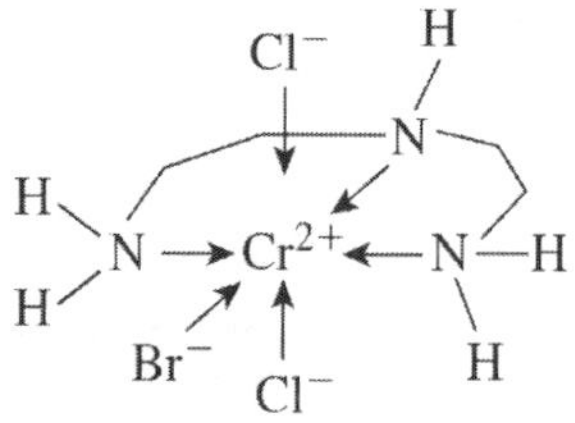   *trans-mer*

These complexes are inequivalent because the middle N atom of dien is inequivalent to the outer two N atoms. In the complex on the left, the bromide is *cis* to this N atom—*trans* on the right.

(c)

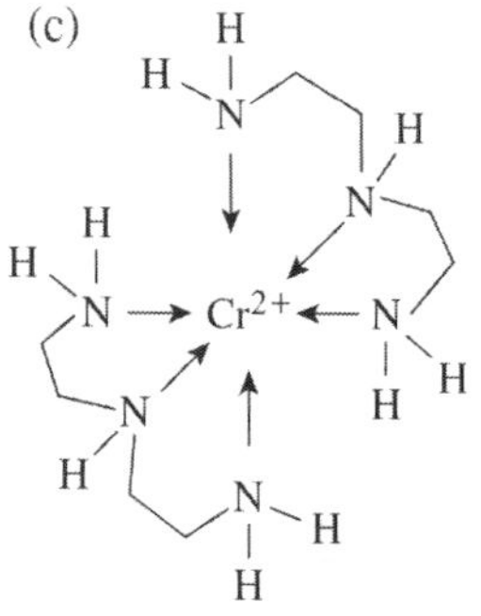

trans-fac 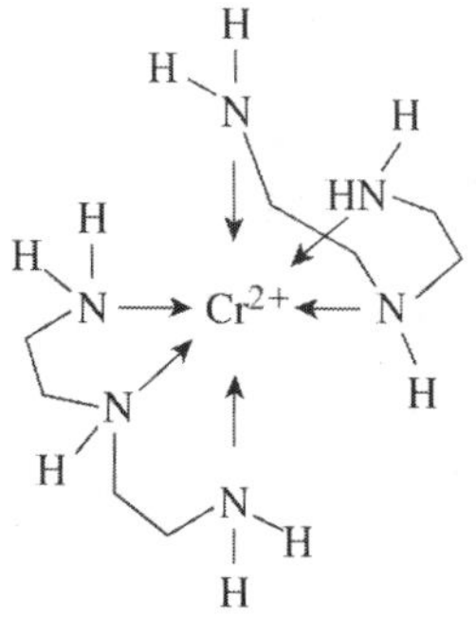   *cis-fac*

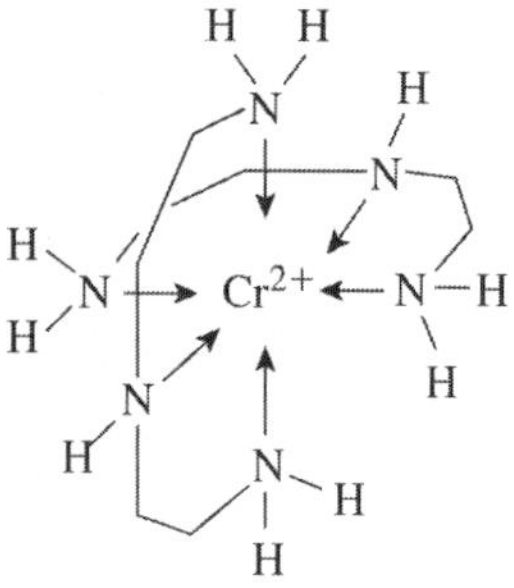

mer

23.35

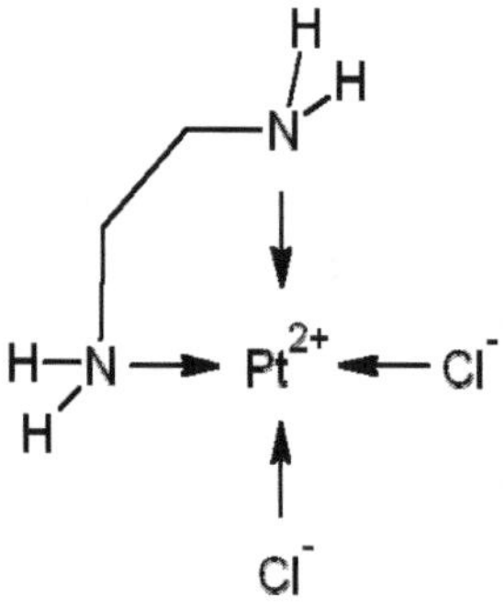

cis

A *trans* isomer would require the N atoms to be on opposite sides of the platinum. The $-CH_2CH_2-$ link between the N atoms is not long enough to allow this.

Section 23.8: Bonding in Coordination Complexes

23.37

(a) The five d orbitals have the same energy. **WRONG**
A tetrahedral arrangement of ligands breaks the spherical symmetry about the metal atom—it is this spherical symmetry about a free metal atom that gives rise to the equal energies of the 5 d orbitals in that case.
(b) The $d_{x^2-y^2}$ and d_{z^2} orbitals are higher in energy than the d_{xz}, d_{yz}, and d_{xy} orbitals. **WRONG**
This is the order of d orbital energies in an octahedral complex.
(c) The d_{xz}, d_{yz}, and d_{xy} orbitals are higher in energy than the $d_{x^2-y^2}$ and d_{z^2} orbitals.
CORRECT
(d) The d orbitals all have different energies. **WRONG**
A tetrahedral arrangement of ligands is still quite symmerical. A less symmetrical arrangement of ligands is required to split the d orbitals into all different energies.

23.39

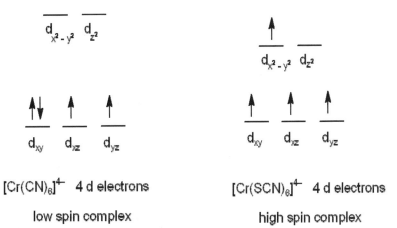

SCN⁻ ions are lower in the spectrochemical series than CN⁻ ions. SCN⁻ ions do not split the d orbital energies as much as cyanide ions do.

23.41

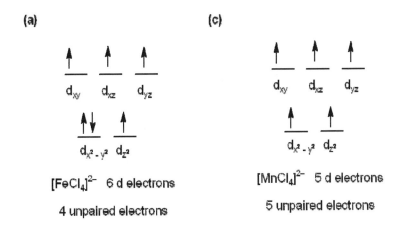

(b)

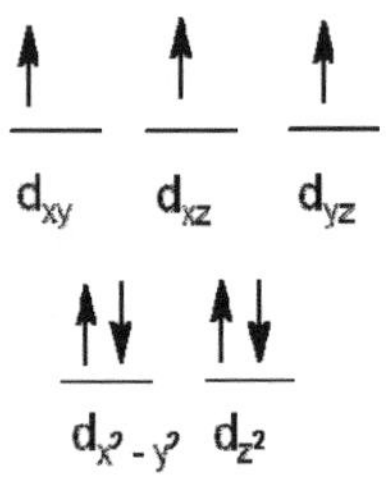

$[CoCl_4]^{2-}$ 7 d electrons

3 unpaired electrons

(d)

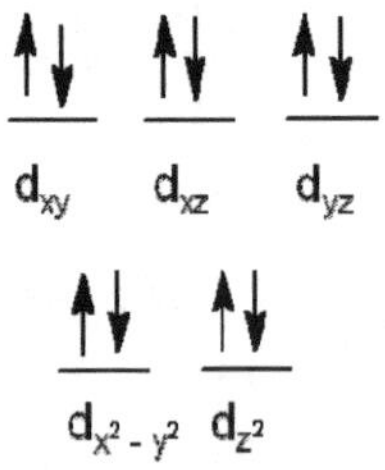

$[ZnCl_4]^{2-}$ 10 d electrons

0 unpaired electrons

23.43

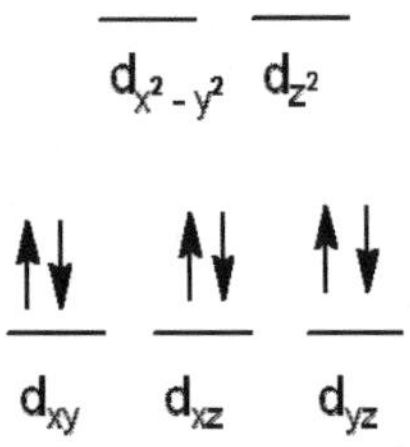

$[Co(en)(NH_3)_2Cl_2]^+$ 6 d electrons

low spin

(a) The coordination number of Co is 6—it has 4 monodentate ligands and 1 bidentate ligand.
(b) The coordination geometry of Co is octahedral—the usual geometry in the case of coordination number 6.
(c) The oxidation state of Co is +3.
(d) There are 0 unpaired electrons.
(e) Because the complex has no unpaired electrons, it is diamagnetic.

(f)

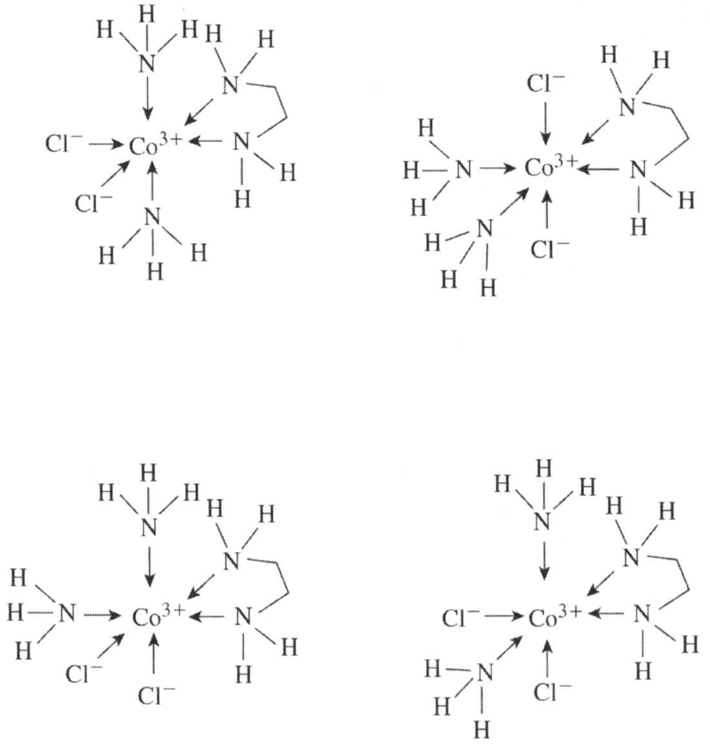

enantiomers

23.45

Hexaaquairon(II) ions, $[Fe(OH_2)_6]^{2+}$, are paramagnetic because the water ligands split the d orbital energies insufficiently to force pairing of the electrons—i.e., it is a high-spin complex with 4 unpaired electrons. When NH_3 is added to the solution the ammonia complex ions $[Fe(NH_3)_6]^{2+}$ are formed. This complex is diamagnetic—it is the low-spin complex with no unpaired electrons—because the d orbital splitting exceeds the energy required to pair the spins. Ammonia is higher in the spectrochemical series than water.

23.47

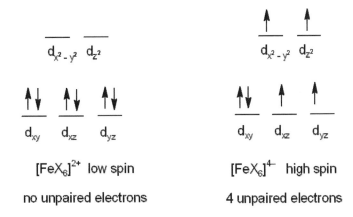

SUMMARY AND CONCEPTUAL QUESTIONS

23.49

cis-tetraamminedichlorocobalt(III) chloride   *trans*-tetraamminedichlorocobalt(III) chloride

23.51

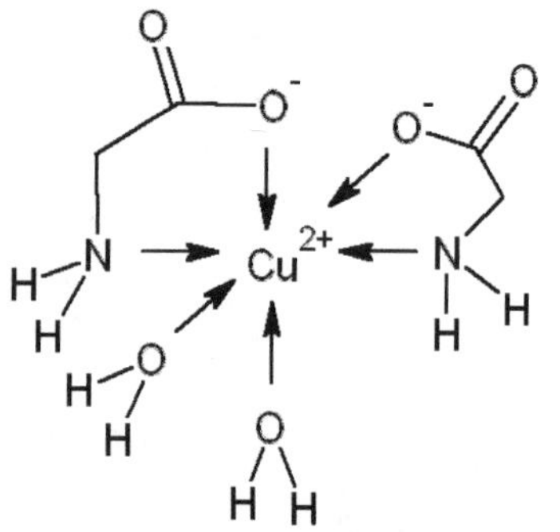

An isomer of $[Cu(H_2NCH_2CO_2)_2(H_2O)_2]$.

(a) The oxidation state of copper is +2.

(b) Copper has coordination number 6.

(c) Copper has 9 d electrons in this complex. Its d electron configuration is shown below.

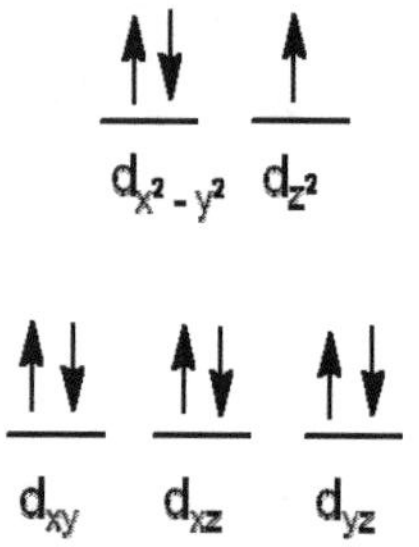

There is one unpaired electron.

(d) The complex is paramagnetic.

23.53

(a) $Ni(s) + 4 CO(g) \rightarrow Ni(CO)_4(g)$

$\Delta_rH° = \Delta_fH°[Ni(CO)_4(g)] - (\Delta_fH°[Ni(s)] + 4 \times \Delta_fH°[CO(g)])$
$= \{-602.9 - (0 + 4 \times (-110.525))\}$ kJ mol^{-1} = -160.8 kJ mol^{-1}

$\Delta_rS° = S°[Ni(CO)_4(g)] - (S°[Ni(s)] + 4 S°[CO(g)])$
$= \{410.6 - (29.87 + 4 \times 197.674)\}$ J K^{-1} mol^{-1} = -410.0 J K^{-1} mol^{-1}

$\Delta_rG° = \Delta_rH° - T \Delta_rS° = -160.8$ kJ mol^{-1} $- 298.15$ K $\times (-410.0$ J K^{-1} mol$^{-1})$ / (1000 J/kJ) = -38.6 kJ mol^{-1}

$K = \exp(-\Delta_rG°/ RT) = \exp(38.6 \times 10^3$ J mol^{-1}/ (8.314 J K^{-1} mol$^{-1} \times 298.15$ K)) = 5.792×10^6

(b) $K \gg 1$. The reaction of Ni(s) and CO(g) is product-favoured.

(c) Because the reaction is exothermic, it is less product-favoured at higher temperatures. Consequently, it is possible to use this reaction to purify nickel metal. Specifically, one would react impure nickel with carbon monoxide at ordinary or even low temperature, where the product is highly favoured. The gaseous product is collected, and then raised in temperature to reverse the reaction and deposit pure nickel. Because the impurities in $Ni(CO)_4(g)$ are other gases, they will not contaminate the solid nickel product. The original solid impurities remain in the solid phase from the first reaction.

CHAPTER 24
The Chemistry of Modern Materials

REVIEW QUESTIONS

Section 24.1: Materials: Ancient and Modern Building Blocks

24.1

There are many possible answers. One example is:
Carbon: most common are graphite (weakly bound sheets of hexagonal lattices, flaky), diamond (tetrahedral, very hard), nanotubes (cylindrical nanostructure, cavity/pocket), and fullerene ("buckyballs"—C_{60} or C_{70}, 5- and 6-membered rings bound to create a large sphere, good electron acceptor, insoluble)

Section 24.2: Metals

24.3

Each Mg atom has one $3s$ and three $3p$ valence shell orbitals. Thus, there are $4N$ molecular orbitals formed from the valence orbitals of N Mg atoms. Each Mg atom contributes 2 valence electrons, while each molecular orbital can accommodate 2 electrons. Thus, ¼ of the available molecular orbitals are filled with electrons.

24.5

(a) Metal atoms that can form a solid solution in iron, as interstitial atoms, must be smaller than iron atoms. Iron atoms have a radius of 156 pm. Aluminum atoms have a radius of 118 pm. These are considerably smaller than Fe atoms, and might occupy interstitial sites in a solid solution with iron.

(b) Metal atoms that can form a solid solution in iron, as substitutional atoms, must be similar in size to iron atoms, and have similar electronegativity. Neighbouring elements in the same period as iron have similar size and similar electronic properties. Thus, Mn (radius = 161 pm) or Co (radius = 152 pm) might be expected to form such solid solutions.

Section 24.3: Semiconductors

24.7

The band gap of GaAs is 140 kJ mol^{-1}. The energy gap for an individual electron is this energy divided by Avagadro's number—i.e., 140 kJ mol^{-1} / 6.022×10^{23} = 2.32×10^{-22} kJ = 2.32×10^{-19} J (per photon)

The maximum wavelength of light that can excite a transition across the band corresponds to the minimum frequency that can do so. This corresponds to photon energy equal to the band gap.

$$E_g = h\nu = hc/\lambda$$

So,

$$\lambda = hc/E_g = 6.626 \times 10^{-34}\,\text{J s} \times 2.998 \times 10^8\,\text{m s}^{-1} / 2.32 \times 10^{-19}\,\text{J} = 8.56 \times 10^{-7}\,\text{m} = 856\ \text{nm}$$

This corresponds to light in the near infrared region of the electromagnetic spectrum.

Section 24.4: Ceramics

24.9

Pyrex, or borosilicate glass, is made by adding boron oxide to the melt. The glass that results upon rapid cooling (slow cooling produces opaque ceramic materials) has a lower thermal expansion coefficient (i.e., for a given increase in temperature its degree of expansion is less) than ordinary glass. Thermal expansion (or contraction upon cooling) causes strains in ordinary glass that can cause it to break. This is a problem for glass that is heated in an oven.

24.11

(a) Hydroplasticity is the property of materials that become plastic when water is added to them. Clays have this property. When wet, they are readily deformed into almost any desired shape. They retain this shape after deformation, allowing the wet clay to be heated in an oven—i.e., fired—to make the shape firm and rigid.

(b) A refractory material can withstand very high temperatures without deformation. They also have low thermal conductivities, making them useful for thermal insulation—especially under high temperature conditions. For example, the space shuttle uses refractory ceramic tiles in its heat shield to protect against the extreme temperature conditions of re-entry. The Columbia space shuttle disaster resulted from a missing heat shield tile.

SUMMARY AND CONCEPTUAL QUESTIONS

24.13

To operate a 700 W microwave oven, we would need N cells of the type described in 24.12, where

$$700\ \text{W} = N \times 0.0925 \times 0.25\ \text{W} = N \times 0.0231\ \text{W},$$

accounting for the 25% efficiency, i.e.,

$$N = 700 / 0.0231 = 30300\ \text{cells required.}$$

The total area of the solar panel $= 30300\ \text{cm}^2 = 3.03\ \text{m}^2$.

24.15

In Table 24.1 of the text, the composition of pewter is given as 91% Sn, 7.5% Sb, 1.5% Cu. To get the density of pewter requires knowing the arrangement of the atoms and appropriate values for the atomic radii. Here, we will assume that the atoms are arranged pretty much the way they are in the bulk elements. We will calculate the density of pewter as a weighted average of the densities of the constituent elements. From www.webelements.com we get the bulk element densities:

$$Sn \quad 7310 \ kg \ m^{-3}$$
$$Sb \quad 6697 \ kg \ m^{-3}$$
$$Cu \quad 8920 \ kg \ m^{-3}$$

From these densities, we get

$$0.91 \times 7310 + 0.075 \times 6697 + 0.015 \times 8920 \ kg \ m^{-3} = 7288 \ kg \ m^{-3}$$

as the approximate density of pewter.

24.17

While hydrogen bonding is usually stronger than dispersion forces (locally), we cannot make a blanket statement that hydrogen bonding is always stronger. Dispersion forces increase with the size of the molecules involved and, more specifically, with the number of electrons. This is why iodine, a non-polar diatomic molecule, is a solid at normal temperature. Iodine atoms have many more electrons than O atoms and especially H atoms, and they are much further from the nucleus so are much more polarizable. The dispersion forces between iodine molecules are consequently substantial—even, based on the evidence of their phases at ambient temperature, more effective than the hydrogen bonding networks in water.

CHAPTER 25
Biomolecules

IN-CHAPTER EXERCISES

Exercise 25.1—Nomenclature

(a) Threose is an aldotetrose.
(b) Ribulose is a ketopentose.
(c) Tagatose is a ketohexose.
(d) 2-deoxyribose is an aldopentose.

Exercise 25.3—Fischer Projections

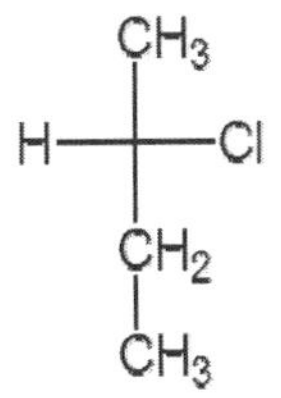

(S)-2-chlorobutane

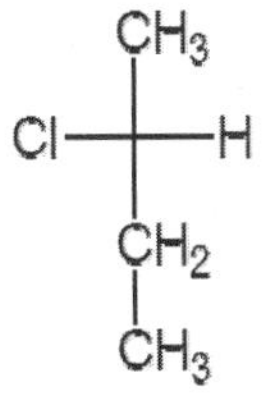

(R)-2-chlorobutane

Exercise 25.5—Stereochemistry

(a)

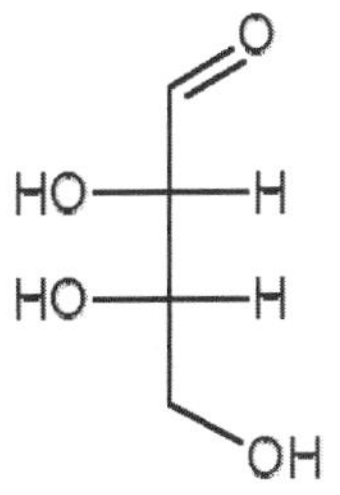

L-sugar

(b)

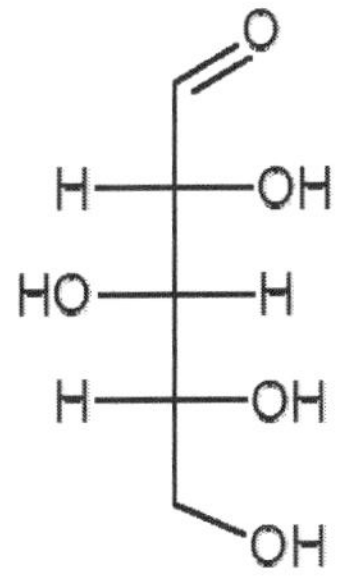

D-sugar

(c)

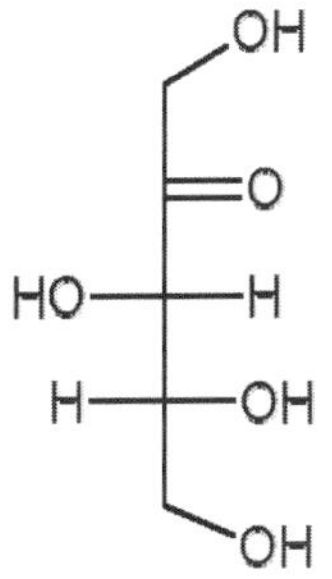

D-sugar

At the bottom stereocentre (furthest from the C=O group), the hydroxyl group is on the right = D-sugar.

Exercise 25.7—Carbohydrate Structure

α-D-galactopyranose:

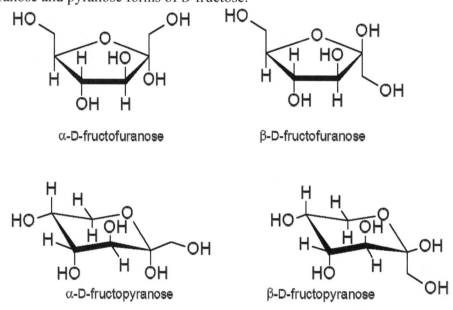

Exercise 25.9—Carbohydrate Structure

Furanose and pyranose forms of D-fructose:

α-D-fructofuranose

β-D-fructofuranose

α-D-fructopyranose

β-D-fructopyranose

Exercise 25.11—Glycosides

Acetal product of the acid-catalyzed reaction of β-D-galactopyranose with ethanol:

Exercise 25.13—Nucleotides

DNA dinucleotide AG:

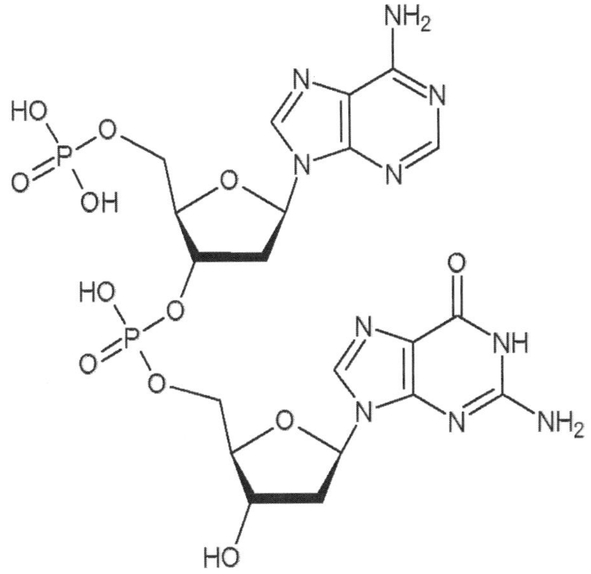

Adenine is at the 5′ end.

Exercise 25.15—Nucleotides

DNA sequence (3′) CCGATTAGGCA(5′) is complementary to GGCTAATCCGT (by default this is 5′ to 3′).

Exercise 25.17—RNA

RNA sequence (5′) CUAAUGGCAU (3′) is complementary to DNA sequence (3′) GATTACCGTA (5′).

Exercise 25.19—RNA

Several different codons can code for each amino acid. The codons are written with the 5′ end on the left.
(a) Ala, alanine, is coded by GCU, GCC, GCA, or GCG.
(b) Phe, phenylalanine, is coded by UUU or UUC.
(c) Leu, leucine, is coded by UUA, UUG, CUU, CUC, CUA, or CUG.
(d) Tyr, tyrosine, is coded by UAU or UAC.

Exercise 25.21—RNA

mRNA base sequence (5′)–CUU-AUG-GCU-UGG-CCC-UAA–(3′) gives rise to tRNA sequence,
(3′)–GAA UAC CGA ACC GGG–(5′)

REVIEW QUESTIONS

Section 25.1: Molecules and Melodies of Life

25.23

The correlation between pitch and polarity used in the pieces allows us to identify regions of high polarity versus those of low polarity, clearly showing the alternating regions of low and high polarity in both proteins. In addition, the transmembrane regions are identified by distinct musical devices (glockenspiel in the piece corresponding to AgOR1 and strings in the piece corresponding to 5-HTT). We can see (hear) that 5-HTT has more transmembrane domains than AgOR1, but the transmembrane domains in AgOR1 are generally longer than those in 5-HTT.

25.25

Primary structure = amino acid sequence. Secondary structure = regions of α-helix, β-sheet, etc. The backbone of calmodulin (or any other protein) corresponds to the chain constructed from the peptide bonds, excluding the amino acid side chains.

Section 25.2: Carbohydrates

25.27

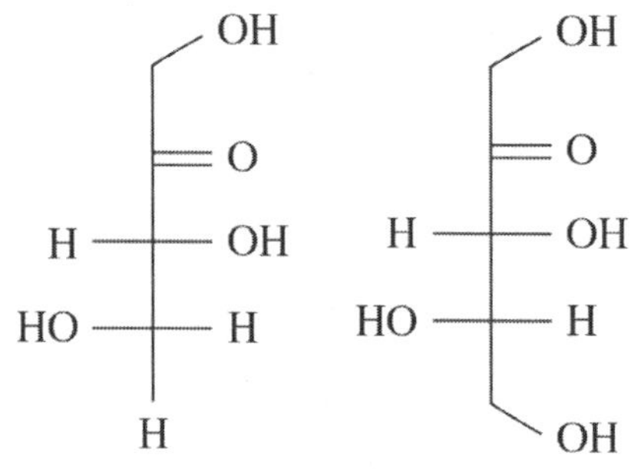

25.29

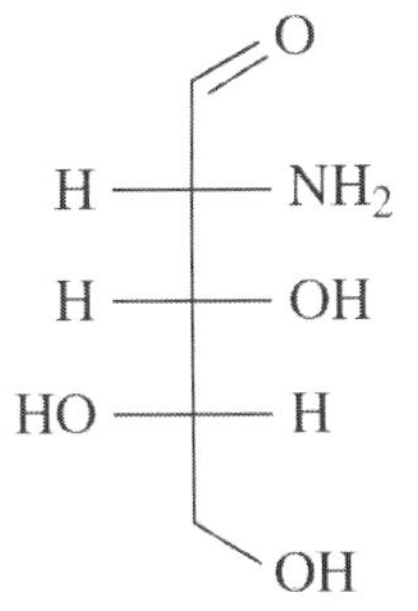

25.31

R or *S* stereochemistry of ascorbic acid:

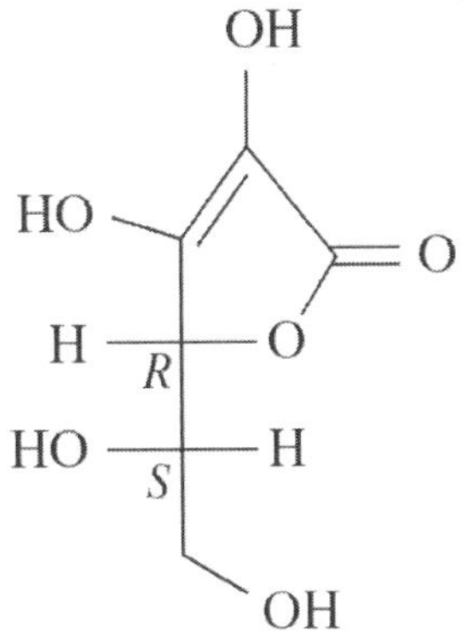

25.33

This is

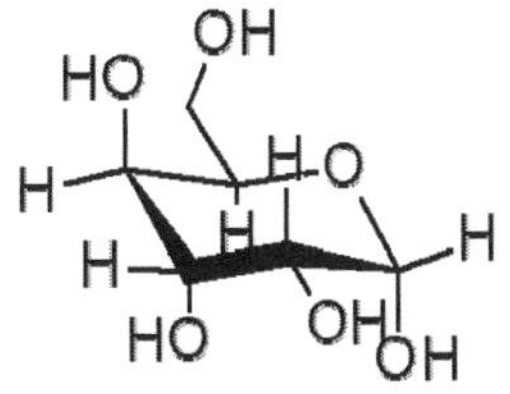

a-D-Gulopyranose

It is the β anomer of the pyranose form, and a D-sugar.

25.35

D-ribose and L-xylose are diastereomers of each other, so would not be expected to have the same physical properties. Their **(a)** melting points, **(b)** solubilities in water, **(c)** specific rotations, and **(d)** densities are all different.

25.37

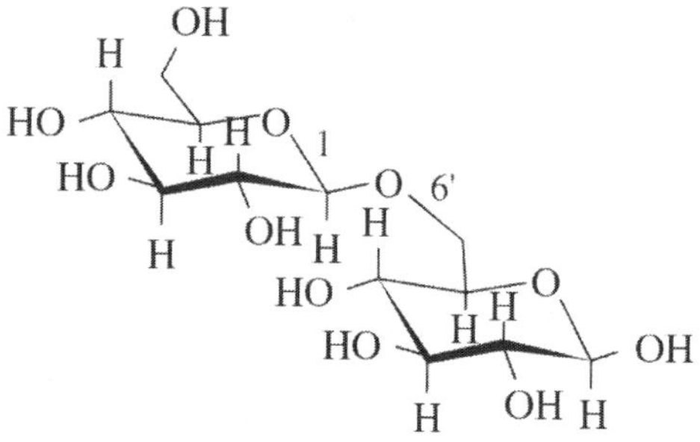

Gentiobiose

Section 25.3: Amino Acids, Peptides, and Proteins

25.39

(a) Amino acid valine:

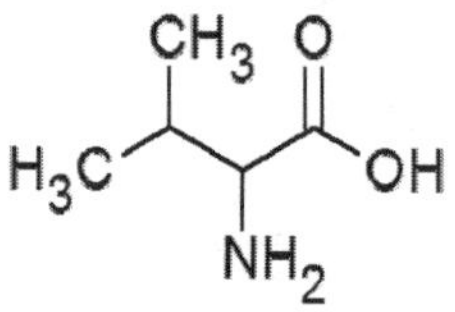

(b) Zwitterionic form of valine:

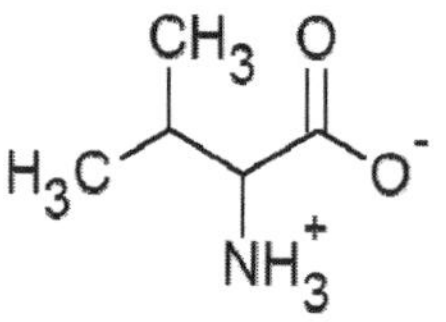

(c) The zwitterion is the predominant form at physiological pH.

25.41

Two possible peptide linkages between glycine and alanine:

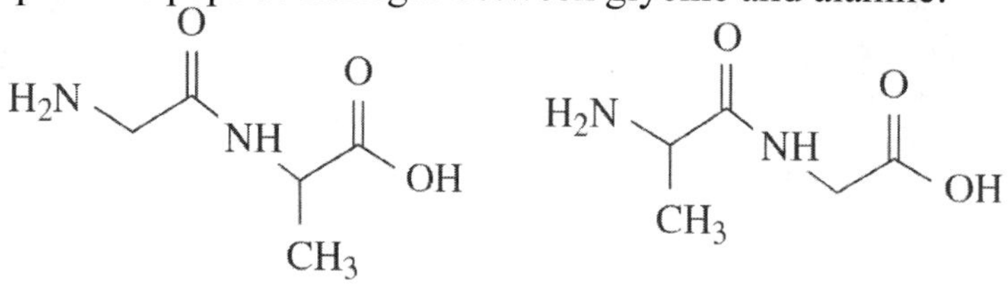

25.43

Dipeptide alanine-isoleucine:

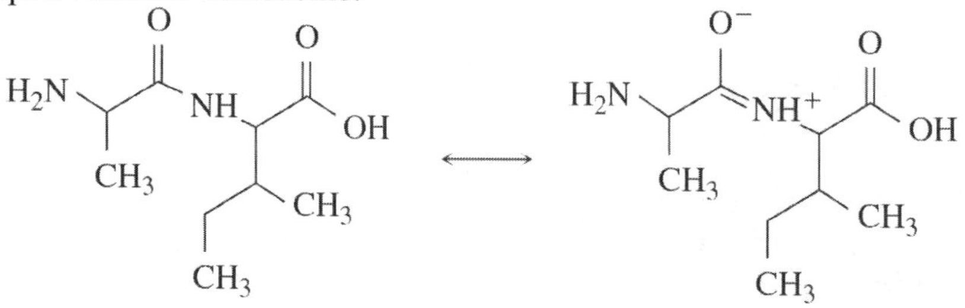

Section 25.4: Nucleic Acids and Nucleotides

25.45

A description of the quaternary structure of reverse transcriptases would show us the way the two subunits are arranged with respect to each other.

25.47

RNA tetranucleotide AUGC:

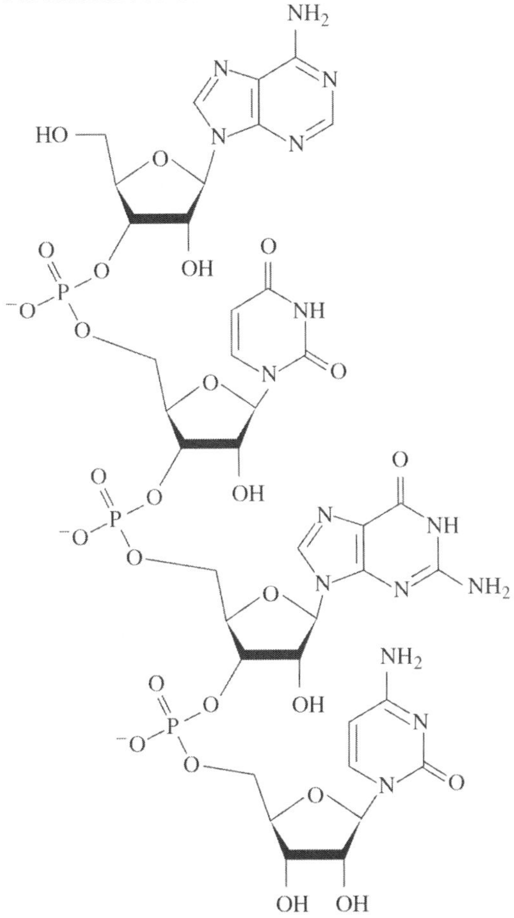

25.49

Watson and Crick noticed that the amounts of adenosine and thymine (and also cytidine and guanosine) were always the same whereas the A to C ratios vary significantly from species to species. They proposed that A pairs with T and C pairs with G in a double stranded structure. They further proposed a double helical arrangement of the strands.

SUMMARY AND CONCEPTUAL QUESTIONS

25.51

Raffinose:

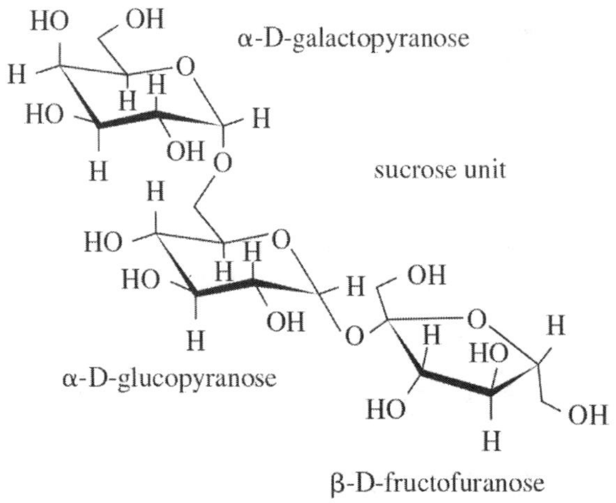

α-D-galactopyranose

sucrose unit

α-D-glucopyranose

β-D-fructofuranose

25.53

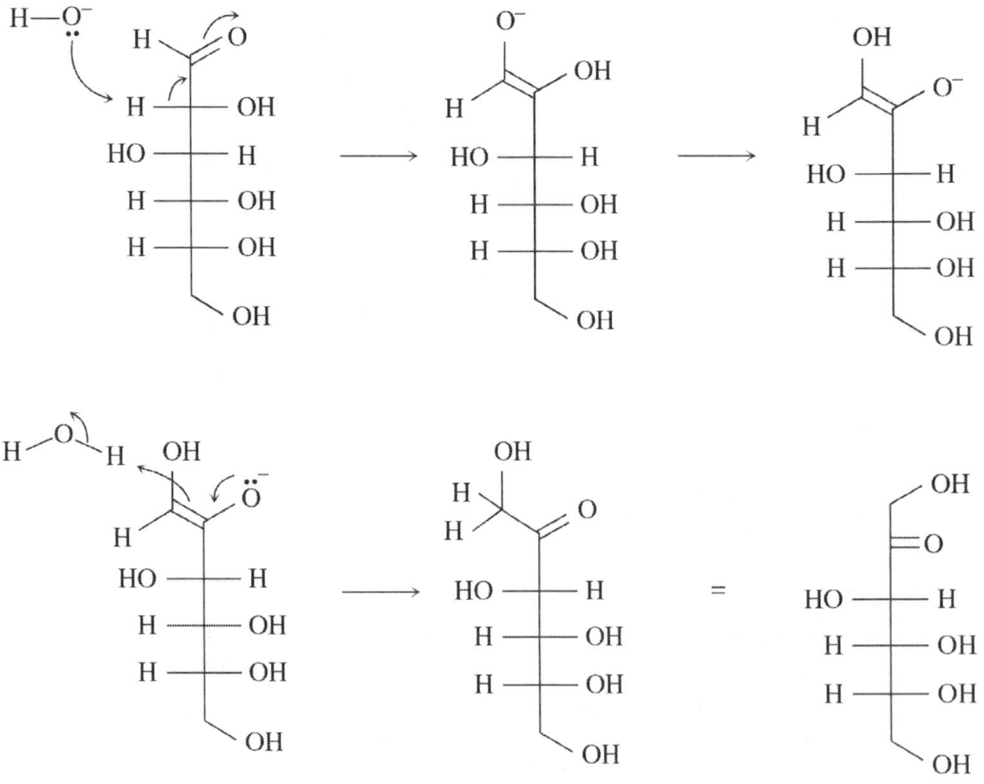

25.55

(a) Hydrogen bonds hold the two strands in place next to each other. Asymmetry of the links causes the strands to bend into helices.

(b) If the two strands were held together by covalent bonds, it would be too difficult to separate the strands in order to "read" the genetic information. The energy required to break covalent bonds linking the strands would be sufficient to break covalent bonds within the strands. In any case, "unzipping" of the double helix would be much slower—in addition to costing more energy—if the strands were linked by covalent bonds. Hydrogen bonds are of intermediate strength, stronger than other intermolecular forces and weaker than the covalent bonds that give the strands their unique structure.

25.57

(a) GAA codes for Glu, glutamic acid.

(b) TTC is the sequence in the original DNA that led to this codon in the mRNA.

(c) A mutation in the DNA such that this sequence becomes TGC leads to the codon, GCA, in the mRNA. This mutated DND leads to the amino acid Ala, alanine.

CHAPTER 26
Nuclear Chemistry

IN-CHAPTER EXERCISES

In these solutions, the symbols $_{-1}^{0}e$ and $_{-1}^{0}\beta$ can be used interchangeably, as can $_{2}^{4}He$ and $_{2}^{4}\alpha$.

Exercise 26.1—Equations for Nuclear Reactions

(a) $_{86}^{222}Rn \rightarrow {}_{84}^{218}Po + {}_{2}^{4}He$

The 86 protons of $_{86}^{222}Rn$ (i.e., its atomic number) are divided between the polonium nucleus—84 protons—and the α particle—2 protons. Similarly, the number of neutrons is conserved, as reflected in the atomic mass numbers—$222 = 218 + 4$.

(b) $_{84}^{218}Po \rightarrow {}_{85}^{218}At + {}_{-1}^{0}e$

The symbols $_{-1}^{0}e$ and $_{-1}^{0}\beta$ can be used interchangeably.

Exercise 26.3—Radioactive Decay

(a)

$_{90}^{232}Th \rightarrow {}_{82}^{208}Pb + 6\,{}_{2}^{4}He + 4\,{}_{-1}^{0}e$

(b)

$_{90}^{232}Th \rightarrow {}_{88}^{228}Ra + {}_{2}^{4}He$

$_{88}^{228}Ra \rightarrow {}_{89}^{228}Ac + {}_{-1}^{0}e$

$_{89}^{228}Ac \rightarrow {}_{90}^{228}Th + {}_{-1}^{0}e$

Exercise 26.5—Radioactive Decay

(a) Silicon-32 has a high neutron-to-proton ratio. It most likely decays by beta emission to form phosphorus-32.

$_{14}^{32}Si \rightarrow {}_{15}^{32}P + {}_{-1}^{0}\beta$

(b) Titanium-45 has a low neutron-to-proton ratio. It most likely decays by positron emission to form

$_{22}^{45}Ti \rightarrow {}_{21}^{45}Sc + {}_{1}^{0}\beta$

(c) Plutonium-239 is beyond—in mass and atomic number—the band of stability. It most likely decays by alpha emission.

$_{94}^{239}Pu \rightarrow {}_{92}^{235}U + {}_{2}^{4}He$

(d) Potassium-42 has a high neutron-to-proton ratio. It most likely decays by beta emission.

$_{19}^{42}K \rightarrow {}_{20}^{42}Ca + {}_{-1}^{0}\beta$

Exercise 26.7—Half-Life

(a) Starting with 1.5 mg of this isotope, after 49.2 years we have
$$1.5 \text{ mg} \times (1/2)^{49.2/12.3} = 1.5 \text{ mg} \times (1/2)^4 = 1.5 \text{ mg} \times (1/16) = 0.094 \text{ mg}$$
(b) A sample of tritium decays to one eighth of its activity in 3 half-lives—i.e., in 36.9 years.
(c) A sample decays to 1.00% of its original activity in n half-lives.
i.e., $\quad 0.0100 = (1/2)^n$
Therefore, $\quad \ln(0.01) = n \ln(1/2) = n(-\ln 2)$
and $\quad n = 6.64$
This corresponds to 6.64×12.3 years $= 81.7$ years.

Exercise 26.9—Kinetics of Nuclear Decay

The activity decreases by the factor $5.00 \times 10^1 / 1.08 \times 10^{11} = 4.63 \times 10^{-10}$ in n half-lives.
$$n = \ln(4.63 \times 10^{-10})/(-\ln 2) = 31.0 = t/200 \text{ years}$$
It will take $t = 31.0 \times 200$ years $= 6200$ years.

Exercise 26.11—Isotope Dilution

The isolated sample consists of the fraction, $20/50 = 0.40$, of the original threonine. Since the isolated sample weighs 60.0 mg, there must have been $60.0\text{mg}/0.40 = 150.0$ mg of threonine in the original sample.

REVIEW QUESTIONS

Section 26.2–26.3: Radioactivity and Nuclear Reaction Equations

26.13

(a) $^{54}_{26}\text{Fe} + ^4_2\text{He} \longrightarrow 2^1_0\text{n} + ^{56}_{28}\text{Ni}$
(b) $^{27}_{13}\text{Al} + ^4_2\text{He} \longrightarrow ^{30}_{15}\text{P} + ^1_0\text{n}$
(c) $^{32}_{16}\text{S} + ^1_0\text{n} \longrightarrow ^1_1\text{H} + ^{32}_{15}\text{P}$
(d) $^{96}_{42}\text{Mo} + ^2_1\text{H} \longrightarrow ^1_0\text{n} + ^{97}_{43}\text{Tc}$
(e) $^{98}_{42}\text{Mo} + ^1_0\text{n} \longrightarrow ^{99}_{43}\text{Tc} + ^0_{-1}\beta$
(f) $^{18}_9\text{F} \longrightarrow ^{18}_8\text{O} + ^0_1\beta$

26.15

(a) $^{111}_{47}\text{Ag} \longrightarrow ^{111}_{48}\text{Cd} + ^0_{-1}\beta$
(b) $^{87}_{36}\text{Kr} \longrightarrow ^0_{-1}\beta + ^{87}_{37}\text{Rb}$
(c) $^{231}_{91}\text{Pa} \longrightarrow ^{227}_{89}\text{Ac} + ^4_2\text{He}$
(d) $^{230}_{90}\text{Th} \longrightarrow ^4_2\text{He} + ^{226}_{88}\text{Ra}$
(e) $^{82}_{35}\text{Br} \longrightarrow ^{82}_{36}\text{Kr} + ^0_{-1}\beta$

(f) $^{24}_{11}\text{Na} \longrightarrow ^{24}_{12}\text{Mg} + ^{0}_{-1}\beta$

26.17

$^{235}_{92}\text{U} \longrightarrow ^{231}_{90}\text{Th} + ^{4}_{2}\text{He}$

$^{231}_{90}\text{Th} \longrightarrow ^{231}_{91}\text{Pa} + ^{0}_{-1}\beta$

$^{231}_{91}\text{Pa} \longrightarrow ^{227}_{89}\text{Ac} + ^{4}_{2}\text{He}$

$^{227}_{89}\text{Ac} \longrightarrow ^{227}_{90}\text{Th} + ^{0}_{-1}\beta$

$^{227}_{90}\text{Th} \longrightarrow ^{223}_{88}\text{Ra} + ^{4}_{2}\text{He}$

$^{223}_{88}\text{Ra} \longrightarrow ^{219}_{86}\text{Rn} + ^{4}_{2}\text{He}$

$^{219}_{86}\text{Rn} \longrightarrow ^{215}_{84}\text{Po} + ^{4}_{2}\text{He}$

$^{215}_{84}\text{Po} \longrightarrow ^{211}_{82}\text{Pb} + ^{4}_{2}\text{He}$

$^{211}_{82}\text{Pb} \longrightarrow ^{211}_{83}\text{Bi} + ^{0}_{-1}\beta$

$^{211}_{83}\text{Bi} \longrightarrow ^{211}_{84}\text{Po} + ^{0}_{-1}\beta$

$^{211}_{84}\text{Po} \longrightarrow ^{207}_{82}\text{Pb} + ^{4}_{2}\text{He}$

Section 26.4: Stability of Atomic Nuclei

26.19

(a) Gold-198 decays to mercury-198.

$^{198}_{79}\text{Au} \longrightarrow ^{198}_{80}\text{Hg} + ^{0}_{-1}\beta$

The atomic number increases by 1, while the mass number does not change. This is beta decay.

(b) Radon-222 decays to polonium-218.

$^{222}_{86}\text{Rn} \longrightarrow ^{218}_{84}\text{Po} + ^{4}_{2}\text{He}$

The atomic number decreases by 2, while the mass number decreases by 4. This is alpha decay.

(c) Cesium-137 decays to barium-137.

$^{137}_{55}\text{Cs} \longrightarrow ^{137}_{56}\text{Ba} + ^{0}_{-1}\beta$

This is beta decay.

(d) Indium-110 decays to cadmium-110.

$^{110}_{49}\text{In} \longrightarrow ^{110}_{48}\text{Cd} + ^{0}_{1}\beta$

This is positron emission.

26.21

(a) Bromine-80m is a metastable state of bromine. It most likely decays by gamma emission to form bromine-80.

$$^{80m}_{35}\text{Br} \rightarrow ^{80}_{35}\text{Br} + n\gamma$$

n is the number of gamma photons emitted.

(b) Californium-240 is beyond—in mass and atomic number—the band of stability. However, its neutron-to-proton ratio is very low. It most likely decays by positron emission.

$$^{240}_{98}\text{Cf} \rightarrow ^{240}_{97}\text{Bk} + ^{0}_{1}\beta$$

Alternatively: $^{240}_{98}\text{Cf} \rightarrow ^{236}_{96}\text{Cm} + ^{4}_{2}\text{He}$

(c) Cobalt-61 has a high neutron-to-proton ratio. It most likely decays by beta emission.

$$^{61}_{27}\text{Co} \rightarrow ^{61}_{28}\text{Ni} + ^{0}_{-1}\beta$$

(d) Carbon-11 has a low neutron-to-proton ratio. It most likely decays by positron emission.

$$^{11}_{6}\text{C} \rightarrow ^{11}_{5}\text{B} + ^{0}_{1}\beta$$

26.23

(a) Of the nuclei ^{3}H ^{16}O ^{20}F and ^{13}N

^{3}H and ^{20}F have high neutron-to-proton ratios and are likely to decay by beta emission.

(b) Of the nuclei ^{238}U ^{19}F ^{22}Na and ^{24}Na

^{22}Na has a low neutron-to-proton ratio and is likely to decay by positron emission.

26.25

For ^{10}B,
$\Delta m = 5 \times 1.00783 + 5 \times 1.00867 - 10.01294$ g mol^{-1} = 0.069560 g mol^{-1} = 6.9560×10^{-5} kg mol^{-1}
$E_b = \Delta m\, c^2$
$= 6.9560 \times 10^{-5}$ kg mol$^{-1} \times (2.998 \times 10^8$ m s$^{-1})^2$
$= 6.252 \times 10^{12}$ J mol$^{-1} = 6.252 \times 10^{9}$ kJ mol^{-1}
Per nucleon,
$E_b / n = 6.252 \times 10^9$ kJ mol$^{-1} / 10 = 6.252 \times 10^8$ kJ mol^{-1} nucleons
For ^{11}B,
$\Delta m = 5 \times 1.00783 + 6 \times 1.00867 - 11.00931$ g mol^{-1} = 0.081860 g mol^{-1} = 8.1860×10^{-5} kg mol^{-1}
$E_b = \Delta m\, c^2$
$= 8.1860 \times 10^{-5}$ kg mol$^{-1} \times (2.998 \times 10^8$ m s$^{-1})^2$
$= 7.358 \times 10^{12}$ J mol$^{-1} = 7.358 \times 10^{9}$ kJ mol^{-1}
Per nucleon,
$E_b / n = 7.358 \times 10^9$ kJ mol$^{-1} / 11 = 6.689 \times 10^8$ kJ mol^{-1} nucleons
We see that ^{11}B is the more stable nucleus—it has a higher binding energy per nucleon.

26.27

For ^{40}Ca,
$\Delta m = 20 \times 1.00783 + 20 \times 1.00867 - 39.96259 \text{ g mol}^{-1} = 0.36741 \text{ g mol}^{-1} = 3.6741 \times 10^{-4} \text{ kg mol}^{-1}$
$E_b = \Delta m\, c^2$
$\qquad = 3.6741 \times 10^{-4} \text{ kg mol}^{-1} \times (2.998 \times 10^8 \text{ m s}^{-1})^2$
$\qquad = 3.302 \times 10^{13} \text{ J mol}^{-1} = 3.302 \times 10^{10} \text{ kJ mol}^{-1}$
Per nucleon,
$\qquad E_b / n = 3.302 \times 10^{10} \text{ kJ mol}^{-1} / 40 = 8.255 \times 10^8 \text{ kJ mol}^{-1}$ nucleons
This result is consistent with Figure 26.6.

26.29

For ^{16}O,
$\Delta m = 8 \times 1.00783 + 8 \times 1.00867 - 15.99492 \text{ g mol}^{-1} = 0.13708 \text{ g mol}^{-1} = 1.3708 \times 10^{-4} \text{ kg mol}^{-1}$

$E_b = \Delta m\, c^2$
$\qquad = 1.3708 \times 10^{-4} \text{ kg mol}^{-1} \times (2.998 \times 10^8 \text{ m s}^{-1})^2$
$\qquad = 1.232 \times 10^{13} \text{ J mol}^{-1} = 1.232 \times 10^{10} \text{ kJ mol}^{-1}$
Per nucleon,
$\qquad E_b / n = 1.232 \times 10^{10} \text{ kJ mol}^{-1} / 16 = 7.700 \times 10^8 \text{ kJ mol}^{-1}$ nucleons

Section 26.5: Rates of Nuclear Decay

26.31

64 h corresponds to 64 h / 12.7 h $= 5.04$ half-lives. At this time, the mass of ^{64}Cu is
$\qquad 25.0 \text{ µg} \times (1/2)^{5.04} = 0.760 \text{ µg}$

26.33

(a) $^{131}_{53}\text{I} \rightarrow {}^{131}_{54}\text{Xe} + {}^{0}_{-1}\beta$
(b) 40.2 days corresponds to 40.2 days / 8.04 days $= 5.00$ half-lives. At this time, the mass of ^{131}I is
$\qquad 2.4 \text{ µg} \times (1/2)^{5} = 0.075 \text{ µg}$

26.35

13 days corresponds to 13 days / (78.25 / 24) days $= 4.0$ half-lives. At this time, the mass of ^{67}Ga is
$\qquad 0.015 \text{ mg} \times (1/2)^{4.0} = 9.4 \times 10^{-4} \text{ mg}$

26.37

(a) $_{86}^{222}\text{Rn} \rightarrow {}_{84}^{218}\text{Po} + {}_{2}^{4}\text{He}$

(b) 20.0% of ^{222}Rn remains after n half-lives.

$0.200 = (1/2)^n$

$n = \ln(0.200) / \ln(1/2) = 2.32$

2.32 half-lives corresponds to 2.32×3.82 days $= 8.87$ days

26.39

0.72 of the ^{14}C remains after n half-lives.

$0.72 = (1/2)^n$

$n = \ln(0.72) / \ln(1/2) = 0.47$

0.47 half-lives corresponds to $0.47 \times 5.73 \times 10^3$ years $= 2700$ years

This is the age of the bone fragment.

26.41

(a) A cobalt-60 source will drop to 1/8 of its original activity after n half-lives.

$0.125 = (1/2)^n$

$n = \ln(0.125) / \ln(1/2) = 3$

3 half-lives corresponds to 3×5.27 years $= 15.81$ years

(b) One year corresponds to $1 / 5.27 = 0.190$ half-lives. At this time, the fraction of ^{60}Co remaining is

$(1/2)^{0.190} = 0.877$

26.43

(a)

$_{11}^{23}\text{Na} + {}_{0}^{1}\text{n} \rightarrow {}_{11}^{24}\text{Na}$

$_{11}^{24}\text{Na} \rightarrow {}_{12}^{24}\text{Mg} + {}_{-1}^{0}\beta$

(b)

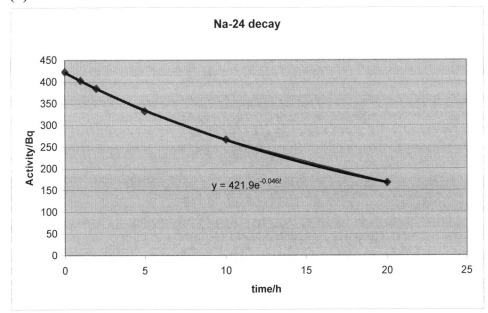

The data is clearly well fit by a simple exponential decay, $421.9\,e^{-0.046\,t}$. The parameters of this fit are well approximated using just the first and last data. $k = 0.046$ corresponds to

$$t_{1/2} = \ln 2 \,/\, k = \ln 2 \,/\, 0.046 \text{ h} = 15 \text{ h}$$

Section 26.6: Artificial Nuclear Reactions

26.45

$$^{239}_{94}\text{Pu} + 2\,^{1}_{0}\text{n} \rightarrow \,^{241}_{94}\text{Pu}$$

$$^{241}_{94}\text{Pu} \rightarrow \,^{241}_{95}\text{Am} + \,^{0}_{-1}\beta$$

26.47

$$^{238}_{92}\text{U} + \,^{12}_{6}\text{C} \rightarrow \,^{246}_{98}\text{Cf} + 4\,^{1}_{0}\text{n}$$

It must be carbon-12.

26.49

(a) $^{114}_{48}\text{Cd} + \,^{2}_{1}\text{H} \longrightarrow \,^{115}_{48}\text{Cd} + \,^{1}_{1}\text{H}$

(b) $^{6}_{3}\text{Li} + \,^{2}_{1}\text{H} \longrightarrow \,^{7}_{4}\text{Be} + \,^{1}_{0}\text{n}$

(c) $^{40}_{20}\text{Ca} + \,^{2}_{1}\text{H} \longrightarrow \,^{38}_{19}\text{K} + \,^{4}_{2}\text{He}$

(d) $^{63}_{29}\text{Cu} + \,^{2}_{1}\text{H} \longrightarrow \,^{65}_{30}\text{Zn} + \gamma$

26.51

$$^{10}_{5}\text{B} + \,^{1}_{0}\text{n} \longrightarrow \,^{7}_{3}\text{Li} + \,^{4}_{2}\text{He}$$

Section 26.8: Nuclear Fusion

26.53

$$_{3}^{6}\text{Li} + _{0}^{1}\text{n} \rightarrow _{1}^{3}\text{H} + _{2}^{4}\text{He}$$

SUMMARY AND CONCEPTUAL QUESTIONS

26.55

Carbon-14 levels in the atmosphere are in steady state—determined by the flux of cosmic rays that produce it in the upper atmosphere. Plants take up carbon-14 (via carbon dioxide) in proportion to its concentration in the atmosphere. When they die, no more carbon-14 is absorbed. The relative amount of carbon-14 (to carbon-12) decays exponentially with the known half-life of carbon-14. Measuring this isotope ratio allows the age of the sample to be determined—via the logarithm of the isotope ratio divided by its initial value (the atmospheric level). The method, as described here, assumes the atmospheric level has remained constant. This level is known to have varied by up to 10% in the past. The method can be corrected for this variation, however. Nevertheless, the method is limited to objects between 100 and 40 000 years old. There has not been enough carbon-14 decay in objects less than 100 years old, while the level of carbon-14 is too small to give an accurate age measurement beyond 40 000 years.

26.57

Radiation can cause transmutation of elements within cells, or can simply ionize or otherwise disrupt cellular molecules. The resulting chemical reactions can significantly affect cellular function and even damage DNA molecules. This can lead to cell death, if the cell is unable to repair the damage at the rate it occurs. Thus, exposure of humans to radiation can cause radiation sickness and even death, at sufficiently exposure levels. The killing of tissue by radiation can be used to treat diseases such as cancer, however. By focusing radiation on a tumour, we can kill the cancer tissue and possibly cure the patient.

26.59

The amount of ^{87}Rb is decreased by the factor,
$$0.951 = (1/2)^n$$
$n = \ln(0.951) / \ln(1/2) = 0.072$ half-lives
This corresponds to $0.072 \times 4.8 \times 10^{10}$ years $= 3.5 \times 10^{9}$ years.

26.61

The amount of ^{235}U decreases by the factor
$$0.72 / 3.0 = 0.24 = (1/2)^n$$
$n = \ln(0.24) / \ln(1/2) = 2.06$ half-lives
This corresponds to $2.06 \times 7.04 \times 10^{8}$ years $= 1.45 \times 10^{9}$ years.

26.63

Amount of ^{235}U = 1000 g / 235.0439 g mol^{-1} = 4.255 mol
Amount of energy released by fission of uranium-235 = 4.255 mol × 2.1 × 10^{10} kJ mol^{-1} = 8.9 × 10^{10} kJ
To obtain the same amount of energy from coal requires
$\quad$ 8.9 × 10^{10} kJ / 2.9 × 10^7 kJ t^{-1} = 3100 t of coal

26.65

The time between the sample injection and the taking of the blood sample is negligible compared to the half-life of tritium. This is a sample dilution problem. A 1.0 mL sample was injected, and a 1.0 mL blood sample was taken. The activities measured are proportional to the sample concentrations. The ratio of the activities equals the ratio of the blood volume to 1.0 mL. The blood volume is just
$\quad$ 2.0 × 10^6 Bq / 1.5 × 10^4 Bq × 1.0 mL = 130 mL

26.67

We could label the methanol with the radioactive isotope, ^{15}O, carry out the reaction, and then sample the water and look for ^{15}O that could only have come from methanol. The same experiment could be carried out with labelled acetic acid to see if acid ^{15}O ends up in the water product.

26.69

The number of atoms in 1.0 mg of ^{238}U is
$\quad$ 6.022 × 10^{23} mol × 1.0 × 10^{-3} g / 238.050782 g mol^{-1} = 2.530 × 10^{18}
From $\Delta N / \Delta t = -kN$
we get (note that 1 Bq = 1 s^{-1})
$\quad k = (\Delta N / \Delta t) / N$
$\quad\quad$ = 12 s^{-1} / 2.530 × 10^{18} = 4.74 × 10^{-18} s^{-1}
$\quad\quad$ = 4.74 × 10^{-18} s^{-1} × (60×60×24×365) s y^{-1} = 1.50 × 10^{-10} y^{-1}
$\quad t_{1/2} = \ln(2) / k = 4.6 × 10^9$ y
This is close to the literature value, and consistent with the uncertainty expected with a measurement of only 12 events.